A.H. Glattfelder and W. Schaufelberger

Control Systems with Input and Output Constraints

With 284 Figures

Springer

A.H. Glattfelder, Dr.sc.techn
W. Schaufelberger, Dr.sc.techn

Automatic Control Laboratory, ETH Zentrum, CH-8092, Zurich, Switzerland

British Library Cataloguing in Publication Data
Glattfelder, A.H.
    Control systems with input and output constraints.
    (Advanced textbooks in control and signal processing)
    1.PID controllers 2.Predictive control
    I.Title II.Schaufelberger, W.
    629.8'31
    ISBN 1852333871

Library of Congress Cataloging-in-Publication Data
Glattfelder, Adolf Hermann.
    Control systems with input and output constraints / A.H. Glattfelder and W. Schaufelberger.
    p. cm. -- (Advanced textbooks in control and signal processing)
    Includes bibliographical references and index.
    ISBN 1-85233-387-1 (alk. paper)
    1.Automatic control. 2. Control theory. I. Schaufelberger, W. (Walter), 1940- II. Title.
    III. Series.
    TJ213.G5185 2003
    629.8--dc21   2003045430

ISSN 1439-2232
ISBN 1-85233-387-1 Springer-Verlag London Berlin Heidelberg
a member of BertelsmannSpringer Science+Business Media GmbH
http://www.springer.co.uk

MATLAB® and SIMULINK® are the registered trademarks of The MathWorks Inc., 3 Apple Hill Drive Natick, MA 01760-2098, U.S.A. http://www.mathworks.com

The use of registered names, trademarks etc. in this publication does not imply, even in the absence of a specific statement, that such names are exempt from the relevant laws and regulations and therefore free for general use.

The publisher makes no representation, express or implied, with regard to the accuracy of the information contained in this book and cannot accept any legal responsibility or liability for any errors or omissions that may be made.

Typesetting: Electronic text files prepared by authors
Printed and bound in the United States of America
69/3830-543210 Printed on acid-free paper SPIN 10764614

To Ruth and Elisabeth

Series Editors' Foreword

The topics of control engineering and signal processing continue to flourish and develop. In common with general scientific investigation, new ideas, concepts and interpretations emerge quite spontaneously and these are then discussed, used, discarded or subsumed into the prevailing subject paradigm. Sometimes these innovative concepts coalesce into a new sub-discipline within the broad subject tapestry of control and signal processing. This preliminary battle between old and new usually takes place at conferences, through the Internet and in the journals of the discipline. After a little more maturity has been acquired by the new concepts then archival publication as a scientific or engineering monograph may occur.

A new concept in control and signal processing is known to have arrived when sufficient material has evolved for the topic to be taught as a specialised tutorial workshop or as a course to undergraduate, graduates or industrial engineers. *Advanced Textbooks in Control and Signal Processing* are designed as a vehicle for the systematic presentation of course material for both popular and innovative topics in the discipline. It is hoped that prospective authors will welcome the opportunity to publish a structured and systematic presentation of some of the newer emerging control and signal processing technologies.

In control engineering the difference between the theories of academic control and the realities of the industrial application lies in the simple phrase "operational constraints". Most control textbooks omit the actuator saturation block and/or the output constraint conditions on the linear system feedback block diagram but there are very few real-world control systems where they are not present. In the toolbox of the industrial control engineer, the antiwindup circuit and the control override module are essential items for successful control.

Dolf Glattfelder and Walter Schaufelberger have long recognised the importance of these two industrial procedures in overcoming input and/or output constraints in control system designs. The outcome of a decade or more of research is this new volume in the *Advanced Textbooks in Control and Signal Processing* series. At last it is possible to study a global presentation of control designs for system constraints in one volume. The first part of the textbook deals systematically with the single loop control problem whilst the second part covers more advanced control system problems. The reader will find strong motivation from the practical industrial problems and examples in the presentation; as such, industrial control engineers should find the approach illuminating and accessible. The extensions to more complex systems will interest the industrial and academic control readership alike. The book may be used as a course-book, a book for self-study or a reference book and is a welcome entry to *Advanced Textbooks in Control and Signal Processing*.

M.J. Grimble and M.A. Johnson
Industrial Control Centre
Glasgow, Scotland, U.K.
April, 2003

Preface

In current industrial process control, PID technology is firmly established as *the* standard. PI control performs best on linear time-invariant plant models of "dominant first order", *i.e.* a first order lag element or open integrator in series with a small delay, such as used in the famous Ziegler–Nichols rules. Typical examples are control of speed, level, temperature, *etc.*

For plant models of dominant higher order, such as position control in mechatronic systems, linear control performance can be conserved within the classic PID technology by an additional derivative action, or by using cascade arrangements. If performance is still not sufficient, one may resort to linear state feedback, and add an observer if needed.

Such linear time-invariant plant models of low dominant order are often a sufficiently precise approximation of the real process response. This is one of the key success factors of PID technology in practice. In most cases it is valid only in a small enough neighborhood of a given nominal steady state operating point. However, if deviations increase, for instance by applying large reference steps, then the actuator will saturate transiently or even permanently. In other words, plant "input constraints" will manifest themselves. The same effect will appear for reduced deviations, if the closed-loop bandwidth is increased by using advanced control design methods. Typical examples of input constraints are mechanical limits on servomotor position, on valve position, on voltage from a power amplifier, *etc.* Such actuator saturations will appear in any "linear" control loop on a real plant.

Furthermore, in many applications, operational limits will be encountered on some additional variables within the plant. This situation is known as plant "output constraints". Typical examples would be limits on armature current or winding temperature in DC-servomotors, on flow and dynamic overpressure in hydraulic systems, and on temperature, or on local temperature gradients in thermal power systems, *etc.* Thus, any "linear" control loop on a real plant will surely have such output constraints, although they may be far off during normal operation.

This outlines the type of *control problem* investigated in this book.

Considering *methodology*, three general approaches are available for the analysis and design of such input and output constraint systems. The first one would be Optimal Control Theory, by applying the Maximum Principle. Or one solves the problem by numerical optimization of the performance index for the trajectory, subject to the constraints from the system response equations and to the nonlinear constraints on input and output. This approach of Predictive Control has been investigated intensively in the last two decades, and there are numerous industrial applications. It has been developed (and is best suited) for complex multi-input systems with many constraints.

The third approach has been developed empirically, mainly for single-loop PI control. Practitioners were always confronted with constraint control problems on their particular processes. So they had to invent some solution, implement it with the equipment available at the time, check it by experiments on the plant, and if performance was not up to specifications, again invent, implement and test. Such iterative development has led to the current "antiwindup" schemes for input constraint problems and to "override" schemes for output constraint problems. Essentially they are nonlinear add-ons to the basic linear control algorithms. They are available as software blocks in most commercial Process Control Systems (PLCs) today. Therefore, they are quite popular with control design engineers, and are in everyday use in industry. The main motives for their use are that the resulting systems are quite simple, intuitive, and that they are known to perform surprisingly well.

However, not much effort was spent by the inventors on transferring the knowhow to other areas, on exploring the limits of applicability in a more systematic way, and on other general aspects, such as on comparing different solutions, looking for any equivalence, or making them easier to understand, to learn and to teach, or on publishing results to make the basic concepts more generally known, *etc.* In short, there is much isolated practical knowhow and experience around, but not much is available on a more scientific level.

This particular situation has motivated us to do research in this area for many years now, and has led up to this book. Our main objective is a deeper understanding of such systems, and on this basis to design more effective control systems within typical industrial time constraints.

The usual approach from theory would be to aim for a solution of the most general problem first, and afterwards look at special cases, such as PI control. Instead of using such a "top-down' (deductive) approach we decided on a "bottom-up' (inductive) one. The first step is to solve the simple cases, which also happen to be the most frequent ones in applications, while taking care that the assumptions agree with most typical cases. The further steps then are to use these results as building blocks to investigate successively more complex cases, *i.e.* to proceed by "extension' rather than by "specialization'. In other words, we have elected an engineering design approach for this book rather than a mathematically oriented one. Also, the focus will be on exploiting available control theory rather than developing new theory. And it is also more on employing and adapting schemes which are currently in use rather than inventing new ones. – Such choices may be disputed, of course. We made them by considering the likings and needs of design engineers we personally know, and also of many of our students.

The investigations will follow a typical design sequence. For *motivation* we often use a typical industrial case, and then move up one level of abstraction to typical process models which cover a wider area of application. As the *problem statement* we use "specifications", both for clarifying assumptions and for defining benchmarks. Then we consider currently used alternatives for controller *structure*, and present generic structures, from where the different alternatives may be instantiated. This will also show some equivalences, which are important in practical design. Then we do simulations for the benchmark situations in order to get typical, comparable *transient responses*. This will serve as verification, as first screening and the findings will provide a clear basis for redesign of structure, if necessary. However, this can provide only a point-wise view on the behavior of such strongly nonlinear systems. Getting an overall view only by simulations is tedious and always incomplete, extrapolation is dangerous, and the underlying relations have to be extracted with much effort from the data. In this situation, nonlinear *stability* analysis offers a framework which is both elegant and fairly easy to use. It provides this overview, gives a better understanding of the relations of causes (structure) and effects (transient response), and it indicates limits of applicability.

This sequence of motivation, specification, structure, transient response, and stability properties is then repeated for a next, more complex case.

Also, an intermediate level of abstraction is used. This allows one to relate directly the cases being investigated to everyday design problems, and it also provides more general insights and results. On the other hand, the links to theoretical concepts are visible, but the assumptions and objectives are oriented to practical design situations.

To summarize, this design approach starts from specific design problems arising frequently in design practice, and uses them as background for a more abstract treatment. It is inductive (starting with simple problems and then generalizing them), it is a stepwise procedure, and it is iterative, until results are within specifications (and not necessarily optimal in the strict sense). It also systematically considers alternatives, and integrates control theory at key design phases.

The *content* of this book is organized as follows. The first part of the book is devoted to the *standard techniques* based on PID technology. First the basic concepts of antiwindup on a PI regulator and of overrides using PI controls are presented and investigated. Then an additional derivative action is introduced, and other forms of disturbance inputs to the loop, such as finite pulses and high frequency measurement disturbances. Also, more realistic actuator models are considered.

The second part of the book is about *advanced techniques*, *i.e.* extensions of the basic methods to more complex problems and also to additional methods of investigation. The extensions are towards process models of dominant higher order, *i.e.* to plant windup and extended antiwindup, generalized overrides, split range controls, and multi-input plants with internal coupling, *i.e.* multivariable systems.

Concerning methods, short extensions are made into optimal control, specifically minimum-time systems, and into the numerical methods (Model Predictive Control). Robust control methods for quantitative performance evaluation are mentioned but not used. A summary of the nonlinear stability tests used here is given in an appendix.

The *examples* in the text are based on typical industrial situations. Also, the case studies offered at the end of chapters have been abstracted from projects the authors have been involved with.

In such a book, the selection of topics and methods is always arbitrary, and surely no complete coverage is possible. Here, we have tried to investigate the most frequent situations in practical design, to be consistent in the methods as far as possible, but not to be encyclopedic.

The *objective* of the book is to enable the reader to
- make informed choices on such structures,
- interpret correctly any observed transient responses,
- understand better and predict the effect of design variations.

The main aim is to make the industrial control design process more efficient and effective, and also to provide an application-based entry to research and development in this area.

Readers should have a good background in standard linear control theory, from a typical one-year course at the university level. An introduction to nonlinear control systems, to the describing function technique and the sector or circle stability criteria, is recommended. The examples and case studies make use of Matlab/Simulink[©].

The files are available for downloading in .zip form (size: 927 KB) at

`http://control.ee.ethz.ch/~glatt/book.htm`

This book has grown out of our practical experience from industrial applications, from research, and from lecture notes, which have slowly evolved in many years of teaching an advanced course for students majoring in Automatic Control at the Mechanical and Electrical Engineering departments at the ETH Zürich. We greatly appreciate the contributions of many people over the years.

Zürich,   *Adolf Hermann Glattfelder*
June 2003   *Walter Schaufelberger*

Contents

Part II Advanced Techniques

1

Introduction

"Antiwindup", "Overrides", and "Bumpless Transfer" are typical nonlinear add-on functions to linear feedback controllers. They cope with typical non-linear effects "in the large", such as plant input constraints by actuator saturations, plant output constraints by operational limits on secondary plant outputs, and interaction between logic control and feedback control loops. They have been invented and refined by practitioners in an intuitive engineering design way for their standard control equipment. They work well for most control tasks, but are not so well understood from a more scientific point of view.

The aim of this book is to provide a structured introduction into this particular area of applied control, both from an engineering design approach and a more analytic one. This may contribute to a better understanding and a more effective use of such add-on features.

Before going into details of the subject matter, the introduction illuminates the *context* of such systems from different aspects, such as by a typical example from power generation control and other current industrial applications, a short historical perspective on their development, the basic control problems to be investigated, the expectations on control performance of such systems from a higher level of overall plant operations, a discussion of available methods and methodology, and finally an overview on the contents of this book.

1.1 A Typical Case

A typical case shall illustrate the technical context of the design methods. Consider the automatic operation of a production plant in the power and chemical industries. Then the following operational tasks will appear almost everywhere. The first one is to bring the plant from standstill status to nominal operation status through a sequence of controlled steps. On this level of abstraction, the system is seen as a finite state machine, and described best by a Petri net. Control design is based on the well-established IEC 1131 standards. On the next level below, then the second group of tasks is feedback control, where individual loops usually are active during one sequence step only.

· This is illustrated on a *hydropower unit control system* in Table 1.1 , in a simplified, typical form, where many details have been omitted.

Table 1.1. Sequence steps and feedback control tasks

Sequence steps	Control loop activated	Design situation
Run-up	Feedforward input stroke limit to run-up opening	Speed control, antiwindup
Speed-no-load	Speed control, PI action	Linear, disturbance suppression
Synchronizing	Speed control, driving relative speed and rotor angle to zero	Linear range, high performance response
Generator switch-in	Switch from speed control to power control (both PI)	Bumpless transfer
Power to grid	Power control with transient water pressure rise	Output constraint control
Generator trip	Power control to speed control	Bumpless transfer
Shutdown	Feedforward input stroke limit to zero	No feedback control, antiwindup for speed regulator

On the next level down, simulation results illustrate the performance of antiwindup and override schemes we are going to investigate here.

Fig. 1.1 shows the response of the speed control loop at the transition from feedforward run-up to the "speed-no-load" operation. For the left hand plot a standard linear PI speed regulator was used in combination with the standard start-up limit on main servomotor position (here set at 10% higher than speed-no-load opening), and no antiwindup. Note that the speed overshoot is excessive. Usually this would produce an overspeed trip.
On the right hand plot one of the current antiwindup features has been added. It effectively eliminates the speed overshoot, and cuts time to synchronization from 150 to 30 seconds.

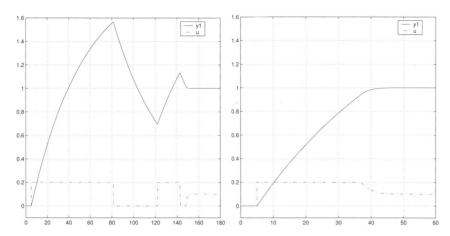

Fig. 1.1. Run-up and control of speed (y_1) on a hydropower unit, with no anti-windup (left) and one of the standard antiwindup schemes (right)

Fig. 1.2 shows the response of the power control loop to setpoint steps with overrides acting from the transient water pressure rise. The non-minimum phase response on y_1 is due to the elastic water hammer effect.

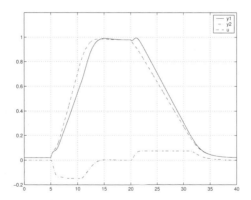

Fig. 1.2. Setpoint step response of power (y_1) control on the hydropower unit with turbine opening u and constraints to -0.15% and to $+0.075\%$ on transient water pressure rise (y_2)

Note that the standard features perform well in this particular case.

In industrial control systems many similar control tasks show up:

- In hydropower systems
  - limits on upstream or downstream reservoir level or
  - generator power limits, if water level difference is substantially higher than nominal
  - *etc.*
- In gas turbines with power control
  - gas turbine turbine inlet temperature limit
  - differential expansions
  - thermoshock limits

  and for the exhaust steam generators in combined cycle plants:
  - steam temperatures at the exit and at critical intermediate positions
  - corresponding thermal gradients and thermoshock in the associated steam turbine
  - *etc.*
- In mechatronic systems with position control
  - speed limits
  - acceleration limits
  - acceleration buildup ("jerk") limits,
  - transient torque limits across flexible couplings
  - *etc.*
- A very frequent case is a servomotor with both slew and stroke limits (saturations on both rate and position), modeled by an integrator with a P controller for linear range positioning and with an input saturation for slew and an output saturation for stroke.

This list of examples can be easily expanded and extended into other areas. It motivates investigating the above-mentioned design techniques in a more general setting.

1.2 A Historical Perspective

Such add-on features go back to mechanical and hydraulic governors, as they were used in power generation control up to the 1960s. Consider for instance the flywheel speed governor for hydropower turbines shown in Fig. 1.3, where the main servomotor provides the integral action. It was equipped with a mechanical blocking of the main servomotor position, in order to produce a smooth run-up without too much upset of the waterway. This mechanical block had to be set manually by the operator before starting the unit and had to be removed after the unit had reached nominal speed and had been synchronized and switched to the grid, but prior to loading the unit. This is a typical input constraint case.

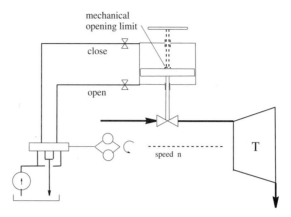

Fig. 1.3. Flywheel speed governor on a hydropower unit, with a mechanical (screw down) opening limit on servomotor opening for run-up

An output constraint situation is sketched in Fig. 1.4, for a steam turbine unit delivering both low-pressure steam and electric power into an insular grid.

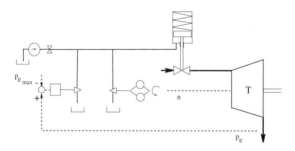

Fig. 1.4. Hydraulic speed/frequency control on a steam-turbine generator unit connected to an insular grid with exit pressure override (schematic)

Here the low pressure may not exceed a high level, regardless of the electric power demand. This is implemented by an override controller, which opens an additional outlet on the control oil circuit to the turbine inlet valves. It thus limits steam inflow to what can be absorbed in the low-pressure subsystem.

In the 1960s pneumatic controllers were the standard in the chemical industries. They essentially consist of a high-gain element (fixed orifice and variable nozzle) and additional positive and negative force feedbacks (by bellows as capacities and adjustable orifices as resistors); see Fig. 1.5.

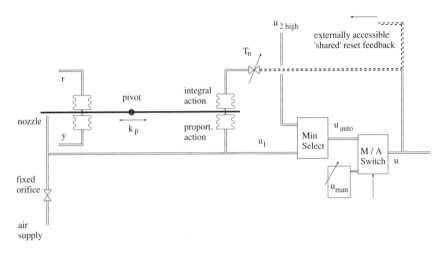

Fig. 1.5. Using standard pneumatic PI controllers with externally accessible reset (T_n) feedback for bumpless manual-to-automatic transfer and override control by minimum selection

Manual-to-automatic transfer was implemented through external access to the feedback capacities in order to equilibrate their pressure rapidly with the externally supplied one. For override control purposes, this switch was replaced by a pneumatic relay selecting and passing through either the minimum or the maximum of two or more input pressures; Fig. 1.5. And finally, actuator limitations inside the regular working range determined by the supply pressure were implemented by applying corresponding constant pressures to the minimum or maximum selectors. Such structures using standard equipment were described by Buckley [21].

In the 1970s, analog electronic PID control hardware became the standard in the power generation industry. Overrides were used such as in the BrownBoveri "Turbomat" for loading of steam turbine units with constraints from thermal gradients within thick-walled casings, and similarly the "Freilastrechner" from Siemens–KWU for temperature load swing constraints in once-through boilers and steam turbines. A list of such techniques was compiled within the Belgian "Projet CHANCE". Plant input saturations were typically handled by limiting diode feedback on the operational amplifier Fig. 1.6(a), and bumpless transfer by external access to the feedback capacitor for integral action Fig. 1.6(b), and overrides by paired diodes for minimum or maximum selection Fig. 1.6(c). Implementation of overrides required additional hardware. It added costs and increased the complexity of the control system for operators and maintenance. So the technique was used sparingly, and only to supplement feedforward measures, such as conservatively slow reference ramping instead of reference steps, by limiting allowable load swings and ramps

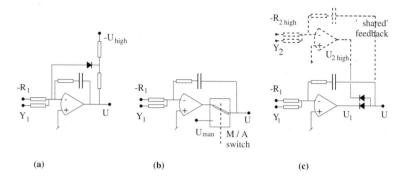

Fig. 1.6. Using analog electronic PI controllers with externally accessible reset feedback for (a) actuator saturation, (b) bumpless manual-to-automatic transfer and (c) override control by minimum selection

severely, and giving conservative instructions to operators. Obviously all this lowers the overall plant productivity.

With the introduction of digital control the restrictions from implementation were much reduced. But the add-ons had to be "re-invented", *e.g.* see [16, 25], because the high-gain loop structures in pneumatic and analog electronic equipment were not directly portable for numerical stability reasons. This has led to both "output" and "input conditioning" techniques [32]. Today, the necessary features are available in all commercial process control systems. So the traditional hardware constraint has been removed, and the way to exploit the design techniques fully is open.

In *research*, antiwindup has become a recognized topic since the 1980s [26, 27], , with many current publications (for instance a special issue of *European Journal of Control*, May 2001), research-oriented books, *e.g.* [13], and at scientific conferences, such as in the proceedings of the recent *American* and *European Control Conferences*. It is interesting to note that such a development has not yet taken place for output constraints and overrides [28, 34].

In *textbooks*, the subject of antiwindup, bumpless transfer and overrides has been present since the mid-1980s [1]; but this has been as a supplement to linear control, and not as a subject matter on its own. To our knowledge, no textbook has appeared recently, that has been devoted exclusively to this area and which covers it both from a practical design and a more abstract analytic point of view. So there is no established way of presenting the material, such as for linear control. Thus, we had to develop a structure, which makes sense to us, but of course may be discussed at will.

1.3 The Control Problem

So far three typical design tasks have emerged: for input constraints, bumpless transfer and output constraints. The aim of this section is to state the associated control problems more clearly.

1.3.1 Control Systems with Input Constraints

Consider the system to be controlled, or "plant" for short, in Fig. 1.7. There are two (scalar) inputs, the manipulated or control variable u and the disturbance input v, and one (scalar) output, the controlled variable y. A typical example would be a tank with level y to be controlled to its reference value r, with persistent outflow v and manipulated inflow u. Thus the value of v determines the operating point of u.

One of the very basic nonlinearities of such simple control loops are *saturations* on the control variable $u(t)$. They are due to the working range of the actuator, which is always bounded by physical reasons to a low and a high limit. A model often used is

$$u = \text{SAT}(u_{lin}) = \begin{cases} u_{low} & \text{if} \quad u_{lin} < u_{low} \\ u_{lin} & \text{if} \quad u_{low} \leq u_{lin}(t) \leq u_{high} \\ u_{high} & \text{if} \quad u_{lin} > u_{high} \end{cases}$$

This is the "input constraint" situation, Fig. 1.7, where "input" refers to the plant, i.e. the system to be controlled, and not to the closed-loop system.

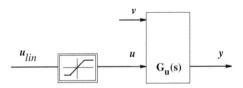

Fig. 1.7. Control system with "input" saturation

A typical example is a valve in the process industries, which can be operated from its low (fully closed) to its high (fully open) limit position.

In normal operation (*i.e.* in the range of operating conditions the feedback loop was originally designed for), the control loop then stabilizes at some equilibrium $\bar{u}$:

$$u_{low} + \Delta u_{low} < \bar{u} < u_{high} - \Delta u_{high} \tag{1.1}$$

situated at a finite distance (Δu) inside the working range limits of u. This allows regulation as specified for sufficiently small-sized loop input changes (to the reference r and/or to the disturbance v), such that $u(t)$ does not touch the limit values. This is the typical situation in control design. Here, the model of the dynamic response may be linearized around the equilibrium, and linear controllers (classical PID, state feedback, *etc.*) are commonly used with much success.

For larger loop input changes, however, the control variable will saturate, either transiently or permanently. In the second case the control function will be permanently interrupted, and control objectives clearly cannot be met any longer. In other words, this is an "ill-posed" control problem, and must be remedied by redesign of the actuator subsystem, such as by an expanded working range or additional actuators. This case shall be considered later. First, the focus will be on the transiently saturating case (the "well-posed" control problem), where a new equilibrium can be established with $\bar{u}$ inside limits; see Eq. 1.1.

In such situations, where the controller output transiently exceeds the saturations on the actuator, the linear loop performance may seriously degrade and become unacceptable; see Fig. 1.1. As we shall see, this is due to excessive run-away or "windup" of state variables both in the controller and the plant. It calls for "antiwindup" measures.

1.3.2 Control Systems with Mode Switch

Another basic nonlinearity in almost all control loops is the "manual-to-automatic" control mode switch on the manipulated variable $u(t)$; Fig. 1.8.

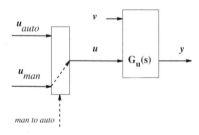

Fig. 1.8. Control system with manual-to-automatic transfer switch on $u(t)$

The input signal to the actuator is to be switched from an external source such as
- a manual input generator manipulated by the operator
- a programmable logic controller output, or
- the output of a second controller
to the output of the linear controller, which so far has been operating in open-loop conditions. Later, the loop usually has to be switched back from

automatic mode to manual mode again.

Then the transfer in both directions has to be "bumpless", meaning that no excessive control transient should be generated in the loop by this switching action. This requires appropriate values of all state variables in all subsystems, that is

- in the plant,
- the automatic controller,
- and in the manual input generator or the second controller.

If this is not the case, then the subsequent transient may move u to the input constraints, or output constraints may be met.

In process control, this assumption usually means that the control system has to be brought close to equilibrium in the vicinity of the target closed loop operation point by an appropriate control movement $u_{man}(t)$. Allowable deviations from the target equilibrium are defined backwards, such that deviations of $u_{auto}(t)$ after the transfer stay within predefined bounds, say ± 5 % of full range of u.

Note however that this is rather conservative. The general case would be automatic switching of u in transients to achieve overall optimal response. Such mixed continuous-discrete predictive control is an interesting new topic in control.

1.3.3 Control Systems with Output Constraints

A third major class of basic nonlinear systems exhibits "output constraints". Consider Fig. 1.9: for large transients, an additional output $y_2(t)$ may exceed operational limits. Then "overrides" are used, see Fig. 1.2. From the list of examples given in Sect. 1.1, it seems that output constraints are quite frequent in practice, and become even more so, as operational limits are fully exploited in order to maximize plant output. In most situations, more than one constrained output is present. Here, cases of one y_2 per u shall be considered first.

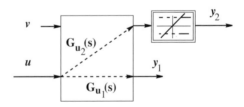

Fig. 1.9. Control system with constraint on the secondary output y_2

The constrained output $y_2(t)$ is proportional to a single state variable of the plant or to a linear combination of those.

Output constraints are operational limits rather than "hard" physical limits, such as the mechanical end stops in an actuator. Transgressing limits is possible, but will advance ageing of process equipment, such as by low-cycle fatigue, and shall be avoided. Very large transgressions will cause spontaneous equipment failure, but this is to be avoided by separate safety functions and shall not be considered here.

Output constraints are typically "soft" constraints:

$$y_{2_{low}} - \Delta y_{2_{low}} < y_2(t) < y_{2_{high}} + \Delta y_{2_{high}} \tag{1.2}$$

i.e. there exists a continuously differentiable model of the response along the constraint values $y_{2_{low}}$ or $y_{2_{high}}$ for small enough deviations (within the span indicated by the $\Delta y_{2_{hi,low}}$ values).

For the control problem to be "well posed", three key conditions must be fulfilled.

1. The constrained output $y_2(t)$ is controllable from $u(t)$.
2. During the transient (run-up, load swing) the $u(t)$ required to drive $y_2(t)$ to its upper limit $r_{2_{high}}$ or its lower $r_{2_{low}}$ limit shall decay to a value within the operating range of u determined by the input constraints. This has to be valid for the specified range of disturbances v. If not, then again the actuator subsystem must be redesigned.
3. At the final steady state of the control system, the steady state value $\bar{y}_2$ of $y_2(t)$ must lie at a finite distance between the upper and lower limits respectively.

$$r_{2_{low}} + \Delta r_{2_{low}} < \bar{y}_2 < r_{2_{high}} - \Delta r_{2_{high}} \tag{1.3}$$

with the Δr_2 values being positive finite. Otherwise, control would not be transferable to the main control loop for y_1 by u near final steady state.

In other words, the output constraints are not reached for small enough deviations from the final steady state of the main control loop. And for large deviations, control of $y_2(t)$ to its limit by $u(t)$ is feasible, *i.e.* the limit values are reachable.

1.4 Expected Control Performance

A third aspect of context is the performance of such control strategies as expected from an overall plant operation view.
Here we shall give some outlines and general directions based on experience.[1]
Note that this is a key element for selecting the appropriate analysis and design tools from control theory.

---

[1] More precise specifications may then be stated for the individual design case.

- For small deviations from steady state, *i.e.* linear conditions, high-performance regulators are to be used, typically including integral action (in order to drive steady state control errors to zero).
  This means that only well designed and well tuned linear regulators for narrow range control will be considered further. In other words, if the linear analysis of the narrow range control loop reveals slow or weakly damped modes, then this has to be remedied first.
- Typical transients considered here are from an initial steady state to a final steady state, where $u(t)$ is to end up inside its linear operating range such as to allow linear narrow range control to be active around the final steady state. Reference (setpoint) steps are applied, while the load is kept constant, and also load (disturbance) steps with constant reference.
  This amounts to equivalent initial conditions responses decaying to the final equilibrium.
- For large deviations, the closed-loop performance to be attained is characterized as follows:
  - Such transitions should be as fast as the process will allow, considering the constraints from actuators and secondary variables.
    In other words, the actuator saturations should be fully exploited, and steady state errors along constraints should decay to zero, such as to exploit the given operational limits fully. Also, de-tuning of feedback to avoid the constraints is not a valid option.
  - No perceptible overshoot on controlled variables $y(t)$ is allowed, both for the main output $y_1(t)$ and the constrained secondary outputs $y_2(t)$.
  - Smooth enough transient of control variable $u(t)$, *i.e.* no "chattering", no jumps exceeding typically $\pm 5\%$ of working range. This is to avoid excessive wear, noise or power consumption.
  - Main plant parameters shall be known *a priori* with an error margin of approximately ± 25 %. This is acceptable for feedback control, but is difficult to absorb by a pure feedforward control.

This means that, for the wide range responses, "minimum-time" behavior is predominant, but it must be sufficiently robust. It is also not intuitively clear how well other widely investigated performance criteria (such as the "minimum quadratic" one) fit in here. And for deviations decaying into the linear range, transition must be made to a linear design, such as by the linear quadratic method, with no discontinuities on $u(t)$.

1.5 Methods

Methods are an important component of context, and must be discussed now as such. Three general approaches are available. So far, the focus has been on the intuitive method of add-ons. However, other methods are available since the 1960s, such as optimal control theory and the numerical optimization approach.

1.5.1 Optimal Control

In optimal control theory, the Maximum Principle [4] applies for large deviations, where constraints are met.[2] An optimal feedforward control function $u^*(t)$ is produced. Converting this into an optimal feedback of states $u^*(x)$, i.e. the synthesis problem, is a separate and nontrivial step. Of special interest for the overall plant operation (as discussed above) is the case of minimum-time transients. They produce a "bang-bang" control sequence $u^*(t)$, and the feedback synthesis leads to "switching hypersurfaces". They are instructive and simple to use in the two-dimensional case, e.g. for the two-open-integrators chain with control saturation only, but unfortunately not for general higher order systems.

The input saturation case is said to be a regular problem, whereas state (output) limits yield a singular problem, which needs additional arguments to solve. It will essentially result in a very high-gain feedback of the constrained state along its constraint value. Also, it is assumed that the endpoint $x = 0$ may be reached by at least one control sequence $u(t)$ from the given initial condition $x_0 \neq 0$. In other words, the reachability is not investigated, but assumed. Also, such minimum-time systems are known to be very sensitive to parameter uncertainty and to persistent unknown inputs. This is due to the basic property of being a feedforward approach. A well-established way out is sliding mode control [10]. Finally, the trajectory is only considered up to the end-time t_E. Afterwards, control has to be transferred to its equilibrium value. If the switching surface concept from "wide range" control is continued for $t > t_E$, then chattering appears on u in the "small range". However, the situation there is linear. Therefore, a common solution is to transfer control to a linear regulator.

To summarize, the optimal control solution may be very helpful in clarifying actual versus ideal performance. As such, it shall be used here.

1.5.2 Numerical Optimization: Model Predictive Control

The numerical optimization approach goes back to the concept of "Dynamic Programming" [3] from the 1960s. It has been developed further in industry for cases where a simple single-loop-based approach would not generate a good enough trajectory. It is best suited for complex multi-input systems with strong couplings and with many constraints. The number of variables for the optimization is a concatenation of the number of input variables u_i, a sufficient number of samples of the individual control inputs, and the number of samples from process outputs $y_j(k)$ required to cover its impulse responses. So, numerical optimization requires a substantial computational effort, making it well suited for slow (thermal) processes and large sampling intervals T_s.

---

[2] In fact, for many theorists, the add-on methods described here are just patches. They are there because the problem was not correctly stated and solved. And if this were the case, they would not be needed any more.

Again, it is assumed that at least one trajectory exists from the initial conditions to the final steady state. Then the control error between actual output and a reference trajectory is considered, which is to converge to near zero for all steady states. Typically, open integrators are also introduced in the algorithm before outputting $u(t)$ to the actuators, such that the inputs $v(t)$ to the integrators also tend to zero for all steady states. Then a performance index is selected. Usually, this is a quadratic one weighing both $e(t)$ and $v(t)$. The performance index built up along the predicted trajectory is then minimized, subject to the constraints from the system response equations and to the nonlinear constraints on input and output. This results in a feedforward discrete time optimal control variable sequence $u^*(kT_s)$.

By applying only the first control move of this sequence, and repeating the whole procedure at the next sampling time (receding horizon) with newly acquired measured variable values, an iterative sort of feedback is implemented. If no nonlinear input and output constraints are present then this converges to the basic discrete time linear quadratic control. – To avoid the numerical optimization in real time, "explicit" model predictive control is currently being developed, where the optimization is performed offline and then a lookup table is used for online operation.

This predictive control approach has received much attention from the research community in the last two decades, and there is a large and quickly growing number of industrial applications. It may have some convergence problems, which tend to exclude it from safety-critical applications. It also may not be cost effective, if the control problem is not of the "very complex and multivariable" type. So there is an ongoing need for simple and reliable methods for the many less complex control design tasks.

1.5.3 The Intuitive Approach

The third general approach shall comprise the intuitive, empirical methods of add-on functions developed basically for the most common single input single output (SISO) linear controllers. It is also known as the *"two-step design procedure"*:

-   The first step is to design a linear control around steady state, assuming that a linearized model is an adequate representation of actual closed-loop dynamics.
-   In the second step, additional nonlinear feedbacks are designed to improve response, until specifications are met.

The procedure will not produce an optimal solution in the strict sense, but from our experience may be iteratively brought close to it.

It is best suited for comparatively simple, but frequent cases, where only standard function blocks are available and the overall engineering effort is strictly limited. It is also well accepted and used by practitioners. This makes the approach attractive to design engineers.

Considering methodology, this is a typical engineering design approach. From the beginning, practitioners were confronted with specific control problems with constraints in their particular plants. So they were forced to invent a solution, implement it with the means available at the time, check it on the real plant and, if performance was not up to specifications, improve on the original design, implement and check again. Not much effort was spent on more general aspects, such as on transfer to other situations, on simplifying the design and making it more transparent, easy to understand and learn (this may even have been avoided for competitive reasons), and on exploring the limits of applicability (often avoided as well), *etc.*

As such solutions have been developed on an empirical basis, a more theoretical analysis is needed to establish their properties more consistently, and also to show the limits for their "good usage".

From a theory perspective, the deductive analytical "top-down" approach would be solving the most general problem first, and only then looking at special cases (such as PI(aw) control). This is, of course, a quite valid alternative, and "abstraction saves time".

But theory in its own interest may assume system properties which seldom hold in practice. This will severely restrict the straightforward application to real-world design problems. A typical example is standard nonlinear stability analysis. It requires, that the open-loop transfer function in the linear range is Hurwitz stable, which is not the case in many practical applications. So this requires some work-around, which then becomes a key factor to practical application. Another element is the inherent conservativeness of nonlinear stability tests. From a theoretical point of view, global asymptotic stability may be the only worthwhile result. However, this is much too restrictive for practical design purposes. In other words, the stability radius becomes a key issue and must therefore be investigated much more closely. A third element is that practical design problems can mostly be broken down into small and simple subproblems, which are relevant only in specific phases of the process operation (see Sect. 1.1). Thus, the simple special cases are very frequent, and the general solution would have to be broken down explicitly first, in order to provide the "missing links" between the general theory and the everyday design problems, and this "takes time".

For all these practical reasons, we shall use an intermediate approach, on a median level of abstraction. We start from physical examples as an introduction and motivation. Then we move up one level of generalization and proceed to classes of processes which are frequent in practice, and choose typical representatives as benchmarks.

We also pick up some currently used regulator structures and present generic structures, from where the different structures may then be instantiated. This will also show possible equivalences, and thus addresses a main need of control engineers. In the past, investigations have often stopped at this point.

We continue by simulations for the benchmark situations in order to get typical, comparable responses. This will serve as verification, as the first screening and the findings will provide a clear basis for redesign of structure, if necessary. However, simulations alone can provide only a point-wise view for such strongly nonlinear systems. Getting an overall view by simulations only is tedious and always incomplete. Extrapolation is dangerous. And the underlying relations have to be extracted with much effort.

Nonlinear stability will provide a tool which is both elegant and easy to use to get this overview, and to understand better the underlying relations of structure and transient response, as well as to perceive the limits of applicability. We shall use the circle and Popov criteria and the describing function method. This procedure is then repeated for the next, more complex case, again for an appropriate class of processes and controls.

So, three views shall be used in the following, *i.e.* "structure", "transient response" and "stability properties". From our experience it is crucial to have a good balance of these three views to obtain a good design result. Note that most discussions of alternatives in practice are about "structure" only, whereas the other views are not taken into account.

On the one hand the median level of abstraction chosen here allows to see the relations to everyday design problems directly. In addition, it provides more general results, including stability and limits of application. On the other hand, the links to theoretical concepts are also directly visible, but the assumptions and aims are oriented to practical design situations.

To summarize, this approach may be characterized as a typical engineering design process, which starts from specific design problems arising often in design practice, and uses this as background for more abstract treatment. It is inductive (starting with simple problems and then generalizing them), it is a stepwise procedure, and it is iterative, until results are within specifications (and not necessarily optimal in a strict sense). It also systematically considers alternatives, and integrates suitable analytical tools at each design phase. It aims at a more efficient and effective design process by the way of better understanding.

1.6 Contents

As stated above, we have opted for an inductive rather than a deductive approach, *i.e.* addressing the basic simple cases first, and extending them step by step until a reasonable coverage of the field from an application point of view is attained. Within each step we shall adhere to the investigation sequence described above, *i.e.* a motivating real-world example, some abstraction to a class of problems, a specification statement, alternative structures, transient response, nonlinear stability properties, and relations to optimal control. This has led us to organize the book into two main parts.[3]

The first part is devoted to **'Standard Techniques'**, meaning those areas of application, where standard PID algorithms perform sufficiently well, and considering ease of implementation, using what is directly available as standard software blocks in today's industrial process control systems.
This implies that we focus on discrete-time (sampled data) algorithms. However, analysis and quantitative design in control is habitually performed in the continuous domain. Therefore, we shall have to discuss continuous-time equivalents as well.
This also implies that we restrict ourselves in this part to plant models of the Ziegler–Nichols type, *i.e.* being of a dominant first-order lag (or one open integrator) in series to a (small) pure time delay, which shall be a conservative approximation of non-modeled fast dynamic modes.

$$G_u(s) = \frac{Y(s)}{U(s)} = e^{-sD} \frac{b}{sT_1 + a} \qquad (1.4)$$

This finally implies that we deal with the basic SISO loop situation, for which the standard antiwindup and override schemes have been developed.

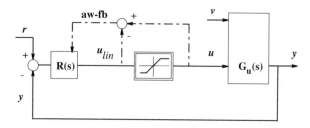

Fig. 1.10. Anti windup control

We shall start with the basic *input constraint* case, *i.e.* with the most common case of P- and PI control with input stroke saturation, Fig. 1.10, demonstrating the "windup effect" and different "anti windup" features (abbreviated in the following to PI(aw) control). We then add a derivative action, *i.e.* we extend to PID(aw) control, and alternatively to a PI(aw)-P cascade. We also expand to other disturbance forms, such as high-frequency measurement noise models and single pulses of finite length.

---

[3] We are aware that any such separation is arbitrary, and may well be disagreed upon. Here, we have selected practical and implementation-oriented allocation criteria.

Then the basic *output constraint* case is addressed, such as using the basic "override" structure shown in Fig. 1.11.

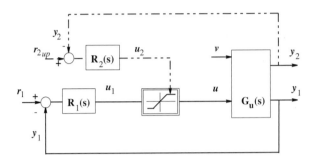

Fig. 1.11. Override control for an upward constraint on y_2 at $r_{2_{up}}$

Next, *combined constraints* on input and output are considered, on the common case of an actuator subsystem with both "slew" (rate) and "stroke" (position) saturations. However, we shall not discuss the "bumpless transfer" situation at length here. It may be considered as a generalized form of input and output constraints implementation.

The second part of the book deals with *'Advanced Techniques'*.
A first attribute of an "advanced case" shall be that the control algorithms are non-standard in current process control systems. It may take some ingenuity to build the additional functions needed from available elementary function blocks, or else a customized module in C-code is required.
A second attribute is that the standard add-on techniques from the first part will not perform sufficiently well and will have to be augmented, *i.e.* additional add-on functions are needed.

First we discuss the *input constraint* case on plants of higher dominant order, where PID control will perform poorly and additional state feedback is indicated, which may be implemented e.g. by multiple cascade arrangements. This will lead to "plant windup", "dynamic antiwindup", trajectory generators with antiwindup, and the use of override techniques for antiwindup purposes. It will also lead one to perceive more clearly the limits of good usage of such add-on techniques. We shall establish the link to MPC here.

The *output constraint* case shall be discussed on higher order plants as well, and also in more complex and multiple constraint situations. Again, this will indicate the limits of good usage for override schemes.

So far, the scalar plant input signal case has been investigated. We now proceed to the split-range situation, where plant response to the control inputs is so different as to require separate control algorithms.

We finally investigate the full multi-input-multi-output (MIMO) case with strong interaction of the local control loops. This will lead to crosscoupling of the antiwindup feedbacks as well.

Concerning examples and problems for the reader, we have tried to be consistent with the general approach of this book, *i.e.* not to ask for proofs of theorems to be worked out, but rather to design and analyze control systems with a visible industrial background. The case studies at the end of chapters have been developed from industrial projects the authors have been involved with.

Readers should have a good background in standard linear control theory, such as from a typical one-year course at university level. An introduction to nonlinear control systems, to the describing function technique and the sector or circle stability criteria is recommended. The examples and case studies make use of Matlab/Simulink$^{©}$. The files used to generate the figures are available for download from http://www. (size 924 KB).

1.7 Objectives

This book has grown out of our practical experience from practical applications, from research, and from lecture notes. They have slowly evolved in many years of teaching a one-semester advanced course in this area for students in Automatic Control at the Mechanical and Electrical Engineering departments of ETH Zürich. We greatly appreciate the contributions of many people during all those years.

Our objective for this book is to enable the reader to

- make informed choices on antiwindup and override add-on structures
- interpret correctly any observed transient responses
- understand better and predict the effect of design parameter variations
- recognize when a given design task is outside the area of "good usage" of such simple add-ons, and when one should resort to more demanding methods.

This should help the industrial control design process to be more efficient and effective. It also may provide an application-based entry to research and development in this area.

Part I

Standard Techniques

PI Control with Input Saturations

We start our investigations with the basic case of PI control, with actuator saturation and on plants of dominant first order. This situation covers a very large percentage of industrial cases. It is also the most simple one and will serve as a building block for more involved cases.

A motivating example is what happened to John.

2.1 John's Case

Many years ago John was a young engineer with the Hydraulic Transient Analysis group of a major producer of hydropower equipment. He knew linear control theory from his studies, but was not intimately familiar with all details of the electronic turbine regulator used by the company at that time. One day he was called to step in to commission such a regulator. It was a retrofit for the previous mechanical governor on a Pelton turbine in a small power station in the Swiss Alps.

The customer wanted him to reproduce the run-up and speed stabilization transient obtained with the mechanical governor in Fig. 2.1 (left). A maximum delay of 30 s was specified from the startup command given at standstill to synchronization.

After a very serious effort at tuning the regulator, all he was able to produce was similar to Fig. 2.1 (right). His verbal comment was that initially he got an oscillatory response where the decay rate was low. He remembered from linear control theory that the gain should be reduced to improve the decay rate, but all he got was an even slower decay. Fortunately, just then, the senior specialist was available again, and by tuning they obtained a result similar to Fig. 2.2 (left). This was within the specification on startup delay, but still was not as good as previously, Fig. 2.1 (left). The following investigation will show that the remaining overshoot-undershoot sequence on speed is due to windup, and it can be eliminated by more precise antiwindup, as demonstrated by Fig. 2.2 (right).

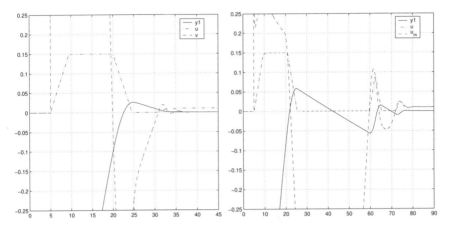

Fig. 2.1. Run-up and control of speed (y_1) on a small Pelton hydropower unit, with the mechanical governor model (left), and with the Electronic Turbine Regulator and the tuning John ended up with (right)

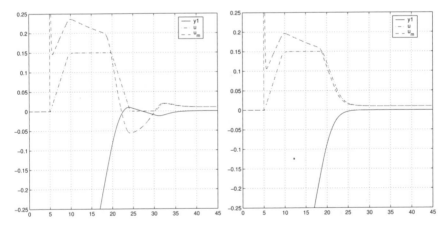

Fig. 2.2. Run-up response of speed (y_1) after tuning by the senior specialist (left), and with an improved antiwindup scheme (right)

2.1.1 Modeling

The plant consists of three subsystems, the turbine wheel and rotor, the waterway with the nozzle, and the servomotor to actuate the nozzle plunger. We shall use simplified models without discussing all details. All variables v are to be in "per units" v/v_R, scaled by their reference ('rated') values v_R.

- The rotor with its inertia Θ and speed ω is subject to a driving torque

$$\frac{T_d}{T_R} = \left(\frac{\omega_{max}}{\omega_R} - \frac{\omega}{\omega_R} \right) \left(\frac{Q}{Q_R} \right); \quad \frac{\omega_{max}}{\omega_R} := 2 \text{ for Pelton turbines} \quad (2.1)$$

with water flow Q. Note that this will lead to a run-away speed of twice nominal and a standstill torque twice the torque at nominal speed.
A small braking torque is due to air and bearing fiction. It is assumed proportional to speed:

$$\frac{T_b}{T_R} = a_f \left(\frac{\omega}{\omega_R} \right) \quad (2.2)$$

and may be observed at coastdown to standstill. A typical value is used: $a_f = 0.01$.
Thus

$$\frac{\Theta \omega_R}{T_R} \frac{d}{dt} \left(\frac{\omega}{\omega_R} \right) = \tau_1 \frac{d}{dt} \left(\frac{\omega}{\omega_R} \right) = \frac{T_d}{T_R} - \frac{T_b}{T_R}$$

$$= \left[2.0 - \left(\frac{\omega}{\omega_R} \right) \right] \left(\frac{Q}{Q_R} \right) - a_f \left(\frac{\omega}{\omega_R} \right) \quad (2.3)$$

which also defines the rotor run-up time coefficient τ_1. Again, a typical value is used: $\tau_1 = 3.0$ s.

- The waterway dynamics are modeled as a rigid mass with inertia, but without elasticity. Linearizing around the steady state operating point $\beta = \bar{A}_n / A_{n_R}$ yields for the flow variation $\delta Q / Q_R$ as a function of nozzle aperture variation $\delta A_n / A_{n_R}$:

$$\frac{1}{2} \beta \tau_w \frac{d}{dt} \left(\frac{\delta Q}{Q_R} \right) = - \left(\frac{\delta Q}{Q_R} \right) + \left(\frac{\delta A_n}{A_{n_R}} \right) \quad (2.4)$$

As typically $\tau_w = 1.0$ s and $\beta = a_f = 0.01$, this leads to a first-order lag with unity gain and a very short time constant compared with the rotor run-up coefficient τ_1. Fig. 2.3 visualizes the model for the two subsystems discussed so far.

- In the servomotor model, in the linear range of operation, the change rate of position h is proportional to oil inflow, which is set proportional to the pilot valve opening v. That is, oil pressure variations due to changing load forces are neglected.

$$\frac{h_R}{v_R} \frac{d}{dt} \left(\frac{h}{h_R} \right) = \tau_2 \frac{d}{dt} \left(\frac{h}{h_R} \right) = \left(\frac{Q_{oil}}{Q_{oil_R}} \right) = \left(\frac{v}{v_R} \right) \quad (2.5)$$

with a typical value of $\tau_2 = 3$ s.

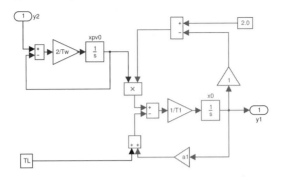

Fig. 2.3. Model of the waterway and rotor dynamics

Two nonlinearities appear in the servomotor for larger input deviations. The first one is the effect of the orifices shown in Fig. 1.3. They limit the oil flow to and from the servomotor and thus the transient water pressure rise. This is a standard safety measure in hydropower plants. The limiting effect is modeled by input saturations,

$$\frac{v_{dn}}{v_R} \leq \frac{v}{v_R} \leq \frac{v_{up}}{v_R} \qquad (2.6)$$

where $\frac{v_{dn}}{v_R} = -0.10$ and $\frac{v_{up}}{v_R} = +0.10$, yielding a travelling time of 30 s for full stroke, which is again typical for such hydropower units.

The second nonlinearity is caused by the mechanical limits imposed on the servomotor position; consult Fig. 1.3. It is modeled by a stiff spring coming into action, if the position exceeds the limit values. To attain equilibrium of forces, the servomotor pressure p_{SM} will rise to the supply pressure p_P, which then reduces the oil inflow to zero for standstill. The relation to use would be $Q_{oil} \sim v\sqrt{\Delta p}$, with $\Delta p = p_P - p_{SM}$. This effect is approximated here by an additive high-gain feedback on the servomotor integrator input, if the servomotor position exceeds the limit values at zero (fully closed position) and at the screw-blocked opening position in Fig. 1.3 for run-up; see Fig. 2.4.

Suppressing all small and fast effects and nonlinearities yields the following approximate model of the dominant plant dynamics, valid within the linear operating range and around steady state at nominal speed:

$$G_{u_d}(s) := \frac{1}{s\tau_1 + a_f} \frac{1}{s\tau_2} \qquad (2.7)$$

This model with $a_f := 0$ shall be used for the design of the feedback parameters next.

2.1.2 The Mechanical Governor

A standard linear PD controller model is used on the plant model, see Fig. 2.4

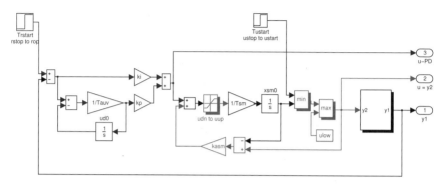

Fig. 2.4. Model of the mechanical governor consisting of a PD controller and the nonlinear servomotor model in the speed control loop with plant model in Fig 2.3

where $e(s) = r(s) - y(s)$

$$G_R(s) = \frac{u(s)}{e(s)} = k_1 + k_2\tau_1\frac{s}{s\tau_v + 1} \quad \text{with} \quad \tau_v = 0.1\tau_1$$

and $G_{R_d}(s) = k_1 + k_2 s\tau_1$ (2.8)

The parameters k_1, k_2 are determined by pole assignment, using the dominant dynamics models G_{u_d} and $G_{R_d}(s)$

$$0 = 1 + G_{u_d}(s)G_{R_d}(s) = 1 + \frac{k_1 + k_2 s\tau_1}{s^2\tau_1\tau_2} \rightarrow s^2 + s\frac{k_2}{\tau_2} + \frac{k_1}{\tau_1\tau_2} \stackrel{!}{=} s^2 + 2\zeta\Omega s + \Omega^2$$

(2.9)

i.e.

$$k_2 = 2\zeta\Omega\tau_2 \quad \text{and} \quad k_1 = \Omega^2\tau_1\tau_2 \tag{2.10}$$

and with $\Omega := 1; \quad 2\zeta = 2.0 \quad \text{then} \quad k_1 = 9; \quad k_2 = 6$

which are typical values for such installations. The simulations with the structure in Fig. 2.4 yield the response shown above in Fig. 2.1

Exercise

- Investigate by simulation the effect of the design parameters Ω and 2ζ.
- And also of the slew and stroke limits on the servomotor.

2.1.3 The Electronic Regulator

The linear controller now consists of a cascade, with a P regulator for servomotor position y_2, and a PI regulator for speed y_1; see Fig. 2.5:

$$u(s) = G_{R_2} e_2(s) = k_2 \left(r_2(s) - y_2(s) \right)$$

$$u_m(s) = r_2(s) = G_{R_1} e_1(s) = \left(k_1 + k_0 \frac{1}{s\tau_1} \right) \left(r_1(s) - y_1(s) \right) \quad (2.11)$$

This provides better disturbance suppression on speed, *i.e.* easier synchronization, than with the mechanical governor.

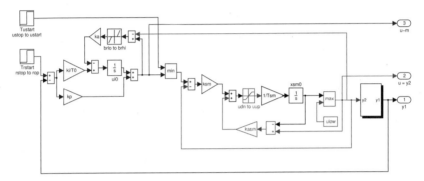

Fig. 2.5. Model of the plant and the electronic turbine regulator ETR

The parameters k_2, k_1, k_0 are again determined by pole assignment

$$0 = 1 + \frac{1}{s\tau_2} k_2 + \frac{1}{s^2 \tau_1 \tau_2} \left(k_1 + k_0 \frac{1}{s\tau_1} \right) k_2 \rightarrow s^3 + s^2 \frac{k_2}{\tau_2} + s \frac{k_1}{\tau_1 \tau_2} + \frac{k_0}{\tau_1^2 \tau_2}$$

$$\overset{!}{=} \left(s^2 + 2\zeta\Omega s + \Omega^2 \right) (s + \Omega) \quad (2.12)$$

yields

$$k_2 = (1 + 2D)\,\Omega\tau_2$$

$$k_1 k_2 = (1 + 2\zeta)\,\Omega^2 \tau_1 \tau_2; \qquad \rightarrow \qquad k_1 = \Omega\tau_1$$

$$k_0 k_2 = \Omega^3 \tau_1^2 \tau_2; \qquad \rightarrow \qquad k_0 = \frac{1}{(1 + 2\zeta)} \Omega^2 \tau_1 \tau_2 \quad (2.13)$$

What John ended up with was equivalent to

$$k_2 = 6.0; \quad k_1 = 3.0; \quad k_0 = 4.5$$

which can be generated from the pole assignment by using $\Omega = 1.0$ as above and $2\zeta = 1$. These parameter values agree fairly well with the usual ones for such plants.

The next step is to model the antiwindup scheme implemented in this particular electronic turbine regulator ETR. It basically consists of a diode feedback as in Fig. 1.6, where the actual position of the servomotor is used for the constraining input $-U_{high}$. The diode breakpoint lets the regulator output stabilize at $+0.05$ above the actual y_2, $i.e.$ there will be a (small) windup, see $u(t)$ and $u_m(t)$ in Fig. 2.1 (right), up to $t \approx 20$ s.

What John was not aware of, but what was well known to the senior specialist, is that there was no such diode feedback in the downward direction, $i.e.$ for the full servomotor closure. This was simply not installed, as with well tuned regulators and turbines with a usual speed-no-load opening, this low limit was never run into for any extended time span, and so "windup" had never been a serious problem along the low limit.

But in John's case it did: a severe mismatch of $u(t)$ and $u_m(t)$ ($i.e.$ "windup") is visible in Fig. 2.1 (right), for $t > 20$ s. After this has died out, for $t > 60$ s and the lower limit is not run into anymore, then servomotor position and speed behave well damped.

The re-tuning by the senior specialist, as shown in Fig. 2.2 (left), was equivalent to

$$k_2 = 8.1; \quad k_1 = 2.7; \quad k_0 = 2.43$$

which can be generated by inserting $\Omega = 0.9$ and $2\zeta = 2.0$. In other words, he increased the closed loop damping 2ζ substantially, and he also decreased the closed-loop bandwidth Ω slightly. Thus, he managed to stay below the specified 30 s without compromising linear closed-loop dynamics unduly. The result is quite sensitive to small parameter changes, as simulations will show.

Exercise
Simulation results are not given here, but you may use this as an exercise.

The next step would be to complete the antiwindup feedback for the fully closed limit, and also to improve it by reducing the diode offsets of 5% from above. In Fig. 2.2 (right), this has been set to 1%. The antiwindup gain has been kept as before at 10.0. This allows to augment Ω to its previous value of 3.0 again, $i.e.$

$$k_2 = 9; \quad k_1 = 3; \quad k_0 = 3$$

and even further, without deteriorating effects on the response. It is now much more robust to small parameter changes. The specified delay of 30 s is now met with a substantial margin.

Exercise
Again investigate this by simulation.

2.1.4 Summary

This section and the example in Sect. 1.1 have shown that:

-   Startup response was well behaved for the PD governor using the servo-motor as an integrator, if both slew saturations (by the orifices in the oil flow) and stroke saturations (by the mechanical end stops) were run into.
-   However, startup response deteriorated substantially if a linear PI controller was used on the servomotor with stroke constraints but no slew constraints; see Fig. 1.1.
-   This was eliminated by using a standard antiwindup in Fig. 1.1.
-   Here, startup transients have again deteriorated with the PI-P controller configuration and incomplete and not sufficiently precise antiwindup.
-   This can be partly compensated by careful tuning, by increasing proportional gain and decreasing integral gain. However, such tuning is very time consuming and the result is very sensitive to small changes.
-   Using a correct antiwindup feedback has given a substantial increase in performance without additional tuning effort and the result is less sensitive.

Note that these findings are related to the specific experiments performed here, and should not be generalized without caution.

2.2 Problem Statement and Test Cases

We focus in the standard input saturation case.

The following set of specifications is used for the control system under study; see Fig. 2.6. It also defines a benchmark for testing controllers with alternate antiwindup forms.

(a) The **plant** is given by its transfer function

$$G(s) = \frac{y(s)}{u(s)} = e^{-Ds}\frac{b}{sT+a} \tag{2.14}$$

This corresponds to the standardized plant model used by the Ziegler–Nichols and Chien–Hrones–Reswick design methods [2], and covers a large percentage of industrial control cases, such as speed, level, pressure, concentration and temperature control. The small delay D shall represent the fast non-modeled dynamics of actuator, process and sensor, and will limit the attainable closed-loop bandwidth to a realistic value.

An additional load input $v(t)$ shall be present as follows:

$$G_v(s) = \frac{y(s)}{v(s)} = -\frac{b}{sT+a} \tag{2.15}$$

Numerical values to be used in the test cases are:

$$b := 1.0; \quad T := 1.0; \quad D := 0.025; \quad a := 1.0 \ \text{or} \ := 0.0 \qquad (2.16)$$

which covers both standard cases "unity gain first-order lag" and "open integrator".

Neglecting the small delay yields the "dominant first-order" plant model

$$G_d(s) = \frac{y(s)}{u(s)} = \frac{b}{sT + a} \qquad (2.17)$$

to be used in the pole assignment design.

(b) As **controllers**, both P and PI(aw) types shall be considered.

The *P controller*

$$u(s) = k_p e(s) + u_{man} \qquad (2.18)$$

usually has an additional offset input u_{man}. Here it is set to zero.

The *PI controller* is in the standard continuous form used in the Ziegler–Nichols rules [2]:

$$R(s) = k_p \left(1 + \frac{1}{sT_i} \right) \qquad (2.19)$$

The integral action $1/sT_i$ often is called "reset action" for historical reasons. For the discrete-time form the continuous integration is replaced by its discrete forward and backward Euler integration equivalent.

$$R(z) = k_p \left(1 + \frac{1}{T_i} T_s \frac{z^{-1}}{1 - z^{-1}} \right) \quad \text{and} \quad R(z) = k_p \left(1 + \frac{1}{T_i} T_s \frac{1}{1 - z^{-1}} \right)$$
$$(2.20)$$

As in almost all applications today, T_s is short compared with the dominant closed-loop time constants. This will not generate a substantial difference in loop response, but the distinction will be relevant in the antiwindup feedback context.

Finally, a one sampling delay z^{-1} is introduced after $R(z)$. In other words, the output values computed from the current input sample will be delivered to the process interface only at the next sampling instant. This covers the finite computation time required by the algorithm, which therefore must be less than T_s. For the test cases, the sampling time T_s is set to

$$T_s = 0.010 \ \text{s} \qquad (2.21)$$

(c) The **PI controller settings** k_p and T_i are obtained by pole assignment using the "dominant first-order' plant dynamics $G_d(s)$ in series to the usual zero-order hold, *i.e.*

$$G_d(z) = b\frac{z^{-1}}{\frac{1-z^{-1}}{T_s}T + az^{-1}} \qquad (2.22)$$

For brevity, we shall directly revert to the equivalent continuous form by substituting

$$\frac{1 - z^{-1}}{T_s} \to s \quad \text{and} \quad z^{-1} \to 1 \qquad (2.23)$$

This implies that the sampling time T_s is to be short compared with the closed-loop dominant time constants (see below).

The closed-loop characteristic equation

$$0 = (sT_i)(sT + a) + k_p b(sT_i + 1) = s^2 + s\frac{a + k_p b}{T} + \frac{k_p b}{T_i T} \overset{!}{=} (s + \Omega)^2 \qquad (2.24)$$

yields

$$k_p b = 2\Omega T - a; \quad \frac{1}{T_i} = \frac{\Omega^2 T}{k_p b}; \quad \text{and for } a = 0: \quad k_p b = 2\Omega T; \quad \frac{1}{T_i} = \frac{\Omega}{2} \qquad (2.25)$$

Setting the closed-loop bandwidth to $\Omega := 5$ rad/s, *i.e.* the dominant closed-loop time constants to $1/\Omega = 0.20$ s, provides a sufficient margin with respect to the sampling time $T_s = 0.010$ s.

Then, for $b = 1$ and $a = 0$: $k_p = 10$; and $\frac{1}{T_i} = 2.5$; *i.e.* $T_i = 0.40$ s

Similarly for the *P controller*

$$k_p b = \Omega T - a \qquad (2.26)$$

and setting the closed-loop bandwidth here to $\Omega = 10$ rad/s yields the same k_p values as for the PI controller.

(d) The following **test sequence** shall be applied to the closed loop in Fig. 2.6, where r is the setpoint and v is the load acting additively to u at the plant input:

-   Initially (index 0) the loop is to be at standstill conditions:
    $r_0 = 0$; and $v_0 = 0$; *i.e.* $\bar{y}_0 = 0$; and $\bar{u}_0 = 0$.
-   At time T_1, a setpoint step to $r_1 = 0.95$ is applied while there is still no load, *i.e.* $v_1 = 0$.
-   Then at time T_2, a setpoint step to $r_2 = 1.0$ is applied while there is still no load, *i.e.* $v_2 = 0$. This will show the small signal (linear) closed-loop response. At equilibrium, a plant input $\bar{u}_2 = (a/b)r_2$ results.
-   At time T_3 a load step $v_3 = 0.90$ is applied, while the setpoint is constant, $r_3 = r_2 = 1$.
-   Finally, at time T_4, a full load reversal is applied $v_4 = -v_3$, again with constant setpoint $r_4 = r_3 = 1$.

No high-frequency measurement disturbance [52] shall be considered.

(f) The **actuator saturation** is specified as follows:

$$u_{low} = -1.0 + \bar{u}_2; \quad u_{high} = +1.0 + \bar{u}_2; \tag{2.27}$$

With $\bar{u}_2$ specified as above, this provides the same "margin to maneuver" for setpoint steps around the speed-no-load condition for both values of a introduced above.

Also, $v_3 = +0.90$ yields a realistic steady state control margin of $\Delta u = 0.10$ for the load step conditions.

Fig. 2.6 visualizes the control loop specified above.

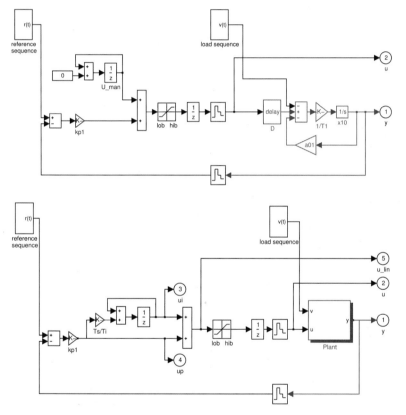

Fig. 2.6. The control loop with P controller (top) and the linear Ziegler–Nichols type PI controller (bottom)

Note that both the Ziegler–Nichols and the Chien–Hrones–Reswick tuning rules would also allow a *derivative* action. In most practical cases, however, the derivative action is set to zero at commissioning. This is to avoid excessive high-frequency movement of the control signal, and thus to reduce actuator wear, noise, *etc.*, if high-frequency measurement disturbances are present. Therefore, the derivative action is set to zero here as well. We shall come back to this later.

2.3 The Reset Windup Effect

The transient response of the system in Fig. 2.6 to the test sequence is shown in Figs. 2.7, 2.8 and 2.9.

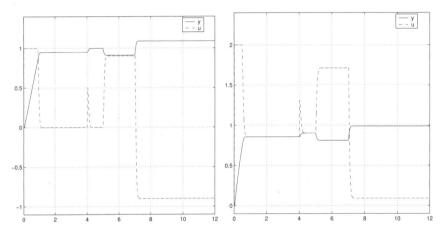

Fig. 2.7. Transient response to the test sequence with P controller for $a = 0$ (left) and $a = 1$ (right), with $u_{man} = 0$

The *P controller* in Fig. 2.7 performs as expected. There is no large overshoot on y, but there is a nonzero steady state control error $\bar{e}$. To suppress this, an integral action (reset action) is inserted.

The *PI controller* in Fig. 2.8 (left) produces an unexpected, large transient overshoot on y in the run-up phase, during which u_{lin} runs into its upper saturation. But $y(t)$ evolves as expected for the small setpoint step, where u_{lin} stays below the saturation. Such overshooting of $y(t)$ also appears in the load swing phases, although it is much smaller.

By inspection of the internal controller signals $u_i(t)$ and $u_{lin}(t)$ in Fig. 2.8 (right), the overshoot is caused by an excessive buildup of the integral action $u_i(t)$ whenever $e(t)$ is large and $u_{lin}(t)$ exceeds its saturation values. This buildup cannot be reduced in time before $e(t)$ approaches its zero crossover, and consequently $y(t)$ will overshoot. This effect is known as **"reset windup"**.

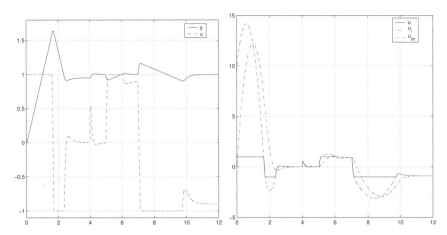

Fig. 2.8. Transient response to the test sequence with PI controller, for $a = 0$

Finally, the case of $a = 0$ seems to be more sensitive to this effect than for $a = 1$; see Fig. 2.9.

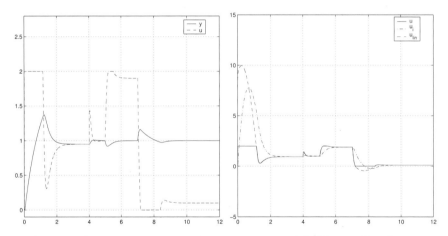

Fig. 2.9. Transient response to the test sequence with PI controller, for $a = 1$

2.4 Antiwindup Structures

Many stratagems have been and are still being invented to alleviate this effect, both for continuous-time and discrete-time forms.

Here, we shall focus on forms where the antiwindup is based on **feedback**, such as shown in Fig. 2.10. Other concepts have been proposed, but these will not be investigated here.

A key issue for design engineers, and the reason for much controversy, is comparative properties of different antiwindup forms, *i.e.* whether they perform identically or not. We shall say that different antiwindup controller structures are **equivalent**, if they produce the same output signal $u(t)$ from the same input signal $e(t) = r(t) - y(t)$. This implies proper initialization of shift elements or integrators. In other words, equivalent structures will produce the same closed-loop response, if all other experimental conditions are the same. Conversely, it is not possible to determine from observing input and output alone which particular form of the set of equivalent forms has been used within. In other words, one is at liberty to choose from the set of equivalent forms by applying other additional design criteria. So, *equivalence* shall be a main topic in the following discussion.

We shall present a generic structure first, from which most of the existing forms with antiwindup feedback can be instantiated. We mainly look at the discrete-time versions.

2.4.1 The Generic Antiwindup Feedback Structure

Here, the focus is on solutions related to the basic antiwindup feedback (awf) structure Fig. 2.10, where R_2 contains the integral action in PI control, or more generally speaking, those parts of the controller where internal states may "run away", *i.e.* "wind up".

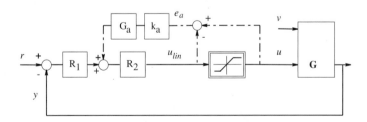

Fig. 2.10. The generic awf controller structure

The basic idea is to monitor the difference e_a of the input and output signal to the saturation block. It is zero within the linear operating range, and nonzero outside, where the integral action (or other such states in the controller) would "wind up". Then, feedback of $e_a = u - u_{lin}$ is used to keep the integral action (or such other states) from running too far off. For more transparency, the transfer function in the awf path is split into the gain k_a and a filter G_a with unity gain.

There are *three design elements* to be selected within the awf path: the gain k_a, the filter dynamics G_a, and the splitting of R into $R_1 R_2$, *i.e.* the location of the summing point for the awf.

G_a is an additional design element to previous proposals for a generic structure [38]. Its usefulness will become clearer later.

By appropriate choice of the three design elements, the most current awf versions in practice can be generated.

Note that Fig. 2.10 also directly generates both the discrete-time and continuous-time implementation by using either z^{-1} or s.

From the generic form in Fig. 2.10, the discrete-time form **A** of the PI controller follows directly from $R = R_1 R_2$, where R is as defined in the previous section:

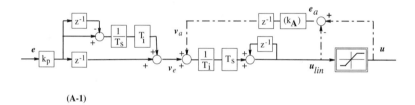

(A-1)

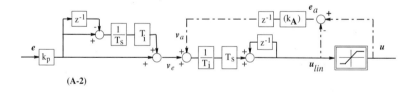

(A-2)

Fig. 2.11. Forms **A** of the PI(aw) controller derived from the generic structure

Form **A-1** uses the forward Euler integration

$$R_1(z) = k_p \left(\frac{1 - z^{-1}}{T_s} T_i + z^{-1} \right) \tag{2.28}$$

and Form **A-2** the backward Euler integration

$$R_1(z) = k_p \left(\frac{1 - z^{-1}}{T_s} T_i + 1 \right) \tag{2.29}$$

and for both

$$R_2(z) = \frac{T_s}{T_i} \frac{1}{1 - z^{-1}} \quad \text{and} \quad G_a = z^{-1} \tag{2.30}$$

Finally, the value of k_A is an open parameter, to be selected in the design process.

In the feedback structure in Fig. 2.11, the awf signal $v_a = k_A G_a e_a$ calculated at each sampling instant may be used only at the *next* sampling instant for addition to the output of R_1. Otherwise an arithmetic loop would result in the controller. So there must be at least one z^{-1} around the awf loop.

The continuous form implementation is

$$R_1(s) = k_p (sT_i + 1); \quad R_2(s) = \frac{1}{sT_i}; \quad G_a = 1.0; \quad \text{and design parameter } k_A$$

Dynamic properties of the awf loop

The output variable of interest is $e_a = -u_{lin} + u$ as a measure of windup of u_{lin} outside u, and the input is e. Then, for the forward Euler case from Fig. 2.11

$$e_a = -u_{lin} + u$$
$$v_e = ek_p \left(\frac{1 - z^{-1}}{T_s} T_i + z^{-1} \right)$$
$$v_a = e_a k_A z^{-1}$$
$$u_{lin} = (v_e + v_a) \frac{T_s}{T_i} \frac{1}{1 - z^{-1}} \tag{2.31}$$

i.e.

$$e_a \left(1 + k_A z^{-1} \frac{T_s}{T_i} \frac{1}{1 - z^{-1}} \right) = u - ek_p \left(1 + \frac{T_s}{T_i} \frac{z^{-1}}{1 - z^{-1}} \right) \tag{2.32}$$

and similarly for the backward Euler case

$$e_a \left(1 + k_A z^{-1} \frac{T_s}{T_i} \frac{1}{1 - z^{-1}} \right) = u - ek_p \left(1 + \frac{T_s}{T_i} \frac{1}{1 - z^{-1}} \right) \tag{2.33}$$

Abbreviating $k'_A := k_A \frac{T_s}{T_i}$, the left side is rewritten as

$$e_a \left(1 + k'_A \frac{z^{-1}}{1 - z^{-1}} \right) = e_a k'_A \frac{1}{1 - z^{-1}} \left[\left(\frac{1 - k'_A}{k'_A} T_s \right) \frac{1 - z^{-1}}{T_s} + 1 \right]$$
$$\text{with} \quad \tau_{A_d} := \frac{1 - k'_A}{k'_A} T_s \tag{2.34}$$

and inserting this into Eqs. 2.32 and 2.33

$$e_a = \frac{1}{k_A} \frac{1}{\tau_{A_d} \frac{1-z^{-1}}{T_s} + 1} \left[u \, T_i \frac{1-z^{-1}}{T_s} - ek_p \left(T_i \frac{1-z^{-1}}{T_s} + g \right) \right] \quad (2.35)$$

where $g = z^{-1}$ for the forward and $g = 1$ for the backward case. In other words the awf loop will be of first order with a time constant τ_{A_d}.
If $k'_A := 1/(n+1)$, then $\tau_{A_d} = nT_s$ is a multiple of the sampling time. And if $n = 0$, then $k'_A := 1$. This amounts to the deadbeat response.

For the *continuous case, i.e.* $T_i \gg T_s$, read directly from Eq. 2.35:

$$e_a = \frac{1}{k_A} \frac{1}{s \frac{T_i}{k_A} + 1} \left[u \, sT_i - ek_p \left(sT_i + 1 \right) \right] \quad (2.36)$$

Later we will need the input e_c to the integral action with the awf loop being closed:

$$e_c = ek_p \left(sT_i + 1 \right) \frac{1}{1 + k_A \frac{1}{sT_i}} + u \frac{k_A}{1 + k_A \frac{1}{sT_i}}$$

$$= ek_p \left(sT_i + 1 \right) \frac{1}{k_A} \frac{sT_i}{s \frac{T_i}{k_A} + 1} + u \frac{sT_i}{s \frac{T_i}{k_A} + 1} \quad (2.37)$$

and also the internal state variable u_i

$$u_i = \frac{1}{sT_i} e_c = ek_p \left(sT_i + 1 \right) \frac{1}{k_A} \frac{1}{s \frac{T_i}{k_A} + 1} + u \frac{1}{s \frac{T_i}{k_A} + 1}$$

$$= \frac{1}{s \frac{T_i}{k_A} + 1} \left[u + ek_p \frac{1}{k_A} \left(1 + sT_i \right) \right] \quad (2.38)$$

There is an equilibration transient within the controller along the saturation with time constant $\tau_{A_c} = \frac{T_i}{k_A}$.

Obviously, the tracking is ideal if $k_A \to \infty$. Then $e_a \to 0$ and the bandwidth of the awf loop $k_A/T_i \to \infty$. However, this is feasible only in the continuous case. In the discrete time forms, there is a limit on k_A from the stability of the awf loop. Its characteristic equation is from Fig. 2.11

$$0 = 1 + k_A z^{-1} \frac{T_s}{T_i} \frac{1}{1 - z^{-1}} = 1 + k_A \frac{T_s}{T_i} \frac{1}{z - 1} = (z - 1) + k_A \frac{T_s}{T_i} \quad (2.39)$$

Using the bilinear transformation $z = \frac{1+w}{1-w}$ yields for w

$$0 = 1 + w - 1 \left(1 - w \right) + k_A \frac{T_s}{T_i} \left(1 - w \right)$$

$$= \left(2 - k_A \frac{T_s}{T_i} \right) w + k_A \frac{T_s}{T_i}$$

$$0 = w + \frac{k_A \frac{T_s}{T_i}}{2 - k_A \frac{T_s}{T_i}} \quad (2.40)$$

Thus for stability

$$0 < k_A \frac{T_s}{T_i} \le 2 \tag{2.41}$$

and for the special case

$$k_A \frac{T_s}{T_i} \overset{!}{=} 1 \quad \to \quad 0 = w + 1 \quad i.e. \quad z = 0 \tag{2.42}$$

there appears a pole at $z = 0$, *i.e.* the deadbeat response.

We shall now look into different forms of PI controllers with such awf. As suggested by Hanus [32], they are put into three groups, *i.e.* "output conditioning", "input conditioning" and "self conditioning". Note that input and output here refer to the controller, and not to the plant, as is now used in most publications. In order to avoid such confusion, we shall use

- "input" and "output" as referring to the plant,
- "control conditioning" instead of "output conditioning", and
- "reference conditioning" instead of "input conditioning".

2.4.2 Control Conditioning

The basic idea of control conditioning is to use e_a to "reset" the integrator such that the controller output $u_{lin}(t)$ stays close to the saturation output u. In other words the control variable $u_{lin}(t)$ is conditioned quite directly. There are several possibilities to implement this.

Forms B

To obtain them, the summing point is moved further downstream across the factor T_s/T_i, see Fig. 2.12 B-1 (a), for the forward Euler case.

By inspection, this structure is equivalent to structure **A-1**, if and only if (iff)

$$k_B \overset{!}{=} k_A \frac{T_s}{T_i} \tag{2.43}$$

Note that for structure **A** the relevant factor for stability of the discrete awf loop was found to be $k_A \frac{T_s}{T_i}$, *i.e.* in fact k_B.

For implementation, the division by the (small) T_s/T_i in the derivative part of R_1 and the subsequent multiplication in the integral part R_2 amounts to poor scaling of internal variables. This can be avoided by moving the factor T_s/T_i into R_1, as shown in Fig. 2.12 (b).

An important special case is shown in Fig. 2.12 (c): for $k_B := 1$ the positive and negative feedback paths from u_{lin} through z^{-1} cancel, and only the positive feedback from u through z^{-1} remains. This produces a particularly simple structure. And as with the analysis of structure **A**, this amounts to the deadbeat response situation.

Finally, note that this structure **B** does not lend itself directly to a continuous form. For this one must go back to structure **A**.

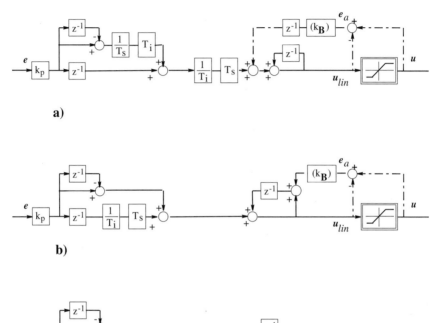

Fig. 2.12. Form **B** of the PI(aw) controller derived from Fig. 2.11:
(a) by shifting the awf summation point;
(b) by subsequent re-scaling of internal variables;
(c) and for the special case of $k_B = 1$

Forms C

The structure in Fig. 2.12 (b) may be described as a PD part for R_1 in series with an I part in backward Euler form with awf. This suggests using a forward Euler integrator as an alternative, Fig. 2.13 (a) and further to (b).

For $k_C := 1$ this may be reduced to Fig. 2.13 (c). Finally, by moving the delay element downstream of the saturation and the branch point produces Fig. 2.13 (d). And this corresponds to Fig. 2.12 (c) with an additional delay on u. This also holds for the general awf case

$$0 \leq k_C \stackrel{!}{=} k_B \leq 1.0 \tag{2.44}$$

In other words the forms **B** and **C** are not equivalent in the strict sense, but very nearly so. The only difference is the one sample delay to be appended to R_B to obtain R_C. It does not influence the awf loop dynamics, but will have a (small) effect on the main loop dynamics.

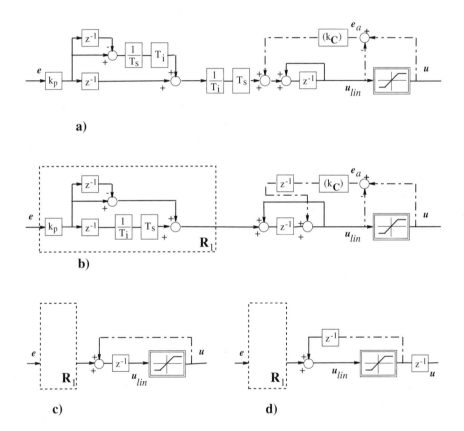

Fig. 2.13. Form **C** of the PI(aw) controller derived from Fig. 2.12 (b)
(a) by shifting the awf summation point,
(b) by subsequent re-scaling of internal variables,
(c) for the special case of $k_C = 1$,
(d) and shifting the delay element

Forms D

Forms **A**, **B**, and **C** are very convenient for theoretical analysis, and shall be used as such later. But, from an implementation point of view, the time derivative should be avoided, because it amplifies any high frequency disturbances. This drawback is remedied by applying the rules for linear block diagram transformations. The summation point of the derivative and proportional path in R_1 is shifted downstream of the integrator in R_2. This yields forms **D-1** with the forward Euler integration and **D-2** with the backward one; Fig. 2.14.

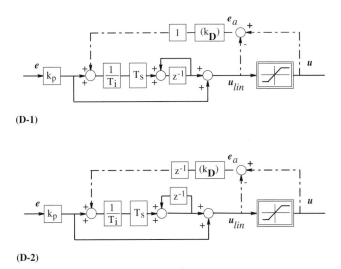

(D-1)

(D-2)

Fig. 2.14. PI(aw) controller derived from Fig.2.10, which avoids the time derivative: **D-1** with the forward Euler integration (top); **D-2** with the backward one (bottom)

Equivalence shall be analyzed as above using the awf tracking response. From Fig. 2.14 (a):

$$e_a = -u_{lin} + u$$

$$u_{lin} = ek_p + (ek_p + e_ak_D)\frac{T_s}{T_i}\frac{z^{-1}}{1-z^{-1}} \tag{2.45}$$

i.e.

$$e_a\left(1 + k_D\frac{T_s}{T_i}\frac{z^{-1}}{1-z^{-1}}\right) = u - ek_p\left(1 + \frac{T_s}{T_i}\frac{z^{-1}}{1-z^{-1}}\right) \tag{2.46}$$

which is identical to Eq. 2.32 for form **A**, if

$$k_D \overset{!}{=} k_A \tag{2.47}$$

i.e. for the discrete awf loop

stability range: $0 < k_D \dfrac{T_s}{T_i} \le 2;$ and deadbeat response for: $k_D \dfrac{T_s}{T_i} = 1$

$$(2.48)$$

Similarly, for the backward Euler case from Fig. 2.14 (bottom):

$$e_a \left(1 + k_D \frac{T_s}{T_i} \frac{z^{-1}}{1 - z^{-1}} \right) = u - e k_p \left(1 + \frac{T_s}{T_i} \frac{z^{-1}}{1 - z^{-1}} \right) \qquad (2.49)$$

which is identical to Eq. 2.35 for form **A**, leading to the same conditions for k_D and to the same awf properties.

Exercise: Connect form **D** to form **B**(b) and investigate the equivalence.

Form E

A very common version (form **E**) is to place the gain k_p in the linear control algorithm downstream of the part $1 + \frac{1}{sT_i}$, whereas it has always been put upstream so far. Then k_p appears in the awf loop.
Using the forward Euler integration:

$$e_a = -u_{lin} + u$$
$$u_{lin} = (e + q)\, k_p$$
$$q = (e + k_E e_a) \frac{T_s}{T_i} \frac{z^{-1}}{1 - z^{-1}} \qquad (2.50)$$

that is

$$e_a \left(1 + (k_p k_E) \frac{T_s}{T_i} \frac{z^{-1}}{1 - z^{-1}} \right) = u - e k_p \left(1 + \frac{T_s}{T_i} \frac{z^{-1}}{1 - z^{-1}} \right) \qquad (2.51)$$

which is identical to Eq. 2.46 for form **D-1**, if

$$(k_p k_E) \overset{!}{=} k_D \qquad (2.52)$$

And correspondingly for the backward Euler case.

2.4.3 Reference Conditioning

So far, the awf idea has been to act on the integral action, which has led to control conditioning. Here the basic idea is to act on the *input* ($e(t)$) of the PI controller, such that $u_{lin}(t)$ does not "wind up". The main consequence is that both the proportional and the integral path inputs are modified by the awf, and not only the integral part as with the "control conditioning".
There are several implementation alternatives, depending on the splitting up of R into R_1 and R_2. The forward Euler algorithm is used.

For Form **F** the awf is added downstream of the gain k_p, *i.e.*

$$R_1 = k_p; \quad R_2 = 1 + \frac{T_s}{T_i} \frac{z^{-1}}{1 - z^{-1}} \tag{2.53}$$

In Form **G** it is moved to the control error $e(t)$,

$$R_1 = 1; \quad R_2 = k_p \left(1 + \frac{T_s}{T_i} \frac{z^{-1}}{1 - z^{-1}} \right) \tag{2.54}$$

The summing point of the awf may be moved even further onto the setpoint $r(t)$, Form **H**. This leads to the concept of the "realizable reference" proposed by Hanus [32].

In any case, the awf loop now contains the proportional action path as well, which will significantly influence its dynamic properties. We shall see that for stability the value of awf gain is restricted to a much lower value, *i.e.* there will be more windup. Intuitively, an additional low-pass filter G_a inserted in the awf path will allow one to increase the awf gain substantially. We shall use a unity gain first-order filter here, with its time constant T_a as an additional design parameter.

$$G_a(z) := \frac{z^{-1}}{\left(\frac{1-z^{-1}}{T_s} \right) T_a + z^{-1}} \tag{2.55}$$

The reasons behind this will become clearer in the following.

Form F

The signal flow diagram is shown in Fig. 2.15, using the forward Euler integration:

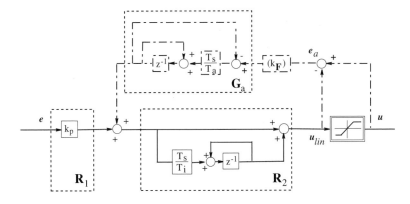

Fig. 2.15. PI(aw) controller with reference conditioning (Form **F**).

Stability of the discrete awf loop

From Fig. 2.15 the characteristic equation is

$$0 = 1 + k_F G_a R_2 = 1 + k_F \frac{z^{-1}}{\left(\frac{1-z^{-1}}{T_s}\right) T_a + z^{-1}} \left(1 + \frac{T_s}{T_i} \frac{z^{-1}}{1 - z^{-1}}\right)$$

$$= 1 + k_F \frac{1}{\left(\frac{z-1}{T_s}\right) T_a + 1} \left(1 + \frac{T_s}{T_i} \frac{1}{z - 1}\right)$$

$$0 = (z-1)^2 + (z-1) \frac{T_s}{T_a} (1 + k_F) + k_F \frac{T_s}{T_a} \frac{T_s}{T_i}$$

$$= z^2 + (a_1 - 2) z + (1 - a_1 + a_0)$$

$$= z^2 + b_1 z + b_0 \qquad (2.56)$$

where for short

$$a_1 = \frac{T_s}{T_a} (1 + k_F) \; ; \quad a_0 = k_F \frac{T_s}{T_a} \frac{T_s}{T_i}$$

$$\text{and} \quad b_1 = a_1 - 2 \; ; \qquad b_0 = 1 - a_1 + a_0$$

Then using the bilinear transformation

$$z = \frac{1 + w}{1 - w}$$

yields as characteristic equation of the transformed continuous system

$$0 = (1 + w)^2 + b_1 (1 + w) (1 - w) + b_0 (1 - w)^2$$

$$0 = w^2 + w \frac{2 (1 - b_0)}{1 - b_1 + b_0} + \frac{1 + b_1 + b_0}{1 - b_1 + b_0} \qquad (2.57)$$

Then

$$0 < (1 + b_1 + b_0) = 1 + (2 - a_1) + (1 - a_1 + a_0) = 4 - 2a_1 + a_0$$

$$= 4 - 2 \frac{T_s}{T_a} (1 + k_F) + k_F \frac{T_s}{T_a} \frac{T_s}{T_i} = 4 \frac{T_a}{T_s} - 2 - k_F \frac{T_s}{T_i} \left(2 \frac{T_i}{T_s} - 1\right)$$

$$\rightarrow \qquad k_F \frac{T_s}{T_i} < 2 \frac{2 \left(\frac{T_a}{T_s}\right) - 1}{2 \left(\frac{T_i}{T_s}\right) - 1} \qquad \text{for the upper bound} \qquad (2.58)$$

and also

$$0 < (1 - b_1 + b_0) = 1 - (2 - a_1) + (1 - a_1 + a_0) = a_0 = k_F \frac{T_s}{T_a} \frac{T_s}{T_i}$$

$$\rightarrow \qquad 0 < k_F \frac{T_s}{T_i} \qquad \text{for the lower bound} \qquad (2.59)$$

There are four interesting cases:

- Let $T_a := T_s$.

Then $\quad G_a(z) = \dfrac{z^{-1}}{(1 - z^{-1})\, 1 + z^{-1}} = z^{-1}$

and $\quad k_F \dfrac{T_s}{T_i} < 2\dfrac{2 \cdot 1 - 1}{2\left(\frac{T_i}{T_s}\right) - 1} = \dfrac{2}{2\left(\frac{T_i}{T_s}\right) - 1} \ll 1.0$

as always $\quad T_i \gg T_s$ (2.60)

- Let $T_a := nT_s$

then $\quad k_F \dfrac{T_s}{T_i}\bigg|_{n>1} = (2n - 1)\, k_F \dfrac{T_s}{T_i}\bigg|_{n=1}$ (2.61)

- Let $T_a := \frac{1}{m}T_i$

then $\quad k_F \dfrac{T_s}{T_i}\bigg|_{m>1} = \left(2\dfrac{1}{m} - 1\right) k_F \dfrac{T_s}{T_i}\bigg|_{m=1}$ (2.62)

- Finally let $T_a := T_i$

then $\quad k_F \dfrac{T_s}{T_i} < 2.0$ (2.63)

This is the same condition as for Form **A**, and hints at possible equivalence.

In other words, the first case $T_a := T_s$ leads to the smallest awf gain k_F, and it increases with increasing T_a, until attaining at $T_a := T_i$ the same level as for the control conditioning forms.

Intuitively, increasing T_a will make the awf loop slower, *i.e.* there will be an increasing transient windup of u_{lin}. It is associated with a decreasing permanent one, as the awf gain can be increased. At this point it is not evident how T_a should be selected with respect to T_i. This shall be answered later.

The main awf stability result from Equation 2.58 can be rewritten as

$$k_F \frac{T_s}{T_a} < 2\, \frac{1 - \frac{1}{2}\frac{T_s}{T_a}}{1 - \frac{1}{2}\frac{T_s}{T_i}}$$ (2.64)

and for $T_a, T_i \gg T_s$

$$k_F \frac{T_s}{T_a} \lesssim 2; \qquad (\text{should not be used for} \quad T_a, T_i \lesssim 10T_s)$$ (2.65)

The same approximate result can be obtained directly from Fig. 2.15 by
- omitting the integral action part in R_2, *i.e.* $R_2 := 1$
- and omitting also in G_a the feedback path around the discrete integrator. This simplified system has the characteristic equation

$$0 = 1 + k_F \frac{T_s}{T_a} \frac{z^{-1}}{1 - z^{-1}} \quad \rightarrow \quad 0 = z - 1 + k_F \frac{T_s}{T_a} = w + \frac{k_F \frac{T_s}{T_a}}{2 - k_F \frac{T_s}{T_a}} \tag{2.66}$$

where the bilinear transformation has been applied, and where $2 - k_F \frac{T_s}{T_a} > 0$ for stability, q.e.d.

Transient response of the awf loop

Consider again the forward Euler algorithm. From Fig. 2.15

$$e_a = -u_{lin} + u \quad \rightarrow$$

$$u_{lin} = \left(ek_p + e_a k_F \frac{z^{-1}}{\frac{1-z^{-1}}{T_s} T_a + z^{-1}} \right) \left(1 + \frac{T_s}{T_i} \frac{z^{-1}}{1 - z^{-1}} \right)$$

$$= ek_p \left(1 + \frac{T_s}{T_i} \frac{z^{-1}}{1 - z^{-1}} \right) + e_a k_F \frac{T_s}{T_i} \frac{z^{-1}}{1 - z^{-1}} \frac{\frac{1-z^{-1}}{T_s} T_i + z^{-1}}{\frac{1-z^{-1}}{T_s} T_a + z^{-1}} \tag{2.67}$$

that is

$$e_a \left(1 + k_F \frac{T_s}{T_i} \frac{z^{-1}}{1 - z^{-1}} \frac{\frac{1-z^{-1}}{T_s} T_i + z^{-1}}{\frac{1-z^{-1}}{T_s} T_a + z^{-1}} \right) = u - ek_p \left(1 + \frac{T_s}{T_i} \frac{z^{-1}}{1 - z^{-1}} \right) \tag{2.68}$$

For the special case $T_a := T_i$ the numerator and denominator polynomials on the left side cancel and

$$e_a \left(1 + k_F \frac{T_s}{T_i} \frac{z^{-1}}{1 - z^{-1}} \right) = u - ek_p \left(1 + \frac{T_s}{T_i} \frac{z^{-1}}{1 - z^{-1}} \right) \tag{2.69}$$

and this is indeed equivalent to form **A**, if

$$k_F \overset{!}{=} k_A \tag{2.70}$$

Similarly, the other special case $T_a := T_s$ yields

$$e_a \left[1 + k_F \left(1 + \frac{T_s}{T_i} \frac{z^{-1}}{1 - z^{-1}} \right) \right] = u - ek_p \left(1 + \frac{T_s}{T_i} \frac{z^{-1}}{1 - z^{-1}} \right) \tag{2.71}$$

If the backward Euler algorithm is used then

$$u - e_a = u_{lin} = \left(ek_p + e_a k_F \frac{z^{-1}}{\frac{1-z^{-1}}{T_s} T_a + z^{-1}} \right) \left(1 + \frac{T_s}{T_i} \frac{1}{1 - z^{-1}} \right) \tag{2.72}$$

that is

$$e_a \left(1 + k_F \frac{T_s}{T_i} \frac{1}{1 - z^{-1}} \frac{\frac{1 - z^{-1}}{T_s} T_i z^{-1} + z^{-1}}{\frac{1 - z^{-1}}{T_s} T_a + z^{-1}} \right) = u - e k_p \left(1 + \frac{T_s}{T_i} \frac{1}{1 - z^{-1}} \right)$$

(2.73)

which is *not equivalent* to the corresponding case of form **A**, due to the element z^{-1} set in bold in Eq. 2.73.

For the *continuous case*, i.e. $T_s \ll T_a$, T_i

$$e_a \left(1 + k_F \frac{1}{sT_i} \frac{sT_i + 1}{sT_a + 1} \right) = u - e k_p \frac{sT_i + 1}{sT_i}$$

$$= e_a \left[\frac{s^2 T_a T_i + (1 + k_F) sT_i + k_F}{sT_i(sT_a + 1)} \right] \quad (2.74)$$

and for the input to R_2 after adding the awf

$$e_c = e k_p + e_a k_F \frac{1}{sT_a + 1}$$

$$= e k_p + \frac{sT_i(sT_a + 1)}{s^2 T_a T_i + (1 + k_F) sT_i + k_F} \left(u - e k_p \frac{sT_i + 1}{sT_i} \right) k_F \frac{1}{sT_a + 1}$$

$$= u \frac{sT_i}{s^2 T_a T_i + (1 + k_F) sT_i + k_F} k_F$$

$$+ e k_p \left[1 - \frac{sT_i + 1}{sT_i} \frac{sT_i}{s^2 T_a T_i + (1 + k_F) sT_i + k_F} k_F \right]$$

$$= u \frac{sT_i}{s^2 T_a T_i \frac{1}{k_F} + \frac{1 + k_F}{k_F} sT_i + 1} + e k_p \frac{1}{k_F} \frac{sT_i(sT_a + 1)}{s^2 T_a T_i \frac{1}{k_F} + \frac{1 + k_F}{k_F} sT_i + 1}$$

$$= \frac{sT_i}{s^2 T_a T_i \frac{1}{k_F} + \frac{1 + k_F}{k_F} sT_i + 1} \left[u + e k_p \frac{1}{k_F} (sT_a + 1) \right] \quad (2.75)$$

and for the integral action output u_i

$$u_i = \frac{1}{sT_i} e_c = \frac{1}{s^2 T_a T_i \frac{1}{k_F} + \frac{1 + k_F}{k_F} sT_i + 1} \left[u + e k_p \frac{1}{k_F} (sT_a + 1) \right] \quad (2.76)$$

Consider first the case **F_a**, $T_a \to 0$. Then

$$u_i = \frac{1}{\frac{1 + k_F}{k_F} sT_i + 1} \left(u + e k_p \frac{1}{k_F} \right) \quad (2.77)$$

there is a slow equilibration within the controller with the time constant

$$\tau_{F_a} = \frac{1 + k_F}{k_F} T_i \geq T_i \quad (2.78)$$

.

Consider now the case $\mathbf{F_b}$, $T_a := T_i$. Then

$$s^2 T_a T_i \frac{1}{k_F} + \frac{1 + k_F}{k_F} s T_i + 1 = \left(s \frac{T_i}{k_F} + 1 \right) (s T_i + 1) \tag{2.79}$$

and

$$u_i = \frac{1}{s \frac{T_i}{k_F} + 1} \left(u \frac{1}{s T_i + 1} + e k_p \frac{1}{k_F} \right) \tag{2.80}$$

Exercise

Compare and discuss Eq. 2.80 with respect to form **A**, Eq. 2.38.

Form G

This form is obtained from the previous one by moving the awf summing point across the k_p block up to the control error e, *i.e.*

$$R_{1G} = 1$$

$$R_{2G} = k_p \left(1 + \frac{1}{T_i} \right)$$

$$G_{aG} = \frac{1}{s T_a + 1}$$

$$k_G \quad \text{to be designed} \tag{2.81}$$

R_2 contains the full control algorithm. This also covers the classic form, where k_p is placed downstream of the $1 + 1/s T_i$ part.

For equivalence to form **F**, the awf closed-loop properties must also be conserved, *i.e.*

$$k_p k_G \overset{!}{=} k_F \tag{2.82}$$

to be replaced in the results obtained above for Form **F**.

Form H

Here, the awf summing point is moved still further to the reference signal $r(t)$, *i.e.*

$$r_c(t) = r(t) + v_a(t) = r(t) + \frac{1}{s T_a + 1} k_H e_a \tag{2.83}$$

It reduces the (large) $r(t)$ to $r_c(t)$ much closer to $y(t)$, such that the windup $e_a(t)$ is small enough. This is the "realizable reference" concept. Then

$$\begin{aligned} e_{c_H} &= r_c(t) - y(t) \\ &= r(t) + v_a(t) - y(t) \\ &= e(t) + v_a(t) \end{aligned} \tag{2.84}$$

In other words, $e_{c_H} = e_{c_G}$, if $k_H = k_G$, and then form **H** is equivalent to form **G**.

2.4.4 Self-conditioning

The concept of self-conditioning is shown in Fig. 2.16.

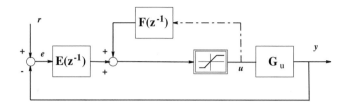

Fig. 2.16. The concept of self-conditioning controllers

There is no extra awf visible, hence the name self-conditioning. The controller transfer function in the linear operating range is

$$R(z^{-1}) = E(z^{-1}) \frac{1}{1 - F(z^{-1})} \qquad (2.85)$$

Note that the transfer functions $E(z^{-1})$, $F(z^{-1})$ in Fig. 2.16 may be the result of advanced design techniques, such as observer-based state feedback. There, the input u to the observer has to be picked up downstream of the saturation, in order to supply to the observer the u which is active on the plant. Otherwise the observer states would run off far away from the corresponding plant states, *i.e.* they would "wind up".

From Fig. 2.16, self-conditioning forms also are related to "Internal Model Control" and similar approaches.

We shall focus on the PI algorithms here and discuss extended forms later.

Forms J

Note that for $F(z^{-1}) = z^{-1}$ and $E(z^{-1}) = k_p \left(T_i \, \frac{1-z^{-1}}{T_s} + g \right)$
this is
- with $g = 1$ equal to Form **B**, with $k_B = 1$, Fig. 2.12 (c), and
- with $g = z^{-1}$ equal to Form **C**, Fig. 2.13 (d).

In other words, the special cases of Forms **B** and **C** are also self-conditioning.

Form K

This is derived from the traditional (and still widely used) positive feedback loop implementation of linear PI controllers, *e.g.* see [2].

Set

$$E(z^{-1}) := k_p \quad \text{and} \quad F(z^{-1}) = \cfrac{1}{1 + \frac{T_s}{T_f}\frac{z^{-1}}{1-z^{-1}}} = \cfrac{z^{-1}}{\frac{1-z^{-1}}{T_s}T_f + z^{-1}} \qquad (2.86)$$

then in the linear operating range

$$R(z^{-1}) = k_p \cfrac{1}{1 - \frac{z^{-1}}{\frac{1-z^{-1}}{T_s}T_f + z^{-1}}} = k_p \cfrac{\frac{1-z^{-1}}{T_s}T_f + z^{-1}}{\frac{1-z^{-1}}{T_s}T_f}$$

$$= k_p \left(1 + \frac{T_s}{T_f}\frac{z^{-1}}{1-z^{-1}}\right) \qquad (2.87)$$

For $T_f \overset{!}{=} T_i$ this is equal to Form **A-1** with the forward Euler algorithm. Note that it is not possible to generate the backward Euler form in this way.[1]

According to Fig. 2.16 this structure is used in saturating operation as well. Note that $F(z^{-1})$ is a unity gain first-order lag element with its input connected always to the saturation output. Therefore, its internal state variable will not increase beyond the saturation values, *i.e.* it will not "wind up". This is the main feature of self-conditioning.

Equivalence is investigated next. We start from Form **F**, Fig. 2.15. The first step is to split up the awf path into two parallel paths, form **L**; see Fig. 2.17.

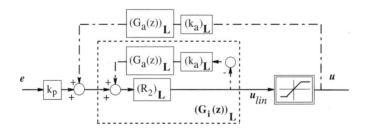

Fig. 2.17. Intermediate structure **L** to investigate equivalence of forms **F** and **K** Note that $(G_a)_L := (G_a)_F$, $(k_a)_L := (k_a)_F$ and $(R_2)_L := (R_2)_F$

Then, if the inner negative-feedback loop has a transfer function $(G_i(z))_L$ equal to one, by inspection form **L** is equivalent to form **K**. That is

---

[1] This will have an algebraic loop.

$$(G_i(z))_L \overset{!}{=} 1 = \frac{(R_2)_L}{1 + (R_2)_L k_L (G_a)_L}$$

i.e. $(R_2)_L = 1 + (R_2)_L k_L (G_a)_L$

and $k_L(G_a)_L = \dfrac{(R_2)_L - 1}{(R_2)_L} = \dfrac{1 + \frac{1}{T_i} T_s \frac{z^{-1}}{1 - z^{-1}} - 1}{1 + \frac{1}{T_i} T_s \frac{z^{-1}}{1 - z^{-1}}} = \dfrac{z^{-1}}{\frac{1 - z^{-1}}{T_s} T_i + z^{-1}}$

If now
- $k_L = 1$, and
- $(G_a)_L$ is a first-order lag (such as used above for $(G_a)_F$),
- and $T_a = T_i$ is inserted,
then forms **F** and **K** are equivalent.
In other words, form **K** is a very special case of form **F**, where *no* free design parameters for the awf are available anymore.

2.4.5 Summary

Concerning methodology, we have derived ten different versions of PI(aw) controllers, **A** to **K**, by systematic variation of a generic form **A**. Note that this does not reproduce the historic development, but helps to get an overview. Note also that this does not cover all existing proposals for awf. Notable exceptions are "feedforward awf", the "back calculation" awf or simply saturating the integral action in **D** on its own.

A second methodic element has been the analysis of the awf-loop response properties, isolated from the rest of the control system. This has produced upper bounds for the gain k_a in the discrete awf loop. It has also shown equivalences, and thus has enabled a reduction from ten versions down to two representative forms
- form **D** for the "output" conditioning subgroup, forms **A** to **E**, and
- form **F** for the "input" and "self"-conditioning subgroup, forms **F** to **K**.
Forms **D** and **F** shall be investigated further. This will also reveal further equivalence properties.

2.5 Transient Responses for the Test Cases

2.5.1 The Effect of Antiwindup Feedback Structure

From the equivalence discussion above, only the two forms **D** and **F** need to be considered further. Their closed-loop performances are compared with the awf gain k_a as the main experimental parameter on the test benchmark from Sect. 2.1. The forward Euler version is used in **D** for consistency with **F** and **K**.

- Fig. 2.18 shows the transient response for form **D**, covering **A** to **E** and **J**,
- Fig. 2.19 for form **F**, covering **G** and **H**, with $T_a < T_i$,
- and Fig. 2.20 for forms **F**, **G** and **H**, with $T_a = T_i$.
- Finally, form **K** is covered by Fig. 2.21 with $k_F = 1$.

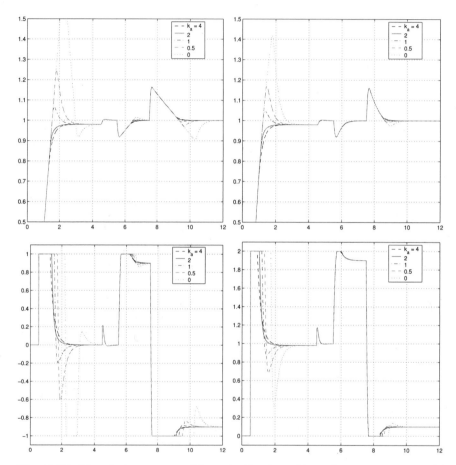

Fig. 2.18. Transient response with controller form **D** to the test sequence with different values of the awf gain k_a: for $a = 0$ (left column) and $a = 1$ (right column). output $y(t)$ (top); plant input $u(t)$ (bottom)

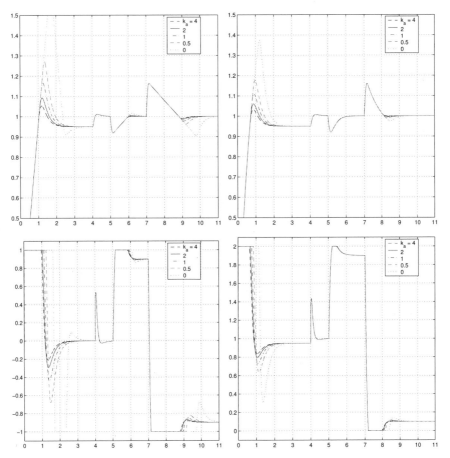

Fig. 2.19. Transient response with controller form **F** with $T_a = T_i/4$ to the test sequence with different values of the awf gain k_a: for $a = 0$ (left column), $a = 1$ (right column).

output $y(t)$ (top); plant input $u(t)$ (bottom)

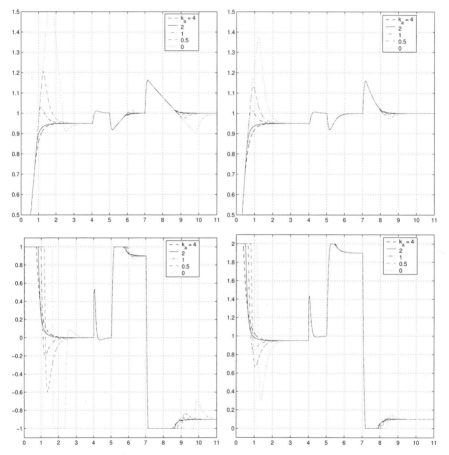

Fig. 2.20. Transient response with controller form **F** with $T_a = T_i$ to the test sequence with different values of the awf gain k_a: for $a = 0$ (left column) and $a = 1$ (right column).
output $y(t)$ (top); plant input $u(t)$ (bottom)

Observe that:

- With all forms, the windup effect is strongly reduced for high enough awf gain values (here $k \geq 1$).
- Again the case $a = 1$ seems less critical than the case $a = 0$.
- From Fig. 2.18, Form **D** eliminates the overshoot for $k_D \geq 2.0$.
- From Fig. 2.19, Form **D** performs better than form **F**.
- From Fig. 2.20, Form **D** performs better also than form **K**.
- The responses for forms **D** and **F** are strikingly similar, if both

$$(T_a)_F := T_i \quad \text{and} \quad k_F := k_D.$$

Note that this holds for arbitrary values of the awf gain, *i.e.* also for $k \neq 2$.

But keep in mind that these findings are subject to the specific experimental conditions defined in this benchmark, and may not be valid in a very different setup.

2.5.2 The Effect of Controller Tuning

John's case has demonstrated that performance of the main loop with awf also depends on how k_p, T_i are tuned.
From the benchmark specification:

$$k_p = 2\zeta \Omega T_1 \quad \text{and} \quad \frac{T_i}{T_1} = \frac{2\zeta}{\Omega T_1} \tag{2.88}$$

In the previous subsection, $2\zeta := 2.0$ has been used.

Now $2\zeta := 1.0$ (as in John's case). The simulations are shown in Fig. 2.21, for the plant case $a = 0$ and for the awf forms **D** and **F** for $T_a = T_i/4$ only.

Considering only $k_a \geq 2$, there is now a considerable overshoot of y even for form **D** in the range of 10% (there was none before), whereas for form **F** with $T_a = T_i/4$ it is now ~ 20 %, *i.e.* twice as large as before.

Exercise
Investigate the other cases not shown here.

2.5.3 Overshoot Analysis

From the general expectations on control performance of such control systems (see Sect. 1.4), they are expected to have "no perceptible overshoot". The simulations shown so far illustrate that this could not be achieved with any arbitrary awf, but requires a well-tuned combination of controller parameters, awf structure and awf gain. So a more detailed investigation is needed. We shall look at the general mechanism first, then turn to our benchmark situation, and investigate the two representative awf forms **D** and **F** separately.

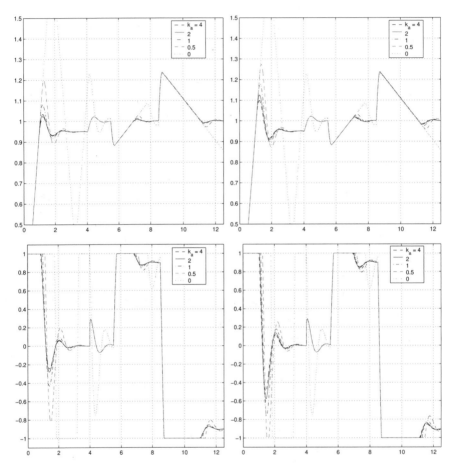

Fig. 2.21. Transient response with de-tuned controller to the test sequence, for $a = 0$, with different values of the awf gain k_a: form **D** (left column) and form **F** $T_a = T_i/4$ (right column).
output $y(t)$ (top); plant input $u(t)$ (bottom)

The General Mechanism

We shall look at the time interval around the last transition from nonlinear operation to linear operation, *i.e.* the final settling to the new steady state. This linear transient starts at time $t = t_T$, where $u_{lin}(t)$ lifts off the respective saturation limit u_{high} or u_{low}. This takes place at the instant where the awf tracking error $e_a(t)$ crosses to zero.

$$\text{for} \quad t < t_T : \quad e_a \neq 0 \quad \text{and for} \quad t > t_T : \quad e_a = 0 \qquad (2.89)$$

The nonlinear motion of the control loop up to this instant may be characterized by the control error $e(t_T)$ and its n time derivatives, where n is the order of the linear system. These values are the initial conditions of the linear system response down to steady state at $e = 0$.

So overshoot depends on the specific combination of values of the initial conditions at the transition time t_T, and on the eigenvalues of the linear closed loop.

Assumptions

We now consider the benchmark from Sect. 2.2 to obtain more specific results.

- We shall look at the plant of dominant first order with a PI controller. So the linear loop is of second order. And we need $e(t_T)$ and $\dot{e}(t_T)$ as initial conditions for the final linear transient.
- From the simulations, the large setpoint step r_1 of the run-up phase is the most critical one for overshoot. So we focus on this particular situation.
- We also focus on the open integrator plant, *i.e.* $a = 0$.
- Only cases of $k_a > 0$ shall be considered further, as for the no-awf case the overshoot is excessive anyhow.
- We shall also focus on the continuous case, because we look at the main control loop here, and where $T_s \ll T_i$ (see Sect. 2.2).

Then, in the nonlinear part of the run-up phase, by inspection of the specific plant structure, the derivative of the control error quickly converges to a constant value

$$\dot{e}(t) = \dot{r_1} - \dot{y}(t) = 0 - \frac{\Delta u}{T} \qquad (2.90)$$

determined by the steady state control margin Δu:

$$\Delta u = u_{high} - \bar{u} \quad \text{or} \quad u_{low} + \bar{u} \qquad (2.91)$$

and assuming a constant r_1.

Then for the control error

$$e(t) = e_0 + \dot{e}\, t = r_1 - \frac{\Delta u}{T}\, t \qquad (2.92)$$

And in the linear part of the transient for $t \geq t_T$, the response e is generated by the superposition of the responses to the initial conditions of the control error $e(t_T)$ and its derivative $\dot{e}(t_T)$ at $t = t_T$, i.e.

$$e(s) = \frac{1}{s^2 + s2\zeta\Omega + \Omega^2} \left[(s + 2\zeta\Omega)\, 1(s)e(t_T) + 1(s)\dot{e}(t_T) \right] \tag{2.93}$$

Analysis for Form D

The transition takes place at the instant $t = t_T$, where $e_a(t)$ hits zero. From Equation 2.46

$$e_a \left(1 + k_D \frac{1}{sT_i} \right) = u - ek_p \left(1 + \frac{1}{sT_i} \right) \quad \rightarrow \quad e_a = -\frac{1}{s\frac{T_i}{k_D} + 1} ek_p \frac{1}{k_D} (sT_i + 1)$$
$$\tag{2.94}$$

if u, i.e. the upper or lower saturation value, is time invariant.

For the special case $k_D = 1$, the numerator and denominator polynomials cancel, and e_a hits zero when the control e error crosses zero, i.e.

$$e(t_T) = 0 \quad \text{while} \quad \dot{e}(t_T) = -\frac{\Delta u}{T} \tag{2.95}$$

In other words an overshoot of $e(t)$ is unavoidable.

And if $k_D \gg 1$, then

$$0 \approx -(sT_i + 1) ek_p \frac{1}{k_D} \quad i.e. \quad e(t_T) = -\dot{e}(t_T)T_i = \Delta u \frac{T_i}{T} \tag{2.96}$$

Now the conditions are favorable to avoid an overshoot of $e(t)$.

Consider now the general case of k_D. In Equation 2.94 there is an additional first-order lag with time constant $\tau_T = T_i/k_D$, which will delay the transition time t_T with respect to the case $k_D \gg 1$. To simplify matters, assume that this first-order lag has attained its steady state before $t = t_T$. This implies that the duration of the run-up phase is at least $\sim 3\tau_T$. Then the time delay Δt_D for transition is equal to the time constant of the first-order lag τ_T, that is

$$\Delta t_D = \tau_T = \frac{T_i}{k_D} \quad \text{and} \quad t_T|_{k_D} = t_T|_{k_D \gg 1} + \Delta t_D$$

$$\rightarrow \quad e(t_T)|_{k_D} = \left(1 - \frac{1}{k_D} \right) \Delta u \frac{T_i}{T} \tag{2.97}$$

which reproduces the previous results for $k_D = 1$ and $\gg 1$.
Note that for $k_D < 1$, $e(t_T)$ will be negative, and overshoot will be even more pronounced than for $k_D = 1$.

Exercise
- Derive Eqs. 2.93 and 2.97.
- Investigate the effect of this assumption on the size of r_1.
- What if the steady state of the first-order lag
  has not been attained at $t = t_T$.

We now turn to the linear transient for $t \geq t_T$.
Introducing $e(t_T)$ from Eq. 2.97 and $\dot{e}(t_T)$ from Eq. 2.90 into Eq. 2.93

$$
e(s) = \frac{1}{s^2 + s2\zeta\Omega + \Omega^2} \frac{\Delta u}{T} \left[(s + 2\zeta\Omega) \left(1 - \frac{1}{k_D} \right) T_i - 1 \right]
$$

$$
= \frac{1}{\Omega^2} \frac{1}{\left(\frac{s}{\Omega}\right)^2 + 2\zeta\left(\frac{s}{\Omega}\right) + 1} \frac{\Delta u}{T} \left[\left(\frac{s}{\Omega} + 2\zeta \right) \left(1 - \frac{1}{k_D} \right) \Omega T_i - 1 \right] \quad (2.98)
$$

from the nominal closed-loop pole assignment to $(s^2 + s2\zeta\Omega + \Omega^2) = 0$ in Sect. 2.2.

If as the first case

$$
2\zeta := 2 \quad \text{and} \quad \Omega T_i = 2\zeta := 2 \quad \text{and if further} \quad k_D := k^* = 2 \quad (2.99)
$$

are introduced into Eq. 2.93, then

$$
e(s) = \frac{1}{\Omega} \frac{1}{\left(\frac{s}{\Omega} + 1\right)^2} \frac{\Delta u}{\Omega T} \left[\left(\frac{s}{\Omega} + 2 \right) \left(\frac{1}{2} \right) 2 - 1 \right] 1(s) = \frac{1}{\Omega} \frac{1}{\left(\frac{s}{\Omega} + 1\right)} \frac{\Delta u}{\Omega T} 1(s)
$$

$$
(2.100)
$$

and in the time domain

$$
e(t) = \frac{\Delta u}{\Omega T} \frac{1}{\Omega} \left[\Omega e^{-\Omega t} \right] = \frac{\Delta u}{\Omega T} e^{-\Omega t} = \frac{1}{2} \Delta u \frac{T_i}{T} e^{-\Omega t} \quad (2.101)
$$

This is the impulse response of a first-order lag, which will not overshoot. This nicely confirms the observation from Fig. 2.18.

Then, using as the second case the controller tuning from Sect. 2.5.2

$$
2\zeta := 1 \quad \text{and} \quad \Omega T_i = 2\zeta := 1 \quad \text{and if} \quad k_D := 2 \quad (2.102)
$$

are put into Eq. 2.93, then

$$
e(s) = \frac{1}{\Omega} \frac{1}{\frac{s}{\Omega}^2 + 1\frac{s}{\Omega} + 1} \frac{\Delta u}{\Omega T} \left[\left(\frac{s}{\Omega} + 1 \right) \left(\frac{1}{2} \right) 1 - 1 \right]
$$

$$
= \frac{1}{\Omega} \frac{1}{\frac{s}{\Omega}^2 + 1\frac{s}{\Omega} + 1} \frac{\Delta u}{\Omega T} \frac{1}{2} \left[\frac{s}{\Omega} - 1 \right] \quad (2.103)
$$

This is the impulse response of a non-minimum phase system, which will strongly overshoot.

And if $k_D \gg 1$ then

$$e(s) = \frac{1}{\Omega} \frac{\frac{s}{\Omega}}{\frac{s}{\Omega}^2 + 1\frac{s}{\Omega} + 1} \frac{\Delta u}{\Omega T} \tag{2.104}$$

which will also produce an overshoot together with its lower damping. This again confirms the observations from Fig. 2.21.

Exercise
Investigate form **B**, Fig. 2.12, with $k_B = 1$ (which also covers form **J**).

Analysis for Form F

We proceed along the same lines. The transition occurs at $t = t_T$, when u_{lin} lifts off the saturation u, *i.e.* when e_a hits zero. Then, from Eq. 2.74

$$e_a \left(1 + k_F \frac{1}{sT_i} \frac{sT_i + 1}{sT_a + 1} \right) = u - ek_p \frac{sT_i + 1}{sT_i} \tag{2.105}$$

$$\rightarrow \quad e_a = -\frac{sT_a + 1}{s^2 T_a \frac{1}{k_F} T_i + \frac{1+k_F}{k_F} sT_i + 1} ek_p \frac{1}{k_F} (sT_i + 1) \tag{2.106}$$

where again the respective upper or lower constraint value u is set as time-invariant.

For the special case of $T_a := T_i$, then

$$e_a = -\frac{sT_i + 1}{(s\frac{T_i}{k_F} + 1)(sT_i + 1)} ek_p \frac{1}{k_F} (sT_i + 1) = -\frac{1}{s\frac{T_i}{k_F} + 1} ek_p \frac{1}{k_F} (sT_i + 1) \tag{2.107}$$

which is equivalent to Eq. 2.94 for form **D**, if $k_F \overset{!}{=} k_D$
and the same overshoot properties hold as discussed there. In particular, if $k_F = 1$ this results in $e(t_T) = 0$, and thus overshoot is unavoidable. We have shown that this special case of form **F** is equivalent to the self-conditioning form **K**.
Then, by inference, one expects the overshoot to be suppressed for $k_F \geq 2$ and for $2\zeta = 2$ in the closed-loop pole assignment.

Another special case of interest is $T_a \rightarrow 0$:

$$e_a = -\frac{1}{s\frac{1+k_F}{k_F} T_i + 1} ek_p \frac{1}{k_F} (sT_i + 1) \tag{2.108}$$

In this case the time constant of the first-order lag element

$$\tau_F = \frac{1 + k_F}{k_F} T_i \quad \text{is} \quad \geq \quad T_i \; \forall \; k_F > 0 \quad \text{and} \quad \lim_{k_F \to \infty} \tau_F = T_i \tag{2.109}$$

i.e. $e(t_T) = 0$ now appears for $k_F \to \infty$ only. Then, by inference, the overshoot will be present for all k_F values, and increase with decreasing k_F entries.

In other words, T_a should be selected large not only for stability of the discrete awf loop (as shown in Sect. 2.4), but also for overshoot suppression.

It is still open, whether T_a should be increased beyond T_i. This can be investigated in the same manner as above. Briefly, a simulation shall give a hint of what may be expected; Fig. 2.22. There, the response is slowed down unduly for $T_a > T_i$, so $T_a = T_i$ seems to be the best selection. Of course this needs to be confirmed by a more general investigation.

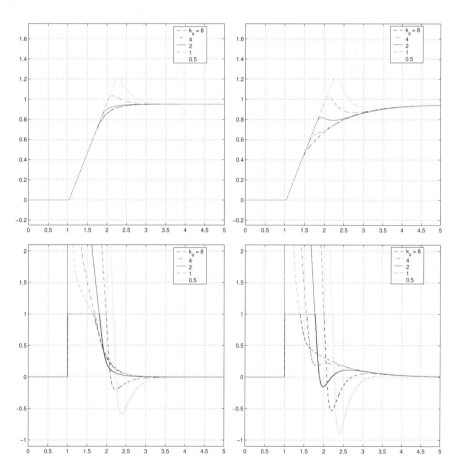

Fig. 2.22. Transient response with nominal controller to the run-up phase of the test sequence, for $a = 0$, with different values of the awf gain k_a and awf form **F**: $T_a := T_i$ (left column) and $T_a := 2T_i$ (right column); output $y(t)$ (top); plant input $u(t)$ and output of the controller u_{lin}(bottom)

Exercise

Investigate form **F** for awf design parameters T_a and k_F in a more general manner, similar to what is presented above for form **D**.

Hint: It may be useful to let $T_a := \tau T_i \quad \forall \ \tau > 0$

2.5.4 The Effect of the Relative Pole Shift ΩT

In the simulation study the relative pole shift from plant bandwidth $\omega = 1/T$ to closed-loop bandwidth Ω has been fixed at $\Omega T = 5$ as per specifications in Sect. 2.2.

Now the results from the overshoot analysis can be used directly to clarify the question, without further simulations.

We again assume that the setpoint step r_1 is of sufficient height, such that the low-pass filter has reached steady state at the transition instant $t = t_T$.

From the nonlinear phase, the transition leaves as initial conditions for the subsequent linear settling transient

$$\dot{e}(t_T) = \frac{\Delta u}{T} \quad \text{and} \quad e(t_T) = \Delta u \frac{T_i}{T} = \Delta u \frac{\Omega T_i}{\Omega T} = \Delta u \frac{2\zeta}{\Omega T} \tag{2.110}$$

i.e. the approaching speed $\dot{e}(t_T)$ does not depend on ΩT, while the lift-off error $e(t_T)$ decreases for increasing ΩT, and also decreases for reduced linear damping ratio 2ζ.

In other words, there is less space for the linear control to reduce the kinetic energy of motion to zero without overshoot.

For the linear transient we consider only the special cases of Form **D** with $k_D = 2$ and $2\zeta = 2$, and for Form **F** we assume further $T_a = T_i$. Then, by inspection of Equation 2.101, the no-overshoot property is conserved. The linear response is smaller by $1/(\Omega T)$ and faster by ΩT.

A word of caution against undue generalization of those neat results: keep the assumptions made earlier in mind.

- The relative pole shift is not a fully free design parameter. It is limited upward by the small delay element approximating the non-modeled fast dynamics in the plant, and downward by the minimum disturbance rejection performance required. For low enough ΩT the full run-up transient will be non-saturating, i.e. linear. Overshoot must then be countered by an appropriate two degree of freedom PI structure (*e.g.* see [2]).
- We have assumed the open integrator plant $a = 0$.
- We also have a comparatively short sampling time T_s.

Exercise

- Confirm the results from above by simulations.
- Investigate the effect of $2\zeta \neq 2$ for variable ΩT on the overshoot.
- Investigate the effect of $a \neq 0$ on overshoot.

2.5.5 Summary

From a methodology point of view, we have started with simulations to obtain a first overview. We then have investigated an area of special concern (the final overshoot) analytically, which has helped us to understand the basic relations and predict the effect of variations of other design parameters.

We have investigated two representative forms: $\mathbf{D}$ for the "control conditioning" and $\mathbf{F}$ for the "reference conditioning" groups of structures, both on the open integrator plant $a := 0$, non-time-varying saturations, and comparatively small sampling time T_s.

For Form $\mathbf{D}$, we have shown that there will be no overshoot in the final settling transient for the awf gain $k_D = 2$ and for the linear closed-loop damping factor $2\zeta = 2$. The main reason is that the initial conditions at the transition point, where u_{lin} lifts off the saturation u, are favorable for the linear settling transient. They are even more so if $k_D > 2$, or $2\zeta > 2$. The relative pole shift ΩT has no significant effect on overshoot.

For Form $\mathbf{F}$, there is an additional awf design parameter T_a. We have shown that, for $T_a/T_i \to 0$, such final overshoot is unavoidable, even for high-gain awf $k_F \to \infty$. We have shown that for $T_a/T_i = 1$ the response of $\mathbf{F}$ is equivalent to $\mathbf{D}$ for all k_F. Finally, we have shown that for $T_a/T_i > 1$ and for all k_F, the transition to the linear operation is advanced further, and a slow mode appears in the run-up response.

Therefore, as initial design values we suggest
- set $2\zeta = 2$ for the main loop
- set $k_F \geq 2$ for the awf loop, and
- for $\mathbf{F}$ set $T_a = T_i$
and then iterate on the process to accommodate modeling deviations and simplifying assumptions.

Also note that the self-conditioning form $\mathbf{K}$, while being very simple in structure, is predisposed to overshooting.

    All statements and conclusions above are subject to the specific experimental setup from Sect. 2.2, and may change for substantially different conditions.

2.6 Stability properties

2.6.1 Motivation

From what we have seen so far in the simulations, stability is not an urgent problem with the control loops as they were specified in the benchmark in Sect. 2.2, as long as a 'reasonably high' awf gain k_a was used. If, however, $k_a \to 0$, as this happened inadvertently in John's case, then large over- and under-shoots of $y(t)$ appear; see also Fig. 2.23.

    The obvious question is whether the control error $e(t)$ still converges and ultimately will end up at the transition instant $t = t_T$ to linear control, or whether this is not the case. This is answered by *nonlinear stability analysis*, as by inspection of Fig. 2.23 the transient of e is substantially shaped by the saturation element.

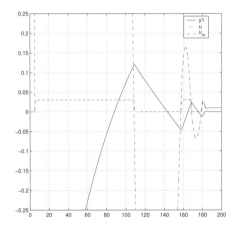

Fig. 2.23. Turbine speed run-up response (see Sect. 2.1) with awf gain $k_a = 0.05$ both for high and low saturations. The opening limit value is lowered to 0.03, to reduce the maximum speed overshoot

At first glance this may seem to be an academic exercise, as in the current particular benchmark situation, stability is not an urgent problem except for the "pathological" case $k_a \to 0$. However, we will see that it becomes important for plants of dominant order higher than one. The results turn out to be of eminent practical value for design. And the simple benchmark case of Sect. 2.2 will establish a useful base line for later investigations.

2.6.2 Methods

In the previous section the focus was on the final phase of the trajectory to the new equilibrium, where $u_{lin}(t)$ lifts off the saturation for the last time, and the successive transient is linear. Here, the focus is on the initial part of the transient, where the saturation has a dominant effect on the trajectory, *e.g.* see Fig. 2.23.

So far, the closed-loop trajectories have been induced by applying setpoint or disturbance (load) steps. In this part, the setpoint or the disturbance steps are interpreted as equivalent initial conditions being applied to the control loop out of the final steady state. In other words, the trajectories are seen as the corresponding *initial conditions responses*. This allows us to use standard nonlinear stability tests, which apply to initial condition responses.

Specifically, the *circle criterion* and the *Popov criterion* shall be used, see *e.g.*

[5, 6, 9, 11]. These deal with convergence of the initial condition response to the new steady state, which obviously is the main interest for practical applications.

The *describing function method* (*e.g.* see [12]) is useful to investigate oscillatory instability, which has been observed on higher order plants (see Chapter 6).

Appendix A provides a short introduction to these nonlinear stability tests.

2.6.3 A Generic Result

The nonlinear system to be investigated is the basic control loop in Fig. 2.6. As shown in Sect. 2.4, the typical implementation forms **A** to **K** may be generated by assigning the appropriate expressions to $R_1(z)$, $R_2(z)$, $G_a(z)$, and k_a and by selecting appropriate parameter values and initial conditions.

To apply the circle criterion, the basic system in Fig. 2.6 must be rearranged into its "canonical form". It has to consist of one nonlinear block and one linear block connected in a negative feedback loop, where the linear block (subsystem) contains all linear time-invariant dynamic elements, and is described by its transfer function; see Fig. 2.24(a).

The circle criterion requires the linear subsystem to be asymptotically stable. In other words, its poles must lie to the left of the imaginary axis for continuous systems, or inside the unit circle for discrete systems, *e.g.*[11].

But this is not the case here, as the PI controller contributes one non-asymptotically stable pole from the integral action, at least for the no-awf case ($k_a = 0$), and others may be contributed by $G(s)$, which contains open integrators or even positive feedback paths in many practical applications. Excluding all such cases from the stability test would seriously diminish its value for practical use.

Fortunately, asymptotic stability of the linear subsystem may be obtained by the "loop transformation" or "pole shifting" technique (*e.g.* see [12, 34] and references therein for more details). Specifically, the saturation nonlinearity is replaced by an equivalent arrangement of a deadspan nonlinearity with breakpoints at u_{low}, u_{high} and unity slope gain, in parallel with a unity gain path; Fig. 2.24(b). Then the unity gain path is joined to the rest of the linear system; Fig. 2.24(c). The new linear part F now has feedback, and thereby is asymptotically stable, as we shall see below.

The closed-loop system of Fig. 2.24(c) is now in its canonical form as required by nonlinear stability tests. Note the sign conventions in Fig. 2.24(c).

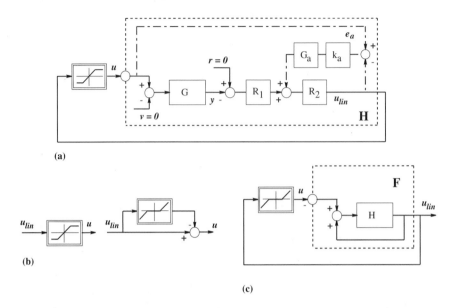

(a)

(b)

(c)

Fig. 2.24. Transformation steps of the nonlinear system of Fig. 2.6 into its canonical form, suitable for the stability test

The nonlinear subsystem

From the specifications of the benchmark in Sect. 2.2, *i.e.* from the control problem being "well-posed", the final steady state value of $\bar{u}_{lin} = \bar{u}$ will be inside the deadspan interval and at a finite distance from the breakpoints u_{low}, u_{high}. Then the origin of the nonlinear characteristic is shifted to $\bar{u}$, and

$$\Delta u_{lin_{max}}\big|_{up} = u_{lin_{max}}\big|_{up} - \bar{u} \quad \text{and} \quad \Delta \bar{u}_{up} = u_{high} - \bar{u} \qquad (2.111)$$

for the upward movement, and similarly for the downward direction

$$\Delta u_{lin_{max}}\big|_{dn} = u_{lin_{min}}\big|_{dn} - \bar{u} \quad \text{and} \quad \Delta \bar{u}_{dn} = u_{low} - \bar{u} \qquad (2.112)$$

For brevity, we shall focus on the more constraining situation, and omit the indexes *up, dn, i.e.*

$$\Delta u_{lin_{max}} = u_{lin_{max}} - \bar{u} \quad \text{and} \quad \Delta \bar{u} = u_{sat} - \bar{u} \qquad (2.113)$$

This implies that the origin is now at the center of the deadspan interval. This is the case for the run-up phase of the test sequence, which from the simulations seems to be the most sensitive one to the saturations. It is not the case for the subsequent load swings.

Exercise

What are the consequences of the assumption to the stability properties?

Also note that, from the specifications, u_{low}, u_{high} are time-invariant, and there is no energy storage (such as by backlash, *etc.*).

Then, to proceed with the test procedure, the upper sector slope from Fig. 2.24(c) is:

$$b_S = \frac{\Delta u_{lin_{max}} - \Delta \bar{u}}{\Delta u_{lin_{max}}} \qquad \text{valid for} \qquad \Delta u_{lin_{max}} > \Delta \bar{u} \qquad (2.114)$$

i.e.

$$\frac{1}{b_S} = 1 + \frac{\Delta \bar{u}}{\Delta u_{lin_{max}} - \Delta \bar{u}} = 1 + \Delta \qquad (2.115)$$

which defines Δ.

The lower sector slope is from Fig. 2.24(c):

$$a_S = 0 \quad i.e. \quad (1/a_S) = \infty \qquad (2.116)$$

The *on-axis* circle for the graphic stability test for both continuous-time systems (e.g. [12]) and for discrete-time systems (e.g.[11]) runs through the points $-1/b_S$ and $-1/a_S$ on the real axis and its center is on the negative real axis. Then, the circle must lie to the left of the Nyquist contour for the linear system part if the frequency is run from 0 to ∞. In other words, there may be no intersections. Then the initial condition response of the nonlinear closed loop will converge to the origin, *i.e.* asymptotic stability is shown.

In our case, $(1/a_S) = \infty$, and the circle degenerates into a vertical straight line positioned at $-1/b = -1.0 - \Delta$ on the real axis.

The *off-axis* circle test and the *Popov* test are applicable if the nonlinear characteristic is time-invariant and non-energy-storing, which is the case here. Then the straight line runs through the same point at $-1 - \Delta$, but it needs not be vertical, it may be inclined. This generally provides a larger region of Δ, where asymptotic stability can be shown, *i.e.* a larger "region of attraction". In other words, the test results will be less conservative.

Furthermore, if the Nyquist contour of the linear subsystem is such that the straight line may be drawn through the point $(-1+j0)$ on the real axis without intersection, *i.e.* $-1/b_S = -1.0$ or equivalently $\Delta \to 0$, then $u_{lin_{max}} \to \infty$ is admissible and the closed-loop response will be *globally asymptotically stable*.

If, on the other hand, $u_{lin_{max}}$ stays below the breakpoint value Δu, then $b_S = 0$ and $1/b_S \to \infty$, *i.e.* the straight line is positioned at $-\infty$. Now the Nyquist contour may evolve in almost all of the complex plane with no intersection, *i.e.* there are no additional restrictions for asymptotic stability of the linear closed-loop system from the nonlinear test.

We shall discuss this situation further in Sect. 2.6.5.

The Linear Subsystem

The second step in the test procedure is to draw the Nyquist contour $F(j\omega)$ for the *linear subsystem*: By inspection of Fig. 2.24(a).

$$H = \frac{k_a G_a R_2}{1 + k_a G_a R_2} - GR_1 \frac{R_2}{1 + k_a G_a R_2} \tag{2.117}$$

and from Fig. 2.24(c)

$$F = \frac{H}{1 - H} = \frac{-GR_1 R_2 + k_a G_a R_2}{1 + k_a G_a R_2 - k_a G_a R_2 + GR_1 R_2} \tag{2.118}$$

inserting $R = R_1 R_2$ yields the main result

$$F + 1 = \frac{1 + k_a G_a R_2}{1 + GR} \tag{2.119}$$

A Further Equivalence

This general result is now used to investigate the alternative forms **D** and **F** from Sect. 2.4 further.

For form **D**:

$$(k_a G_a R_2)_D = (k_a)_D \; 1 \; \frac{1}{T_i} T_s \frac{z^{-1}}{1 - z^{-1}} \tag{2.120}$$

and for form **F**:

$$(k_a G_a R_2)_F = (k_a)_F \; \frac{z^{-1}}{\frac{1-z^{-1}}{T_s} T_a + z^{-1}} \left(1 + \frac{1}{T_i} T_s \frac{z^{-1}}{1 - z^{-1}} \right)$$

$$= (k_a)_F \; \frac{z^{-1}}{\frac{1-z^{-1}}{T_s} T_a + z^{-1}} \left(\frac{1 - z^{-1}}{T_s} T_i + z^{-1} \right) \left(\frac{1}{T_i} T_s \frac{1}{1 - z^{-1}} \right)$$

$$= (k_a)_F \; z^{-1} \left(\frac{\frac{1-z^{-1}}{T_s} T_i + z^{-1}}{\frac{1-z^{-1}}{T_s} T_a + z^{-1}} \right) \left(\frac{1}{T_i} T_s \frac{1}{1 - z^{-1}} \right) \tag{2.121}$$

If both

$$T_a \overset{!}{=} T_i \quad \text{and} \quad (k_a)_F \overset{!}{=} (k_a)_D \tag{2.122}$$

then

$$(k_a G_a R_2)_F = (k_a G_a R_2)_D \tag{2.123}$$

and, as we have from the design of the linear loop

$$(1 + GR)_F \overset{!}{=} (1 + GR)_D \tag{2.124}$$

then finally

$$(F)_F = (F)_D \tag{2.125}$$

that is, the two forms **D** and **F** are *equivalent* also from the point of view of nonlinear stability of the complete loop and not only from the awf loop alone.[2]

Analysis of the Benchmark Cases

The next step is to discuss the general shape of the Nyquist contour of $F(z)$ for the benchmark loop. For brevity, we shall use $G_d(s)$ instead of $G(z)$ with the continuous PI(aw) controllers. Afterwards we shall check the design numerically with the full $G(z)$ and $R(z)$.

(a)   The P controller case
Ideal cancelation of steady state control errors due to persistent setpoints $r \neq 0$ and loads $\bar{v} \neq 0$ is assumed by an appropriate value of u_{man}. Then

$$k_a = 0$$
$$G_a(z) \to 1$$
$$R_1(z) = z^{-1}k_p \to k_p$$
$$R_2(z) = 1$$
$$\text{and} \quad G_d(z) \to \frac{b}{sT + a}$$

where the factor z^{-1} in R_1 accounts for the one sample delay to cover the calculation time requirements.
Then

$$F(s) + 1 = \frac{1 + k_a G_a R_2}{1 + G_d R} = \frac{sT + a}{sT + (a + k_p b)} = \frac{(s/\Omega) + (a/T\Omega)}{(s/\Omega) + 1} \tag{2.126}$$

where $\Omega T = a + bk_p$. For $a > 0$ (and as usually $k_p > 0$), the Nyquist contour of $F(j\omega)$ evolves exclusively to the right of the vertical straight line at -1. Therefore, the closed-loop response will be *globally asymptotically stable*.

And for $a = 0$

$$F(s) + 1 = \frac{(s/\Omega)}{(s/\Omega) + 1} \tag{2.127}$$

the contour starts at $(-1; j0)$. Thus, the closed-loop response can no longer be shown as being globally asymptotically stable, but only up to a bounded $u_{lin_{max}}$. However this upper bound is very large compared to $\Delta \bar{u}$, *i.e.* the loop will be *asymptotically stable in the (very) large*.

---

[2] Note that this nicely explains the striking similarity of the transient responses in Figs. 2.18 and 2.20.

This correlates well with the observations in Fig. 2.7.

(b)   The PI(aw) controller case

For the form **D** in Fig. 2.14, with k_a as parameter, $G_d(s)$ as above, and

$$G_a(z) = z^{-1}$$

$$R_1(z) = z^{-1} k_p \left(z^{-1} + T_i \frac{1 - z^{-1}}{T_s} \right)$$

$$R_2(z) = \frac{1}{T_i} \left(\frac{T_s}{1 - z^{-1}} \right)$$

where the computation time delay again is placed at the entry of R_1, to keep it from interfering with the awf. Then the transition to the continuous form yields

$$G_a(z) = 1$$

$$R_1(z) = k_p \left(1 + T_i s \right)$$

$$R_2(z) = \frac{1}{s T_i}$$

i.e.

$$F(s) + 1 = \left[\frac{s + k_a/T_i}{s} \right] \left[\frac{s^2 + s(a/T)}{s^2 + s(a + k_p)/T + (k_p/T_i T)} \right]$$

$$= \left[\frac{s/\Omega + (k_a/T_i \Omega)}{s/\Omega} \right] \left[\frac{(s/\Omega)^2 + (s/\Omega)(a/T\Omega)}{[(s/\Omega) + 1]^2} \right] \qquad (2.128)$$

If now

$$k_a = k_a^* := T_i \Omega \qquad (2.129)$$

then one of the closed-loop poles is canceled by the zero in $F + 1$ from the awf loop. Therefore, k_a^* is denoted as "compensating" awf gain. And

$$F(s) + 1 = \frac{(s/\Omega) + (a/T\Omega)}{(s/\Omega) + 1} \qquad (2.130)$$

which is the same as for the P controller and, therefore, with the same stability properties as discussed above. In other words one can expect very similar responses. Note that this is independent of a, *i.e.* a need not be zero.

For other ("non-compensating") values of the awf gain, $k_a \neq k_a^*$

$$F(s) + 1 = \frac{(s/\Omega) + (k_a/\Omega T_i)}{(s/\Omega) + (k_a^*/\Omega T_i)} \frac{(s/\Omega) + (a/T\Omega)}{(s/\Omega) + 1} \qquad (2.131)$$

Observe that the second fraction is the shape for $k_a = k_a^*$ and the first fraction is a series lead-lag element. For $k_a > k_a^*$, its phase contribution is negative.

This will increase the real part of $F+1$ in the relevant ω-range, and therefore improve the stability properties.

If, however $k_a < k_a^*$, then its phase contribution will be positive in the relevant ω-range, and the minimum real part of $F+1$ will eventually become negative. And for the "no awf" case $k_a = 0$, its phase contribution at $\omega \to 0$ is $+\pi/2$ and the Nyquist contour will start horizontally to the left. Then with the on-axis circle criterion, asymptotic stability can only be shown for much smaller $u_{lin_{max}}$ than for $k_a \geq k_a^*$.

For $a = 0$, and from the controller design given above, then: $T_i = 2/\Omega$

$$k_a^* = 2 \quad \text{and} \quad F(s) + 1 = [(s/\Omega) + 1] \frac{(s/\Omega)}{[(s/\Omega) + 1]^2} = \frac{(s/\Omega)}{(s/\Omega) + 1} \qquad (2.132)$$

Finally, for form **F** of Fig. 2.15, the low-pass $G_a(s)$ will have a similar effect on the stability properties, as the characteristic polynomial of the awf is the numerator of $F + 1$

$$1 + (k_a G_a R_2)_F = 1 + (k_a)_F \frac{1}{sT_i} \frac{sT_i + 1}{sT_a + 1} \qquad (2.133)$$

and thus produces a positive phase shift to the Nyquist contour, which is zero for $T_a := T_i$ and increases with decreasing T_a. In other words, a decreasing T_a means less favorable stability properties and one would expect more overshoot. This correlates well with the difference of response from e.g. Fig. 2.19 to 2.18.

Reverting to the *discrete-time* case is straightforward. As $1/T_s \ll \Omega$, the shape of F is not affected in the region $0 \leq \omega \leq \Omega$, which will be relevant for the circle test. The same argument holds for the small plant delay D, which has been neglected so far. Note that this assumes that the limits on both T_s and D for discrete feedback stability given in Sect. 2.4 are respected.

2.6.4 Stability Analysis of the Test Cases

The Nyquist contours of $F(z)$ corresponding to the test transient responses of Sect. 2.5 are shown in Fig. 2.25 for form **D**, and in Fig. 2.26 for form **F** with $T_a/T_i = 1/4$.
Note that from the *equivalence* of form **F** at $T_a = T_i$ to form **D** and from the equivalence of form **K** to form **F** at $k_a = 1.0$, no further plots are needed.

Clearly, the shape of the Nyquist contour of $F+1$ is much more favorable in the graphic stability test for the cases $a = 1$ than for $a = 0$, especially for $k_a = 0$. It is also more favorable for $T_a = T_i$ than for $T_a \ll T_i$, and finally more favorable for $k_a = 2$ than for $k_a = 1$, i.e. for form **D** and **F** than for form **K**.

Again, this corresponds well to the simulation findings.

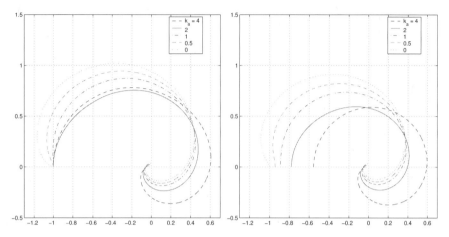

Fig. 2.25. Nyquist contours with controller form **D** to the test system with different values of the awf gain k_a, for $a = 0$ (left) and $a = 1$ (right).
Also covers form **F** with $T_a = T_i$, as well as form **K**, if $k_a = 1$

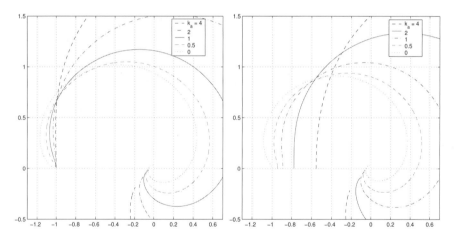

Fig. 2.26. Nyquist contours with controller form **F** to the test system with $T_a = 0.25T_i$ and with different values of the awf gain k_a, for $a = 0$ (left) and $a = 1$ (right)

2.6.5 Estimating the Range of Attraction

In nonlinear stability analysis often only "global asymptotic stability" is considered, implying that the shape of the Nyquist contour of F can only be such that $\Delta_{st} \to 0$ holds, (with Δ_{st} defined in Eq. 2.115, where the index st stands for "stability test").

However, from our experience, in many practical applications the shape of F will be such, that only a finite but small $|\Delta_{st}|$ results. Then $|\Delta u_{lin_{max}}|$ will be large but still bounded. However, all nonlinear system trajectories generating $|\Delta u_{lin}(t)|$ below this upper bound $|\Delta u_{lin_{max}}|$ will converge to the steady state. They evolve in a compact area, which contains the final steady state, *i.e.* within the "region of attraction". This leads to the concept of "stability in the large" rather than "in the whole".

In the following we shall distinguish

- stability *in the small* for $u_{lin_{max}} \leq \Delta u$, *i.e.* responses in the linear range;
- stability *in the large* for $u_{lin_{max}} - \Delta u \geq 1$, *i.e.* for nonlinear transients, with large but still bounded overrun of the saturation ($\Delta > \epsilon$, where $0 < \epsilon \ll 1$);
- stability *in the whole* for $(u_{lin_{max}} - \Delta\bar{u})/\Delta\bar{u} \to \infty$, *i.e.* for nonlinear transients with infinite overrun ($\Delta \to 0$). Note that we exclude the case of $\Delta\bar{u} = 0$ as being "ill posed", because this means that linear closed-loop control is no longer possible around the final steady state.

Note that the concept of stability *in the large* is very suitable for real physical systems, where all states and signals necessarily have both high and low bounds, and where the controller states are bounded by awf gains $k_a > 0$.

The main theoretical difficulty is to determine the precise region of initial conditions for trajectories where this Δ_{tr} (with Δ_{tr} defined as in Eq. 2.115, where the index tr indicates "trajectory") stays above the Δ_{st} from the stability test without actually generating all those trajectories by simulations. An algorithm given by Khalil [12] generates appropriate Ljapunow functions, *i.e.* a conservative approximation.

Another approach is to look for a first estimate instead of a strict upper bound.[3] Such an estimate can be obtained by considering typical transients where the nonlinear effects are most pronounced. From the simulations this would be in particular the large setpoint step responses (for run-up) in the benchmark test sequence (*e.g.* [34, 47]). Then, a first estimate of the maximum Δ_{tr} can be obtained using the maximum setpoint step from the specifications. This value should be greater than Δ_{st}. We shall now go into more detail.

First note that

$$\Delta u_{lin_{max}} - \Delta\bar{u} := -e_{a_{max}} \tag{2.134}$$

and from Sect. 2.4

$$e_a(s) = -e(s)k_p\frac{1}{k_a}G_f(s) \tag{2.135}$$

where $\Delta\bar{u}$ is assumed as time invariant and where $G_f(s)$ is a unity gain filter with coefficients depending on the particular form of antiwindup (**A** through **K**) and of the awf gain value k_a.

---

[3] Note that the following argument is not at all conclusive from the theoretical side, it cannot produce more than an indicative value.

In the following, we exclude the "no-awf" case, and also the cases where (from Sect. 2.5) an overshoot cannot be avoided for the final linear settling transient, i.e.

$$k_a \geq 1 \tag{2.136}$$

If the particular case of **D** with $k_a = 1$ is considered, then $G_f(s) = 1$, and similarly for **F** with $T_a = T_i$ and $k_a = 1$. Then

$$e_a(t) = -e(t)k_p\frac{1}{k_a} \tag{2.137}$$

which makes it particularly simple to determine $e_{a_{max}}$:

$$e_{a_{max}} = -e_{max}k_p\frac{1}{k_a} \tag{2.138}$$

where by inspection of the transients in Sect. 2.5, Figs. 2.18 to 2.21, the control error is largest at the moment T_1 the setpoint step is applied, and then decreases monotonously during the plant run-up, i.e.

$$e_{a_{max}} = -e_{T_1}k_p\frac{1}{k_a} \tag{2.139}$$

In all other cases there is an additional filter, which deforms $e(t)$, and determining $e_{a_{max}}$ analytically is more involved. To avoid this we generate a conservative estimate of $e_{a_{max}}$ by slightly modifying the run-up sequence phase 1 of the benchmark:

We insert a *phase 1a* where the controller is switched on with the reference r preset at r_1, but the process is kept at standstill ($y_{T_1} = 0$) by stepping u_{high} down to zero. The input e to the awf loop then is constant at e_{T_1}:

$$e_{T_1} := r_1 - y_{T_1} = r_1 \tag{2.140}$$

The duration ΔT_{1a} of phase 1a is such that the awf loop reaches its steady state, i.e. approximately three times the dominant time constant of G_f. For form **D**:

$$\Delta T_{1a} \geq \quad \approx 3\frac{T_i}{k_a} \tag{2.141}$$

and correspondingly for the other forms. Using

$$\lim_{s \to 0} G_f(s) = 1 \quad \text{then} \quad \bar{e}_a = -\ e_{T_1}\frac{k_p}{k_a} \tag{2.142}$$

Then in *phase 1b* the process is run up by stepping u_{high} to its specified value. Then $e_a(t)$ starts at $\bar{e}_a$ and afterwards will decrease monotonously, i.e.

$$\bar{e}_a > e_{a_{max}} \tag{2.143}$$

and finally

$$1 + \Delta_{st} := \frac{\Delta \bar{u}}{e_{T_1}} \frac{k_a}{k_p} \tag{2.144}$$

We shall now develop the result for the Nyquist contour further. Having set $k_a > 0$:

$$F(s) + 1 = \frac{1 + k_a G_a R_2}{1 + R\,G} = \frac{[s/\Omega + k_a/(\Omega T_i)]\,[s/\Omega + a/(\Omega T)]}{(s/\Omega)^2 + 2\zeta(s/\Omega) + 1}$$

$$= \kappa_a \frac{[s/(\kappa_a \Omega) + 1]\,[s/\Omega + a/(\Omega T)]}{(s/\Omega)^2 + 2\zeta(s/\Omega) + 1}$$

where $\kappa_a := \dfrac{k_a}{\Omega T_i} = \dfrac{k_a}{k_a^*}$

and as $\Omega^2 := \dfrac{k_p}{T_i T}$; i.e. $\Omega T_i = \dfrac{k_p}{\Omega T}$ from the pole assignment

finally $\kappa_a = \dfrac{k_a}{k_p}\,\Omega T$ \hfill (2.145)

Now both $1 + F(s)$ and $1 + \Delta_{st}$ have the common multiplier $k_a/k_p > 0$, which can be eliminated in both for the graphic test. And the multiplier ΩT is transferred from the Nyquist contour side $1 + F(s)$ to the nonlinear subsystem side $1 + \Delta_{st}$. The test is then performed with

$$(F(s) + 1)' = \frac{[s/(\kappa_a \Omega) + 1]\,[s/\Omega + a/(\Omega T)]}{(s/\Omega)^2 + 2\zeta(s/\Omega) + 1}$$

$$\text{and } (1 + \Delta_{st})' = \frac{\Delta \bar{u}}{e_{T_1}} \frac{1}{\Omega T} \tag{2.146}$$

Further insight is provided by discussing the end points of the Nyquist contour $(F(s) + 1)'$.

$$\lim_{\omega \to 0} (F(s) + 1)' = a \frac{1}{\Omega T} \tag{2.147}$$

which is on the real axis, on the positive side for $a > 0$, at the origin for $a = 0$, and on the negative side for $a < 0$, i.e. for an unstable plant (we have not considered this case so far).
Also

$$\lim_{\omega \to \infty} (F(s) + 1)' = \frac{1}{\kappa_a} = \frac{1}{\Omega T} \frac{k_p}{k_a} = \frac{1}{\Omega T} \frac{2\zeta \Omega T - a}{k_a} = \frac{1}{k_a}\left(2\zeta - a \frac{1}{\Omega T}\right) \tag{2.148}$$

As $k_p > 0$ and as $1 < k_a < \infty$, the Nyquist contour will always end on the positive real axis. Therefore, the area of interest for the stability properties is at $\omega \to 0$.
To avoid an intersection of the straight line and the Nyquist contour there, then

$$- \left(1 + \Delta_{st}\right)' < \left(1 + F(s)\right)' \Big|_{s=j\omega \to 0}$$

$$i.e.: \qquad \frac{\Delta \bar{u}}{e_{T_1}} \frac{1}{\Omega T} < a \frac{1}{\Omega T}$$

$$\text{if } a < 0: \qquad e_{T_1} < \frac{\Delta \bar{u}}{|a|}$$

$$\text{and if } a \geq 0: \qquad e_{T_1} \to \infty \tag{2.149}$$

Note that this is independent of the awf gain $k_a \geq 1$, and of the linear loop design parameter ΩT. For $2\zeta = 2$ the on-axis circle criterion can be used; for lower 2ζ one may have to resort to the off-axis version.

Exercise
- What is the margin introduced by using $\bar{e}_a$ instead of $e_{a_{max}}$?
- Investigate the case for $-0.5 \leq a \leq +0.5$ by simulations.
- How well is the actual stability limit predicted by the circle criterion?
- What is the physical interpretation of the result in Eq. 2.149?

We shall now very briefly apply the same procedure to the *P controller* case. There is no awf here; Fig. 2.6. By inspection of Fig. 2.7, the control error $e(t)$ is largest at $t = T_1$, just after the reference step r_1 has been applied. It then decreases monotonously.
Therefore

$$u_{lin_{max}} = k_p e_{T_1} \tag{2.150}$$

i.e.

$$(1 + \Delta) = \frac{\Delta \bar{u}}{k_p e_{T_1} - \Delta \bar{u}} = \frac{1}{\Omega T} \frac{1}{1 - a/(\Omega T)} \frac{\Delta \bar{u}}{e_{T_1}} \frac{1}{1 - (\Delta \bar{u})/(k_p e_{T_1})} > \frac{1}{\Omega T} \frac{\Delta \bar{u}}{e_{T_1}} \tag{2.151}$$

and for the Nyquist contour

$$(1 + F(s)) = \frac{sT + a}{sT + a + k_p} = \frac{s/\Omega + a/(\Omega T)}{s/\Omega + 1} \tag{2.152}$$

Exercise
- Discuss this case further in the time and frequency domain.
- Look for any equivalence with the PI(aw) case from above.

2.6.6 Summary

A generic result has been derived for the stability test

$$1 + \Delta = \frac{\Delta\bar{u}}{u_{lin_{max}} - \Delta\bar{u}} = \frac{\Delta\bar{u}}{e_{a_{max}}} \quad \text{and} \quad 1 + F(s) = \frac{1 + k_a G_a R_2}{1 + RG}$$

which is not restricted to the dominant first-order plants considered here.

The general shape of the Nyquist contour $1 + F(j\omega)$ for the control loops considered here is such that asymptotic stability in the very large (for $a = 0$) or in the whole (for $a > 0$) can be shown, if the awf gain value k_a is above a threshold k_a^*.

This gain value k_a^* coincides with the value where the overshoot in the final linear settling transient is suppressed (for $a = 0$).

A conservative estimate has been given for the position of the straight line originating from the nonlinear subsystem in the graphic stability test, on the basis of the run-up reference step input, $i.e.$ e_{T_1}.

$$|e_{a_{max}}| < |\bar{e}_a| = |e_{T_1}| \frac{k_p}{k_a} \quad \text{and} \quad 1 + \Delta > \frac{\Delta\bar{u}}{|e_{T_1}|} \frac{k_a}{k_p}$$

This produces a "region of attraction" ($i.e.$ asymptotic stability in the large), even for plants with positive feedback ($a < 0$), for reference steps generating an initial control error up to $e_{T_1} = \frac{\Delta\bar{u}}{|a|}$.

So, nonlinear stability analysis has brought about a much better and more compact understanding of such control loops than what simulations alone can provide.
On the other hand, some simulations are still needed to check the final performance of a stability-based design, due to the inherent conservativeness of the tests.

2.7 Relations to Optimal Control

2.7.1 Motivation

The aim of this section is to explore the relations of the awf control loops investigated so far with Optimal Control. This will help to estimate what is lost by the intuitive design in this respect. It will also establish a base line for later investigations of higher order plants with awf control.

The key decision for Optimal Control is the choice of an appropriate performance index, $i.e.$ the function to be optimized over the trajectory.
Looking at the simulation results suggests separating into a linear end part and a strongly nonlinear initial part.

The control for the *linear part* has been designed by constant state feedback as a structure and pole assignment as a design method. The same controller may be obtained by linear quadratic control on an infinite time horizon and by choosing appropriate weights for the state and control variables. In this respect, the controller used so far is already an optimal one.

For the *nonlinear part*, several approaches are in use, see Sect. 1.5. A very popular one is to go on using the quadratic cost function from the linear part, and inserting additional inequality constraints on control variables and output variables, such as needed. Then there is no closed-form solution as for the linear case, and the optimal control sequence $u^*(kT_s)$ must be determined by numerical optimization. The advantage of this method is that basically the same control algorithm is applied over both parts of the trajectory; there is no method switching. A disadvantage is that it is not clear at the beginning, what shape $u^*(kT_s)$ will have typically.

This type of approach has been investigated much, and many results exist. We shall not go into it here, but refer the reader to the respective literature.

Instead we shall use the minimum transition-time approach. The main reason is that this is basically what practical specifications will want, though not in a mathematically rigorous form. For instance, the speed control of a water turbine should attain the shortest transition time from "standstill" to "ready for synchronization" in order to be of best use for covering demand peaks in the power grid. The alternative would be to operate the unit at speed-no-load, which would continuously draw from stored hydropower. The situation is even more pronounced with gas turbines as prime movers. One disadvantage of this approach is the "method switch", as we shall see.

2.7.2 The Exact Solution

Optimal Control theory assumes that the transients to be optimized start at t_0 at $\mathbf{x}(t_0) = \mathbf{x}_0$ and end at t_1 in the state space origin $\mathbf{x}(t_1) = \mathbf{0}$ at $\bar{u} = 0$. Note that the end condition amounts not only to the control error being zero but all its time derivatives as well, *i.e.* steady state. In other words, this is again the initial condition situation from the stability analysis in Sect. 2.6. It implies that there exists at least one such trajectory from $\mathbf{x}_0$ to $\mathbf{0}$ which then can be optimized. It also implies that $\bar{u} = 0$ lies at a finite distance inside the closed control saturation interval $[u_{low}, u_{high}]$, for reachability. These conditions are familiar and have been complied with in the previous sections.

Optimal Feedforward Control

The cost integral is

$$J = \int_{t_0}^{t_1} 1 \, dt = t_E - t_0; \text{ and for the optimum: } u^*(t)\big|_{t_0}^{t_1} \text{ such that } J_{u^*} \to min$$

$$(2.153)$$

In other words a control sequence u^* from t_0 to t_1 is sought, which minimizes the transition time $t_1 - t_0$ to the state space origin. This amounts to feedforward control.

The Maximum Principle states for the solution $u^*(t)$ that:

- u^* is always at one of the limits of the control interval.
  It is either u_{high} to move upward or u_{low} to move downward.
- In the transition time interval t_0 to t_1, there are at most $n - 1$ switchings between both limits, if the linear system part (the plant) is of order n and has no conjugate complex eigenvalues.

Considering only the dominant dynamics G_d of the benchmark case, then $n = 1$ and there is one real eigenvalue. Thus there will be no switching between u_{low} and u_{high}, but u^* will stay at one limit from t_0 to t_1.

Optimal Controller Synthesis

The aim is to replace $u^*(t)$, a function of time, by a function of plant state $u^*(\mathbf{x})$, such that the required $u^*(t)$ will be generated automatically whilst the system moves along its minimum-time trajectory $\mathbf{x}(t)$ in state space. This is also called solving the synthesis problem, or designing the optimal feedback $u^*(t) = f(\mathbf{x}(t))$.

In the general case this leads to switching hypersurfaces in state space, where $u^* = u_{high}$ on one side, and $u^* = u_{low}$ on the opposite side. In this particular case, obviously

$$u^* = \begin{cases} u_{high} & \text{if } e > 0 \\ u_{low} & \text{if } e < 0 \end{cases} \tag{2.154}$$

Note that u^* is not defined for $e = 0$. It is also not defined outside the optimization time interval, i.e. for $t > T_1$. However, time goes on and some control must be provided there as well.

2.7.3 Ongoing Control

A simple idea would be to let the above switching control algorithm continue after T_1. In practical implementations there are always additional small delays. This then leads to high-frequency oscillations of small amplitude (limit cycles) around the origin. This is known as "sliding mode" control, and there is a considerable literature on this subject. The drawback of the method is noise and wear in the actuator subsystem.

There are several work-around concepts. In the following we discuss two alternatives.

Form I: Suboptimal Control

A popular concept is to substitute the ideal switching characteristic in Equation 2.154 by the well-known saturation characteristic $u_{lin}(t) \to u(t)$, where $u_{lin}(t) = ke(t)$:

$$u' = \begin{cases} u_{high} & \text{if } u_{lin} \geq u_{high} \\ u_{lin} & \text{if } u_{low} < u_{lin} < u_{high} \\ u_{low} & \text{if } u_{lin} \leq u_{low} \end{cases} \tag{2.155}$$

Here, $k \to \infty$ would lead to $u' \to u^*$, *i.e.* to minimum-time control.

For finite k values this amounts to P control with $k_p := k$, with k small enough, such that there are no limit cycles, and the control loop is sufficiently damped in the linear range.

Next we investigate how much is lost against minimum-time by resorting to such P control.

Using the open integrator plant ($a = 0$) and neglecting the small delay $T_t = 0$, then the minimum-time for the run-up phase is

$$t_1 = \frac{\Delta\bar{u}}{|e_{T_1}|}T \tag{2.156}$$

and for finite $k = k_p$ values, lift off of Δu_{lin} from $\Delta\bar{u}$ occurs earlier by τ_1

$$\tau_1 = \frac{1}{k_p}T \quad i.e. \quad \frac{\tau_1}{t_1} = \frac{\Delta\bar{u}}{|e_{T_1}|}\frac{1}{2\Omega T} \tag{2.157}$$

Allowing $3\tau_1$ for decay leads to an overall lengthening of the transient by Δt_1

$$\frac{\Delta t_1}{t_1} > \frac{\Delta\bar{u}}{|e_{T_1}|}\frac{1}{\Omega T} \tag{2.158}$$

The next step is to insert an integral action to suppress steady state offsets, of course with awf. Setting $k_a = k_a^* = 2$ leads to the same response without overshoot from lift-off onwards, and thus to the same lengthening of the transient. This is documented in Fig. 2.28(a), where the small delay and the finite sampling time from Sect. 2.2 are used. The settling time is about 25% longer than the minimum.

Form II: Transfer of Control

The aim is to reduce this lengthening further. The basic idea is shown in Fig. 2.27.

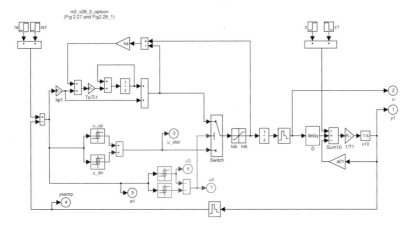

Fig. 2.27. Structure for "transfer of control'

A generator for the minimum-time control sequence $u^*(t)$ is arranged in parallel with the PI(aw) controller. A switch transfers control $u(t)$ between both as follows

$$\text{switch to} \quad u^*(t) \quad \text{if} \quad |e(t)| \geq 2\frac{\Delta \bar{u}}{k_p}$$

$$\text{and back to} \quad u_{PI(aw)} \quad \text{if} \quad |e(t)| \leq 0.25\frac{\Delta \bar{u}}{k_p} \qquad (2.159)$$

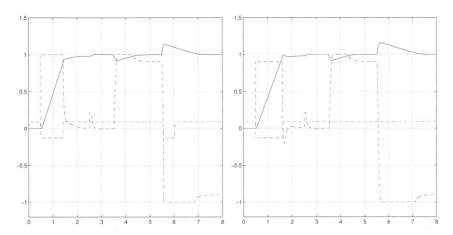

Fig. 2.28. Transient response to the test sequence with "ongoing" linear PI(aw) control (left); with transfer to $u^*(t)$ in the run-up phase only (right)

This allows the transfer to $u^*(t)$ only in the run-up phase, and the load swings later in the test sequence will be handled as before by the PI(aw) feedback.

The settling time in Fig. 2.28(b) is not at the minimum, but about 10% longer due to the small delay and the finite T_s.

Exercise
- Improve this further (hints: reduced T_s, PI(aw) output tracking, *etc.*).
- Investigate the influence of plant feedback $(-0.5 \leq a \leq +0.5)$.

2.8 Case Study:
Temperature Control in a Chemical Batch Reactor

The aim of the design case studies is to provide a framework for active learning of the material presented so far. The idea is not only to do the calculations and simulations, but to check all assumptions carefully as well, and to evaluate the results of the analysis and design critically from a plant operation view.

The example has been very much abstracted from a control engineering project that one of the authors has been involved with.

This case study will be continued at the end of the next chapter.

2.8.1 The process and the main control task

Fig. 2.29 is a much-simplified sketch of the batch reactor.

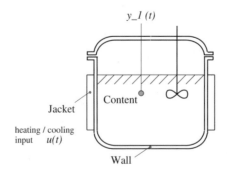

Fig. 2.29. Sketch of the plant

- The *main control* task is to keep the fluid temperature at a given setpoint. Overshoots are to be avoided, because over-temperatures may generate unwanted secondary products from the reactions. Under-temperatures will prolong the batch production time, as reaction rate goes down. Thus, high-performance temperature control is required.

- Temperatures are denoted by ϑ $[^{\circ}C]$, and heat flows by I^* $[W_{th}]$

- Available *instrumentation and operation subsystems*: the fluid content is continuously stirred. Its temperature ϑ_c is measured by a temperature sensor (y_1) in the fluid. The wall temperature and the jacket temperature are not measured.

  The jacket is a source or sink for thermal energy flowing to or from the wall to the content fluid. The response of the jacket and wall subsystems is approximated by a pure delay e^{-sD} with unity gain, where D is short compared with the main time constant τ of the fluid content.

  This delay shall cover the dynamic responses of the fluid temperature sensor and of the heating/cooling control valves as well.

- A *chemical reaction* shall take place in the content fluid, which starts at a given temperature threshold value ϑ_s below the nominal content temperature ϑ_r .

  The *heat flow* from or to this reaction I_R^* shall be modelled by

$$I_R^* = -k_R(\vartheta_c - \vartheta_s) \quad \text{for} \quad \vartheta_c \geq \vartheta_s$$
$$I_R^* = \quad\quad 0 \quad\quad \text{for} \quad \vartheta_c < \vartheta_s$$

  where

$$k_R \quad \begin{array}{ll} < 0 & \text{for exothermal reactions (heat source)} \\ = 0 & \text{for inert reactions} \\ > 0 & \text{for endothermal reactions (heat sink)} \end{array}$$

  The actual value of k_R is not known, but upper and lower bounds are specified.

- The *test sequence* consists of

  – starting from standstill (all variables at zero)
  – step up to the contents temperature setpoint r_1 to nominal, *i.e.* $r_1 = 1.0$;
  – after reaching the steady state, apply a small setpoint step to $r_1 = 1.025$
  – then introduce a small heat flow disturbance step I_d^* to the content. This is to model a warmer reactant flow infused into the content.

  The last two test steps will show the linear loop performance.

- No *measurement disturbances* shall be considered here.

- *Input constraints* from actuator saturation only shall be considered $[-1.0 \ \cdots \ +1.0]$

2.8.2 The Plant Model

In this model, only one (well stirred) main storage domain for thermal energy of the fluid content is considered, resulting in one ordinary first-order differential equation.

Considering the range $\vartheta > \vartheta_s$ around the steady state at $r_1 = 1.0$ and introducing 'per unit' (p.u.) variables, (the index r standing for 'rated')

$$
\frac{\vartheta_1}{\vartheta_r} := x_1 \qquad \frac{\vartheta_s}{\vartheta_r} = x_s;
$$

$$
\frac{I_u^*}{I_r^*} := u; \qquad \frac{I_d^*}{I_r^*} := v; \qquad \frac{I_R^*}{I_r^*} := -\frac{k_R \vartheta_r}{I_r^*}\left[\frac{\vartheta(t)}{\vartheta_r} - \frac{\vartheta_s}{\vartheta_r}\right];
$$

$$
\tau := \frac{c_c m_c \vartheta_r}{I_r^*}; \qquad a := \frac{k_R \vartheta_r}{I_r^*} \tag{2.160}
$$

yields

$$
\tau\frac{d}{dt}x_1 = -a\left(x_1(t) - x_s\right) \qquad + u(t) \quad + v(t)
$$

$$
\text{and} \quad y_1 = x_1
$$

$$
\tag{2.161}
$$

This leads to the dominant denominator polynomial $d(s) = s\tau + a$ and the transfer functions

$$
G_u = \frac{y}{u} = e^{-sD}\frac{1}{s\tau + a}
$$

$$
\text{and} \quad G_v = \frac{y}{u} = \frac{1}{s\tau + a} \tag{2.162}
$$

where the "fast" (non-modelled) dynamics are represented by a series delay element e^{-sD} with $D \ll \tau$.

2.8.3 Parameter Values

$$
\begin{aligned}
\tau &= 900. \text{ s} \\
x_s &= 0.90 \qquad &&\text{reaction threshold temperature, in p.u.} \\
a &= -8.0\cdots+8.0 \qquad &&\text{reaction enthalpy gain, in p.u.} \\
D &= 20. \text{ s} \qquad &&\text{sum of all "non-modelled time constants"} \\
u &= -1.0\cdots+1.0 \qquad &&\text{actuator saturations, in p.u.} \\
v &= +0.025 \qquad &&\text{warm reactant inflow, in p.u.}
\end{aligned}
$$

2.8.4 The Controller

To avoid any overshoot in the linear range, a "two-degrees-of-freedom" linear PI algorithm is suggested for the y_1 loop:

$$u_1(s) = k_{0_1}(r_1 - y_1)\frac{1}{s\tau_0} - k_{1_2}y_1 \qquad (2.163)$$

2.8.5 The Design Program

It is suggested to proceed in three phases. Here are some additional hints:

1. *Linear design*
   Design the controller settings by pole assignment as function of Ω_1.
   What options are available to incorporate the variation of a?
   Select T_s.
   Build a suitable 'Simulink/Matab' file pair with a discrete-time implementation of the controller.
   Check its performance for the test sequence starting from steady state at $\bar{y}_1 = 1.0$, *i.e.* without the run-up phase.

2. *Saturation on $u_1(t)$*
   Start with the stability analysis.
   Determine first for the range of a the steady state $\bar{u}$ relative to the saturation values.
   What sizes of setpoint step deviations δr_1 and step disturbances $\bar{v}$ are admissible for a "well-posed" stability problem?
   Draw the Nyquist contour for the "reduced order and continuous" loop
   - for a high closed-loop bandwidth Ω
   - and different awf gains
   and then for the actual control loop (with $D > 0$ and $T_s > 0$).
   What is the predicted "area of attraction"?

   Check the transient response by simulation for the full test sequence.
   What is the actual "area of attraction"?

3. *Tuning of R*
   Demonstrate the effects of not properly tuning R.
   Hint: implement this by variation of pole assignment in $s^2 + s2\zeta\Omega + \Omega^2$ by
   - decreasing Ω for de-tuning
   - and inserting either $2\zeta \ll 2$ or $2\zeta \gg 2$ for mis-tuning.

2.8.6 Typical Results

Figs. 2.30 and 2.31 may help you to check your solutions.

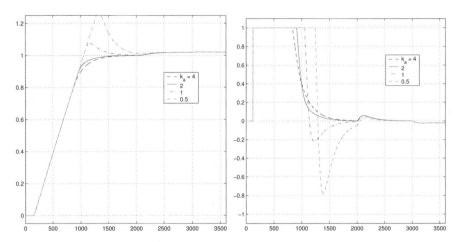

Fig. 2.30. Response to the test sequence with input constraints, $\Omega_1 T_1 = 6$, $a = 0$

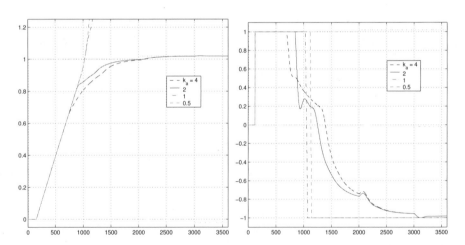

Fig. 2.31. Response to the test sequence with input constraints, $\Omega_1 T_1 = 6$, $a = -8$

2.9 Summary

The basic control loop with a plant of dominant first-order, input saturation and PI(aw) control has been investigated. The performance is satisfactory from an engineering design point of view.

As to *structure*, several simple alternatives are available. They are instantiations of a generic structure. And equivalence properties have been given.

Considering *transient response* on y and u, they perform well (some better than others) for both large setpoint steps and large load swings. Note that the "well-posed" control problem has been considered, where the ultimate steady state control variable $\bar{u}$ is at a finite distance inside of the saturation values. The response approaches that of the minimum-time one, with smooth transition to linear control and with no perceptible overshoot.

Nonlinear *stability analysis* applied to this basic loop shows that the initial condition response will be globally asymptotically stable to the ultimate steady state, if the internal feedback in the plant is negative. If the plant is unstable, the same methods apply and a finite region of attraction will result.

Some areas within the standard PI(aw) technology have not been investigated so far, such as

- the combination of derivative action **D** and awf
- how to implement antiwindup on a PII^2 controller for cases where the tracking error to a setpoint ramp must be brought to zero
- bumpless transfer
- and disturbances other than steps spaced which are widely enough that the loop may settle in between.

However, the implementation of *Output Constraints* is a much more fundamental question, and also a key element for industrial design. So it shall be addressed next, and the other areas of antiwindup just mentioned shall be looked into afterwards.

3

PI Control with Output Constraints

The motivation for investigating this area and its historical background have been briefly discussed in Chapter 1.

As for the input constrained case, many different solutions to output constrained control loops have been developed by practitioners decades ago and are in current use today. The aim here is to introduce selected basic concepts, investigate their dynamic performance in a particular test case, and analyze their nonlinear stability properties.

3.1 Arthur's Case

This case is again from the hydropower control field. Besides motivation, the aim is to demonstrate the basic ideas of output constraint control and also to illustrate how complex practical design problems may be compared with the simplified situations we shall investigate here.

About 20 years ago, Arthur (the senior control specialist from John's case in Sect. 2.1) was asked to do a control feasibility study for a customer operating a pumped storage plant. The large pump was driven by a synchronous motor. Switching such a large motor to the AC grid at standstill would create large overcurrents and torques and is not feasible. So the customer had to use one of his turbine-synchronous generator sets for run-up. It is coupled electrically to the pump motor at standstill. Then the turbine is run up slowly (for instance with approximately constant acceleration), and its generator delivers variable-frequency AC power to the synchronous motor and thereby pulls up the pump unit. This technique is known as "back-to-back" run-up.

In contrast to John's case, the load torque on the turbine is not near zero. At very low speed it is small, and then it increases approximately with the square of rotor speed, while the flow is zero. At about 75% of nominal speed, the pressure delivered exceeds the penstock pressure, and the main pump valve may be opened. Then, the flow up the penstock increases sharply with rotor speed and reaches its nominal value at nominal speed.

After speed stabilization there, the two-unit back-to-back system is synchronized with the grid, and then the AC power supply is switched over to the grid (at full power flow). Note that this amounts to a full load rejection on the turbine. Finally the turbine is shut down.

Note that the whole sequence should be run through quickly for economic reasons, and also because operating the pump at zero flow generates strong vibrations and heat and, if prolonged, may damage the pump. Also, the gradients should not be too large to avoid excessive pressure and flow transients in the waterways. A typical specification would be no more than 60 s from standstill to ready-for-synchronization.

First, we shall extend the plant model from Sect. 2.1 appropriately, then look into Arthur's solution, and finally show the results of a design that exploits the techniques we are going to present in the sequel.

3.1.1 Extending the Plant Model

From what we have developed for John's case, the following extensions have to be discussed; see also Fig. 3.1.
- the rotor acceleration signal as derived from the speed transducer
- the load torque produced by the pump as a function of speed
- the dynamics of the electrical back-to-back coupling, and
- the dynamics of the water columns,

all other elements being unchanged. Details are given below for simulations and exercises.

Water Column Dynamics

The low inertia, incompressible model used so far is a valid approximation for low head units at low flow. In this case, however, the elasticity and inertia effects are such that they have to be included in a useful model. The penstock dynamics are described by a partial differential equation of the telegraph type, which we shall approximate by a lumped model of second order showing the lowest one-end-open resonance mode of the penstock.

From the conservation of momentum of the mass content $m_L = \rho_L L_L A_L$ of the penstock at speed w_L and with nominal head H_r, pressure at turbine p_L and friction pressure loss Δp_f

$$\frac{d}{dt} m_L w_L = F_{drive} - F_{brake} \quad \rightarrow \quad \rho_L L_L \frac{d}{dt} w_L = \rho_L g H_r - p_L - \Delta p_f \quad (3.1)$$

Using p.u. notation with rated values (index r) yields

$$\frac{L_L w_{L_r}}{g H_r} \frac{d}{dt} \left(\frac{w_L(t)}{w_{L_r}} \right) = T_w \frac{d}{dt} \left(\frac{w_L(t)}{w_{L_r}} \right)$$

$$= 1.0 - \left(\frac{p_L(t)}{p_r} \right) - \frac{\Delta p_{f_r}}{p_r} \left| \frac{w_L(t)}{w_{L_r}} \right| \left(\frac{w_L(t)}{w_{L_r}} \right) \quad (3.2)$$

which defines T_w, and where the head is considered constant at rated and the friction pressure loss Δp_f proportional to $|w_L| w_L$ (note that this is to cover any reverse flow situations).

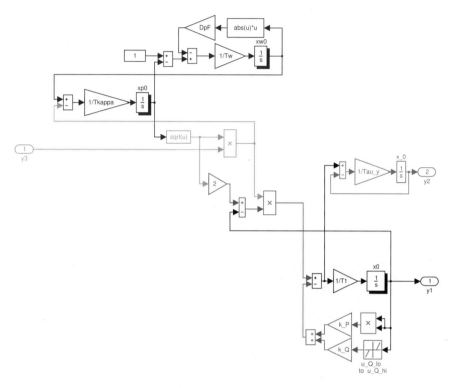

Fig. 3.1. Signal flow diagram (Simulink) of the plant submodel for Arthur's case

For relation of volume change $\Delta V_L(t)$ to pressure $p_L(t)$ with inflow $Q_L = w_L A_L$ and outflow to the turbine Q_T

$$\frac{d}{dt}\Delta V_L = Q_L(t) - Q_T(t) \tag{3.3}$$

and

$$p_L(t) = \frac{1}{\kappa}\frac{\Delta V_L(t)}{A_L L_L} \tag{3.4}$$

i.e.

$$\Delta V_L(t) = \kappa_L A_L L_L p_L(t) \quad \text{and} \quad \Delta V_{L_r} = \kappa_L A_L L_L p_r \tag{3.5}$$

Again in p.u. notation

$$\frac{\Delta V_{L_r}}{A_L w_{L_r}} \frac{d}{dt}\left(\frac{\Delta V_L}{\Delta V_{L_r}}\right) = \frac{\kappa_L \rho_L L_L A_L g H_r}{A_L w_{L_r}} \frac{d}{dt}\left(\frac{p_L(t)}{p_r}\right)$$

$$= T_\kappa \frac{d}{dt}\left(\frac{p_L(t)}{p_r}\right) = \left(\frac{w_L(t)}{w_{L_r}}\right) - \left(\frac{Q_T(t)}{A_L w_{L_r}}\right) \quad (3.6)$$

which defines T_κ.

For $Q_T := 0$ this yields a harmonic oscillator with zero damping and resonance frequency ω_1:

$$\omega_1^2 = \frac{1}{T_w T_\kappa} = \frac{1}{\kappa_L \rho_L L_L^2} \quad (3.7)$$

and with the speed of sound a_L in the penstock as $a_L := 1/\sqrt{\kappa_L \rho_L}$, then

$$\omega_1 = \left(\frac{a_L}{L_L}\right) \quad (3.8)$$

where typically $a_L \approx 1000$ m/s. This is used to determine T_κ, such that the lowest resonance mode ω_1 is at L_L, being a quarter wavelength.

Numerical values for the simulations shall be:

$$L_L = 500\ m; \quad w_{L_r} = 5\ m/s; \quad H_r = 250\ m; \quad \text{that is } T_w = 1.0\ s$$

from $\quad 4L_L = 2000\ m \rightarrow f_1 = 0.5\ Hz; \quad \omega_1 = 3.14\ rad/s \rightarrow T_\kappa = 0.10\ s$

and $\ \Delta p_{f_r}/p_r = 0.05$ $\hspace{6cm}$ (3.9)

Finally the outflow $Q_T(t)$ to the turbine is modeled by

$$Q_T(t) = k_v A_T(t)\sqrt{p_L(t)} \quad (3.10)$$

where the turbine opening A_T is proportional to the servomotor stroke $y_3(t)$; that is, in p.u. notation:

$$\frac{Q_T(t)}{Q_r} = \frac{y_3(t)}{y_{3_r}}\sqrt{\frac{p_L(t)}{p_r}} \quad (3.11)$$

As $p_L(t) \approx 1.0$, *i.e.* $p_L(t) > 0\ \forall\ t$, no exception handling is needed for the square root.

The Back-to-back Coupling

The two rotor inertias are coupled by the electric torque, which is approximately proportional to the difference between both rotor positions, the pole wheel angle. This produces a model from input torque to speed consisting of a harmonic resonator in series with an open integrator. The resonance frequency is typically at 2...3 Hz, *i.e.* a factor of five higher than the hydraulic resonance of the penstock. If the controller output is designed smooth enough, then the pole wheel resonance will not be excited, and thus it is excluded from the model here for simplicity. Then

$$\frac{\Theta\omega_r}{T_r}\frac{d}{dt}\left(\frac{\omega_T(t)}{\omega_r}\right) = T_1\frac{d}{dt}\left(\frac{\omega_T(t)}{\omega_r}\right) = \left(\frac{T_T(t)}{T_r}\right) - \left(\frac{T_P(t)}{T_r}\right) \qquad (3.12)$$

defining the rotor run-up time T_1.

The *numerical value* for the simulations is set to a typical value:

$$T_1 := 5 \text{ s} \qquad (3.13)$$

Driving and Load Torques

The driving torque T_T at the turbine is as in Sect. 2.1:

$$\frac{T_T(t)}{T_r} = \frac{Q_T}{Q_r}\left(2\sqrt{\frac{p_L(t)}{p_r}} - \frac{\omega_T(t)}{\omega_r}\right) \qquad (3.14)$$

where $\sqrt{p_L(t)/p_r}$ introduces the pressure influence on water inflow speed to the turbine.

The load torque T_P of the pump shall be modeled intuitively by

$$\frac{T_P}{T_r} = \frac{T_{P_p}}{T_r} + \frac{T_{P_Q}}{T_r}$$

$$\frac{T_{P_p}}{T_r} = k_P\left(\frac{\omega_T(t)}{\omega_r}\right)^2 \quad \text{for } 0 \le \frac{\omega_T(t)}{\omega_r} \lesssim 2$$

$$\frac{T_{P_Q}}{T_r} = k_Q\left(\frac{\omega_T(t)}{\omega_r}\right)$$

$$\text{with} \quad k_Q = 0 \qquad\qquad \text{for} \qquad 0 \le \frac{\omega_T(t)}{\omega_r} < \frac{\omega_{T_Q}}{\omega_r}$$

$$\text{and} \quad k_Q = \frac{1 - k_P}{1 - \frac{\omega_{T_Q}}{\omega_r}} \qquad \text{for} \qquad \frac{\omega_{T_Q}}{\omega_r} \le \frac{\omega_T(t)}{\omega_r} \lesssim 2$$

$$\text{such that} \quad T_P/T_r = 1 \qquad\qquad \text{at} \qquad \omega_T(t)/\omega_r = 1 \qquad (3.15)$$

Also ω_{T_Q}/ω_r is the pump speed where pressure in the pump at zero flow attains the penstock pressure $p_L/p_r = 1$. And torques due to losses are neglected.

The *numerical values* for the simulations are set to

$$k_P = 0.5; \quad \frac{\omega_{T_Q}}{\omega_r} = 0.75 \quad i.e. \quad \left.\frac{T_P}{T_r}\right|_{\omega=\omega_{T_Q}} \approx 0.281 \qquad (3.16)$$

The Rotor Acceleration Signal $y_2(t)$

In the model, $y_2(t)$ shall be picked up directly from the difference of turbine and pump torques by way of a first-order filter with unity gain and $\tau_{y_2} = 0.1$ s. On the real plant this will have to be obtained as the (filtered) time derivative of the speed transducer output.

The Servomotor Positioning Loop, $y_3(t)$

The structure from Fig. 1.3 in Sect. 2.1 is used with the following *numerical values*:

$$T_{sm} = 1.0\ s;\quad u_{up} = +0.05;\quad u_{dn} = -0.05;\qquad (3.17)$$

that is, the typical 20 s run time for full stroke. All other values are the same as before.

3.1.2 Arthur's Solution

The main restriction he had to comply with was staying as closely as possible to the proven structure of the standard analog electronic turbine regulator. Any add-on functions had to come from the catalog of available boards, and were restricted in numbers.

His basic idea was to use the opening limit setpoint on the standard controller (which was constant during run-up in John's case), and adjust it by an appropriate feedback loop such that rotor acceleration is approximately constant. Here is what he came up with in some more detail; Fig. 3.2.

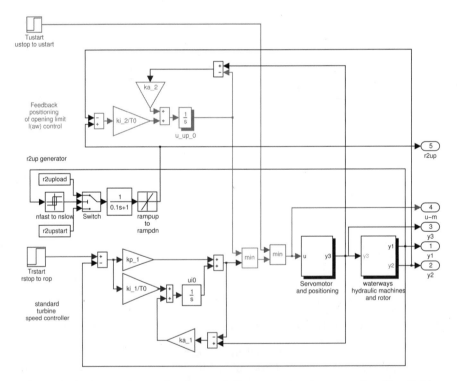

Fig. 3.2. Arthur's solution for the controller extension (converted to Simulink)

- The feedback adjustment of the opening limit is done by a standard substraction module,
- forming the feedback control error $e_2 = r_{2_{up}} - y_2$,
- followed by a standard integrator with adjustable k_{i_2}/T_0,
- and with awf to keep it from running off the actual u.
- The reference $r_{2_{up}}$ is produced by a standard ramp generator,
- which is set to zero at standstill and then ramps up to its preset value.

As the pump torque requirement for constant rotor acceleration would increase steeply after the pump starts delivering flow into the penstock, he also decided
- to lower the setpoint $r_{2_{up}}$, if $\omega_T \geq \omega_{TQ}$,
- such that the turbine opening $u(t)$ does continue at about the same gradient as just before
- in order to avoid upsetting the waterway dynamics excessively.

The controller parameter entries for the simulation have been calculated by pole assignment for a much reduced order model, containing only the servomotor and the rotor dynamics as open integrators with run-up times T_1 and T_{sm} only, and neglecting all other effects, *i.e.*

for the speed regulator

$$\Omega_1 T_1 = 5; \quad k_{sm} = 3\Omega_1 T_{sm} = 3;$$
$$k_{p_1} = 3\Omega_1^2 T_{sm} T_1 / k_{sm} = 5;$$
$$k_{i_1} = \Omega_1^3 T_{sm} T_1 T_0 / k_{sm} = 8.333; \quad \text{where} \quad T_0 := T_1$$
$$\text{and} \quad k_{a_1} = \Omega_1 = 1.0 \tag{3.18}$$

for the acceleration regulator

$$\Omega_2 T_1 = 5; \quad k_{i_2} = \Omega_2 T_0 = 5;$$
$$k_{a_2} = \Omega_1 = 1.0$$

$$\text{start phase setpoint} \quad r_{2_{up}} = 0.16$$
$$\text{loading phase setpoint} \quad r_{2_{up}} = 0.08$$
$$\text{and} \quad r_{2_{up}}, r_{2_{dn}} \text{ ramp rate at} \quad \pm 0.025/s \tag{3.19}$$

The simulation results are shown in Fig. 3.3.

Observe that

- Transitions are executed without bumps or windup effects.
- The run-up time specified ($\leq 60\ s$) is complied with, and this with a feasible control action $u(t)$. Note that r_2 setpoints for both start and loading phases have been tuned accordingly.
- Run-up gradients on $u(t)$ before and after flow into the penstock starts are as prescribed.
- The controller output $u(t)$ seems smooth enough not to excite pole wheel oscillations (see above).

- Both closed-loop bandwidths Ω_1 and Ω_2 entries are near the upper limit, which is imposed by the waterway dynamics. They cannot be increased substantially without generating strong oscillatory modes.
- But, by inspection of Fig. 3.3, Ω_1 is at the low limit regarding closed-loop performance for synchronization. Note that this is to be done under full power, and this puts tight specifications on performance of the speed control loop.
- Finally, there is a very substantial residual error e_2, especially in the pump flow phase, where near the end e_2 attains around 50% of the setpoint value. Note that this happens even with the integral controller R_2. This makes tuning of the setpoints r_2 a tedious task.

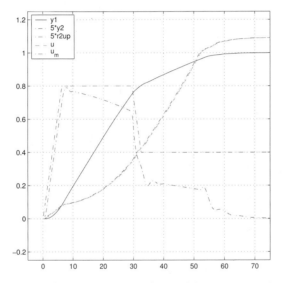

Fig. 3.3. Arthur's solution: the model run-up transient

3.1.3 An Advanced Solution

The control system of Fig. 3.2 has been designed within the implementation restrictions, as they were imposed by the analog regulator equipment available to Arthur at that time. It functions correctly, and the results may be considered acceptable, but they are not very good. Further performance improvement would be very welcome, *i.e.*

- both Ω_1 and Ω_2 should be improved by at least a factor of two,
- and the residual error e_2 should be reduced at least by a factor of five,
    to $\leq 10\%$ of r_2.

This can be attained without losing the smoothness properties on $u(t)$; see
Fig. 3.4,
- if the implementation restrictions from above are put aside,
- and if state feedback is added to both loops for active damping
  of the penstock oscillation,
- and, finally, if the override techniques we shall investigate later
  are put to use.

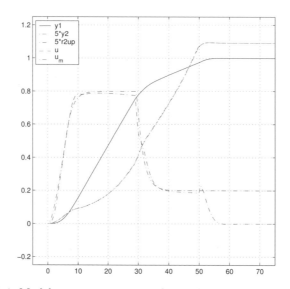

Fig. 3.4. Model run-up transient with an advanced control structure

Comments

On current digital turbine control systems, function blocks may be configured
more freely, so this is now a realistic assumption.

Active damping feedback allows the closed-loop bandwidth to be moved up
to $\Omega_1 = \Omega_2 = 3$ rad/s in the model; see Fig. 3.4. On the real plant, $p_L(t)$
may be measured with low expense, whereas $Q_L(t)$ will have to be estimated
by an observer. Note that such additional feedback has been implemented
successfully in a number of hydropower stations, but it is not yet routine. Note
that this add-on measure will be unavoidable for plants where the penstock
resonance frequency is lower than in our case.

Exercise

  - Use the data given as the basis of a "case study".
  - Design your own control system.
  - And improve on what is shown in Fig. 3.4 !

3.2 The Basic Concept

3.2.1 Problem Statement

Consider the control problem of Fig. 3.5, see also Sect. 1.3.

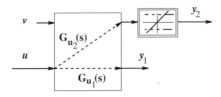

Fig. 3.5. Control system for y_1 by u with upper and lower constraints on the secondary output y_2

The main controlled variable is y_1. Associated with it is the feedback controller R_1 with setpoint r_1. Then a second output variable y_2 is to be constrained to operational limit values $y_{2_{lo}}$, $y_{2_{hi}}$.

A typical case is a DC-servomotor, where y_1 is the rotor speed, y_2 is the armature current (which is to be limited), and u is the voltage supplied by the power amplifier (which is not to be constrained here). Another typical case is position control of a rigid mass, where y_1 is its position, y_2 is speed, and u is force or torque applied, which again shall not be constrained.

The final steady state $\bar{y}_2$ of the second output shall lie at a finite distance inside the limits $y_{2_{lo}}$, $y_{2_{hi}}$, such that linear control of y_1 around r_1 is feasible for small but finite deviations δr_1, δv without y_2 interfering with its limits $y_{2_{lo}}$, $y_{2_{hi}}$. In other words, control shall not be stuck to the constraint limit permanently. This would be considered as an "ill-posed" problem, to be remedied by structural changes in the plant.

Also, no input constraints (on u) shall be active here. In other words, only transients will be considered, where the input constraints are not met.[1] This also implies sufficiently bounded disturbance inputs v. Therefore, the investigations are clearly restricted to a subset of all operational transients. A more complete problem setup with both output and input constraints being active will be considered in Chapter 4.

All input-output paths indicated in Fig. 3.5 are linear and time invariant, described by their transfer functions.

In particular, the response of y_2 to u is to be linear across the limits $y_{2_{lo}}$,

---

[1] Note that input constraints are always present in control loops due to physical actuator saturations.

$y_{2_{hi}}$, within a sufficiently large region around the limits to accommodate the control trajectories to be expected.

Then

- The first basic element is to modify u_1 delivered by the controller R_1 prior to applying it to the plant as u, such that y_2 does not exceed its respective limit value $y_{2_{lo}}$ or $y_{2_{hi}}$, but runs along it, and as closely as possible.

  Intuitively this will produce the least possible constraint on u versus u_1, and thereby the weakest interference with the y_1-loop, and thus the least performance degradation in the y_1 transient, which is still consistent with respecting the output constraint.

- The second basic element is that this modification on u_1 is done by *feedback*, *i.e.* feedback controllers $R_{2_{lo}}, R_{2_{hi}}$, rather than by *feedforward*, for better rejection of parameter changes and disturbances. A feedforward scheme would amount to putting appropriate saturation clamps on stroke and/or slew of u. In practice, often a combination of feedforward and feedback schemes is used.

- The third basic element concerns easy setup and tuning. The reference values $r_{2_{lo}}$, $r_{2_{hi}}$ to be used on the feedback controllers should be equal to the respective limit values $y_{2_{lo}}$, $y_{2_{hi}}$. In other words, the corresponding control errors $e_{2_{lo}}$, $e_{2_{hi}}$ should converge to zero, when the system trajectory evolves along the constraints.

  This leads to integral action in the constraint feedback controllers $R_{2_{hi}}$, $R_{2_{lo}}$, the same as in the main controller R_1.

3.2.2 Typical Controller Structures

Three typical implementations of these basic ideas are given in Fig. 3.6.

- The *Nonlinear-Additive* form was introduced with hydraulic-mechanical regulator equipment more than 50 years ago. Note that it will function properly only if there is *no* integral action upstream of the summing point, *i.e.* neither in R_1 nor in R_2. In other words, there will always be nonzero control errors e_2 along the constraints, as $u(t)$ is a linear combination of $e_1(t)$ and $e_2(t)$. Also, the gain k_2 in the y_2-path must be several times higher than the gain on y_1 in R_1 to produce any useful constraining effect. In order to attain a sufficient phase margin in the y_2-loop for such a high gain k_2, some compensation is usually required. This is indicated in R_2'. Note that an integral action is admissible downstream of the summing point to obtain $\bar{e}_1 \rightarrow 0$ at least.

- The *Cascade-Limiter* form is very popular in drive systems, such as for current limiting within speed control loops, or for speed limiting in position control loops. It is applied in other contexts as well, but, as we shall see, it is not as versatile as the bottom form Fig. 3.6(c). Note that both R_1 and

R_2 must contain integral action to have control errors e_1 and $e_{2_{lo}}$, $e_{2_{hi}}$ converge to zero. Then R_1 needs awf from the $r_{2_{lo}}$, $r_{2_{hi}}$ saturation.

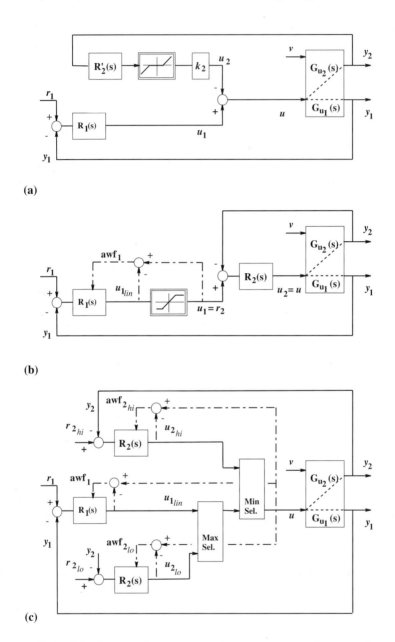

(a)

(b)

(c)

Fig. 3.6. Typical implementation structures for output constraint control *Nonlinear-Additive* (top) *Cascade-Limiter* (center) *Selectors* (bottom)

- The *Selectors* form has been developed in the 1960s for electronic analog and pneumatic equipment for control in the power generation and chemical industries; see also Sect. 1.2. It is also known as the "lowest wins" strategy [26]. This suggests that u is constrained in the "up"-direction only.

  Here it is extended for both "up" and "down" directions: u_1 is modified into u by a Maximum Selection with $u_{2_{lo}}$ for control along the lower constraint $r_{2_{lo}}$, which is followed by a Minimum Selection with $u_{2_{hi}}$ for control along the upper constraint $r_{2_{hi}}$. All controllers have integral action with awf from u.

Note that in all three cases the output constraint feedback manipulates the control variable $u(t)$, *i.e.* they use the *"control conditioning"* approach. We shall investigate them in more detail now using the following benchmark where necessary. In Chapter 7 we shall also consider structures using the *"reference conditioning"* approach.

3.3 The Benchmark Test

Consider the positioning control task mentioned above.

(a) The **plant** shall be described by the two first-order differential equations for speed s and position p:

$$\frac{d}{dt}p = s$$

$$\frac{d}{dt}(ms) = +F_d - F_b$$

where the mass m is constant and the driving and braking force F_d and F_b are both independent of speed s and position p. Introducing rated values for all variables produces a "per unit' notation

$$\text{for position} \quad x_1 = \frac{p(t)}{p_r} \quad \text{with} \quad \tau_1 = \frac{ms_r}{F_r}$$

$$\text{for speed} \quad x_2 = \frac{s(t)}{s_r} \quad \text{with} \quad \tau_2 = \frac{p_r}{s_r}$$

$$\text{and} \quad b_u = \frac{F_{d_f}}{F_r}; \quad b_v = \frac{F_{b_f}}{F_r}$$

$$\begin{bmatrix} \tau_1 & 0 \\ 0 & \tau_2 \end{bmatrix} \frac{d}{dt} \begin{bmatrix} x_1 \\ x_2 \end{bmatrix} = \begin{bmatrix} 0 & +1 \\ 0 & 0 \end{bmatrix} \begin{bmatrix} x_1 \\ x_2 \end{bmatrix} + \begin{bmatrix} 0 & 0 \\ +b_u & -b_v \end{bmatrix} \begin{bmatrix} u & 0 \\ 0 & v \end{bmatrix} \quad (3.20)$$

where F_{d_f}, F_{b_f} are the respective full-scale values.

Letting $y_1 := c_1 x_1$ and $y_2 := c_2 x_2$ with sensor gains c_1, c_2 and introducing again a small delay D to cover the non-modeled fast dynamics yields the transfer functions to u

$$G_{u_2} = \frac{y_2}{u} = b_u c_2 \frac{1}{s\tau_2} e^{-sD}$$

$$\text{and} \quad G_{u_1} = \frac{y_1}{u} = b_u c_1 \frac{1}{s^2\tau_1\tau_2} e^{-sD} = \frac{c_1}{c_2} G_{u_2} \frac{1}{s\tau_1} \quad (3.21)$$

Setting $D := 0$ yields the dominant dynamic response G_{d_1} and G_{d_2}. Similarly, the response to the load force input v is modeled by:

$$G_{v_2} = \frac{y_1}{v} = - b_v c_2 \frac{1}{s\tau_2} \quad (3.22)$$

$$\text{and} \quad G_{v_1} = \frac{y_2}{v} = - b_v c_1 \frac{1}{s^2\tau_1\tau_2} = \frac{c_1}{c_2} G_{v_2} \frac{1}{s\tau_1} \quad (3.23)$$

Numerical values for the test case are entered as follows:

$$b_u = b_v := 1; \quad \tau_1 := 5; \quad \tau_2 := 1; \quad D := 0.025; \quad c_1 = c_2 := 1 \quad (3.24)$$

(b) **Controller Structures**
Consider first both feedbacks of y_1 and y_2 separately.

Controller R_1
Applying basic linear design rules leads to a standard linear cascade arrangement for the main controlled variable y_1, with a P controller with gain k_{s_1} in the inner loop for the speed signal y_2, and a standard PI(aw) controller in the outer loop for y_1, *i.e.*

$$R_1 = \frac{u_1}{e_1} = k_{p_1}\left(1 + \frac{1}{sT_{i_1}}\right)k_{s_1} \quad (3.25)$$

with the awf form **E** of Sect. 2.4.2 and gain k_{a_1}.

Controllers R_2
Applying standard design rules for the output constraint feedback of y_2 to $r_{2_{lo}}, r_{2_{hi}}$ leads to two standard PI(aw) controllers $R_{2_{lo}}$, $R_{2_{hi}}$,

$$R_{2_{lo}} := R_{2_{hi}} = \frac{u_2}{e_2} = k_{p_2}\left(1 + \frac{1}{sT_{i_2}}\right) \quad (3.26)$$

with awf of form **D** and gains k_{a_2}. In other words, both controllers $R_{2_{lo}}$, $R_{2_{hi}}$ shall have the same parameters k_{p_2}, T_{i_2}.
Note that this is directly applicable to the *selectors* form of Fig. 3.6. The other two forms require further modifications.

For the *nonlinear additive* form the controllers must be split into an linearly equivalent series arrangement with an integrator $1/(s\tau_0)$ used by all loops, and

$$u_1 = k_{p_1}\frac{\tau_0}{T_{i_1}}(sT_{i_1} + 1)e_1 - s\tau_0 k_{s_1} y_2$$

$$\text{and} \quad u_{2_{lo,hi}} = k_{p_2}\frac{\tau_0}{T_{i_2}}(sT_{i_2} + 1)e_{2_{lo,hi}} \quad (3.27)$$

The *cascade limiter* form will require the inner loop controller to be of PI type in order to suppress persistent errors e_2 along the constraints. The outer loop controller may then be either of P type (which is often the case in mechatronics) or of PI(aw) type. In this case it turns out to be more convenient for pole assignment design if the structure of the inner loop controller is modified, such that its setpoint $r_2(t)$ is only active on the integral part and not on the proportional path as well. This slightly nonstandard structure is tolerated here, as it is often available in industrial process control systems.

(c) The **controller parameters** are determined by pole assignment using the "dominant dynamics" models G_{d_1} and G_{d_2}.

Controller R_1
For the *selector* form, in a first step the state feedback gains $k_{1_0}, k_{1_1}, k_{1_2}$ of the main controller R_1 are determined. From the characteristic equation

$$0 = s^3 \tau_0 \tau_1 \tau_2 + s^2 \tau_0 \tau_1 (a_2 + b_u c_2 k_{1_2})$$
$$+ s \tau_0 (a_1 + b_u c_1 k_{1_1}) + b_u c_1 k_{1_0}$$
$$\stackrel{!}{=} (s + \Omega_1)^3$$

$$i.e. \quad b_u c_2 k_{1_2} = 3\Omega_1 \tau_2 - a_2$$
$$b_u c_1 k_{1_1} = 3\Omega_1^2 \tau_1 \tau_2 - a_1$$
$$b_u c_1 k_{1_0} = \Omega_1^3 \tau_0 \tau_1 \tau_2 \tag{3.28}$$

where in this case $\tau_0 := \tau_2; \quad a_2 = a_1 := 0; \quad b_u c_2 = b_u c_1 := 1;$

and for the standard PI P cascade structure, Eq. 3.25

$$k_{s_1} = k_{1_2}; \quad k_{p_1} = k_{1_1}/k_{1_2}; \quad T_{i_1} = k_{1_1}/k_{1_0} \tag{3.29}$$

Controllers R_2

$$0 = s^2 \tau_0 \tau_2 + s \tau_2 (a_2 + b_u c_2 k_{2_2}) + (a_1 + b_u c_2 k_{2_1})$$
$$\stackrel{!}{=} (s + \Omega_2)^2$$

$$\text{yielding} \quad b_u c_2 k_{2_2} = 2\Omega_2 \tau_2 - a_2$$
$$b_u c_2 k_{2_1} = \Omega_2^2 \tau_0 \tau_2 - a_1 \tag{3.30}$$

with $\tau_0 := \tau_2; \quad a_2 = a_1 := 0; \quad b_u c_2 := 1;$

and for the standard PI structure

$$k_{p_2} = k_{2_2}; \quad T_{i_2} = k_{2_2}/k_{2_1} \tag{3.31}$$

with *numerical values* $\Omega_1 := 3$, $\Omega_2 = 2\Omega_1 = 6$, and $T_s = 0.010$.

For the *cascade-limiter* form specified above, a very simple and convenient approach is to design both controllers in one step to the same pole locations at $-\Omega$. The inner loop controller is given as

$$u_{lin} = \frac{k_{2_1}}{s\tau_0}(r_2 - y_2) - k_{2_2}y_2 \tag{3.32}$$

Then the characteristic equation is:

$$0 = s^4\tau_0^2\tau_1\tau_2 + s^3\tau_0^2\tau_1(a_2 + b_uc_2k_{2_2})$$
$$+ s^2\tau_0^2(a_1 + b_uc_2k_{2_1}) + s\tau_0(b_uc_1k_{1_1}k_{2_1}) + b_uc_1k_{1_0}k_{2_1}$$
$$\stackrel{!}{=} (s + \Omega)^4$$

i.e.
$$b_uc_2k_{2_1} = 4\Omega\tau_2 - a_2$$
$$b_uc_2k_{2_2} = 6\Omega^2\tau_1\tau_2 - a_1$$
$$b_uc_1k_{1_1} = 4\Omega^3\tau_0\tau_1\tau_2/k_{2_1}$$
$$b_uc_1k_{1_0} = \Omega^4\tau_0^2\tau_1\tau_2/k_{2_1} \tag{3.33}$$

with $\tau_0 := \tau_2$.

The *numerical value* $\Omega := 3$ is suitable for $D = 0.025$ from above and for setting $T_s := 0.010$ s.

(d) The **test sequence** shall be basically the same as in Sect. 2.2 for the input saturation case. Some refinements are added to model a typical startup sequence in more detail.

After starting the simulation, $r_1 := 0$ until time $T = 0.5$ s and then $r_1 := 0.98$.

Also, a shutdown input constraint is inserted on u. It forces $u := 0$ until time $T_{on} = 5.0$ s and only then transfers u to the controller output.

The control system then has to run up along the $r_{2_{hi}}$ constraint, which will take approximately 10 s (see numerical values in items (a) and (e)). During this phase, a load step $v_1 = +0.90$ is applied at time $T_{on} + 6$ s, to check the performance of the output constraint loop (similar to what appeared in Arthur's case).

After settling at $y_1 = r_1 = 0.98$, a small setpoint step to $r_1 = 1.00$ is applied as in Sect. 2.2, and then a large reverse load swing to $v_2 = -0.90$, to check the performance of the y_1-loop.

Again no high-frequency measurement disturbance shall be considered.

e) The **constraint setpoints** are set to

$$r_{2_{hi}} = +0.5 \quad \text{and} \quad r_{2_{lo}} = -0.5 \tag{3.34}$$

that is, to a minimal transition time of 10 s from 0 to 100% of position.

3.4 Structures for Output Constraint Control

The next step is to investigate candidate structures for output constraint control of the *"control conditioning"* class more closely and check them on the benchmark test if necessary.

We shall start with the nonlinear additive idea, which seems to be the most intuitive, then go on to the selector concept and finally investigate the cascade-limiter arrangement.

A broad sweep will be made, ending up in a "generic structure". Its transient response properties will be discussed in more detail in the following section. Typical forms of the *"reference conditioning"* class shall be discussed separately later.

3.4.1 The Nonlinear-Additive Concept

Designing the separate linear controllers in Sect. 3.3, item (b) has led to a series arrangement of an integrator downstream of the nonlinear addition point in Fig. 3.6(a), which is used by all feedback loops, and corresponding PD elements upstream of it.

Form A

This leads to Fig. 3.7. Note the shutdown input constraint switch, which is put within the discrete integrator loop, and thus conforms to the awf structure of Fig. 2.12(c). Note also that the setpoint step r_1 is only applied to the proportional path and not differentiated, to avoid undue upset of u and thus of the output constraint feedback. Finally, note that in R_1 the $d^2 y_1/dt^2$ path is replaced by using dy_2/dt with appropriate weight, for numerical reasons.

The response to the benchmark test sequence in Fig. 3.8, shows that
- the y_1-loop performs well,
- but the y_2 constraint feedback is far from working as specified.

By inspection of the structure in Fig. 3.7 this must be due to the *interaction* of the main control signal v_1 with the constraint feedback output v_2 along the constraint.

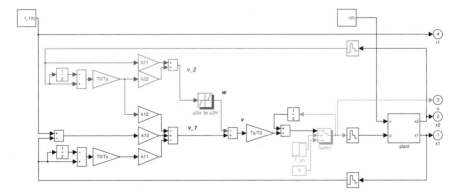

Fig. 3.7. Structure of the controller form **A**

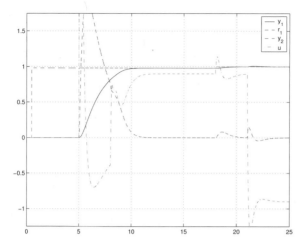

Fig. 3.8. Response to the benchmark test sequence with form **A** load step applied at $t = 8$ s

Form B

This defect can be suppressed very easily; see Fig. 3.9.
The input w_i to the deadspan nonlinear block is augmented by v_1, *i.e.*

$$w_i := v_1 + v_2$$
$$\text{and} \quad v = v_1 - w$$
$$\text{and also} \quad u_{2_{hi}} = k_{2_1} r_{2_{hi}}$$

$$\text{Now if} \quad w_i \leq u_{2_{hi}} \quad \text{then} \quad w = 0 \quad \text{and} \quad v = v_1$$
$$\text{but if} \quad w_i > u_{2_{hi}} \quad \text{then} \quad v = w_i - u_{2_{hi}}$$

$$\text{i.e.} \quad v = v_1 - w_i + u_{2_{hi}}$$
$$= v_1 - v_1 - v_2 + u_{2_{hi}}$$
$$v = -v_2 + u_{2_{hi}} \tag{3.35}$$

and similarly along the lower constraint $u_{2_{lo}}$.
Now v depends only on v_2, as $u_{2_{hi}}$ is a constant offset which is determined by the output constraint setpoint $r_{2_{hi}}$; see above. In other words, the interaction of v_1 on v_2 is now *eliminated*.

This is confirmed in the simulation in Fig. 3.10, where now $e_2 \to 0$ for the run-up trajectory along $r_{2_{hi}}$, as per specification.

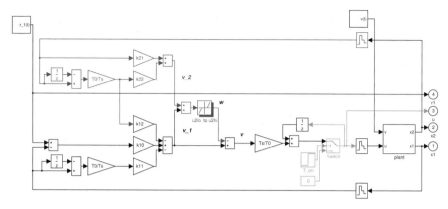

Fig. 3.9. Structure of the controller form **B**

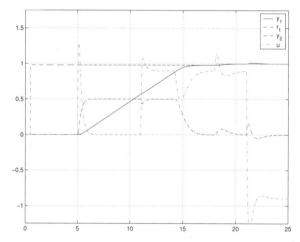

Fig. 3.10. Response to the benchmark test sequence with form **B** load step v applied at $t = 11$ s

Form C

There may be applications where the integral action is not needed nor wanted for some other reasons. This leads to the very simple structure of P cascade for y_1 and P control for y_2; Fig. 3.11. The controller parameters have been determined by pole assignment to the same locations Ω_1, Ω_2 as above. Note that in the response in Fig. 3.12 there are now persistent control errors on both e_2 and e_1 (from v_1, v_2 being nonzero).

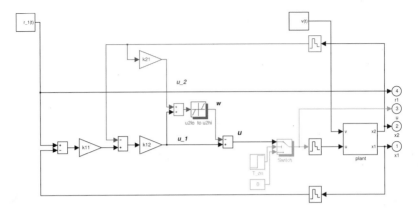

Fig. 3.11. Structure of form **C** with P controllers

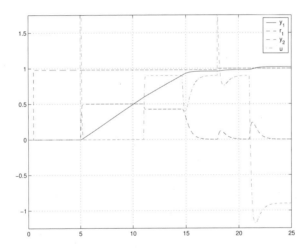

Fig. 3.12. Response to the benchmark test sequence with form **C**

Form D

The aim of developing form **B** further into **D** is to avoid the differentiation, make the form more flexible, and make it implementable by standard software modules on current process control systems. This needs several small manipulations; see Fig. 3.13 for illustration.

-   Fig. 3.13(a): The output constraint feedback through the deadspan non-linear element is split into two separate branches for $u_{2_{hi}}$ and $u_{2_{lo}}$.

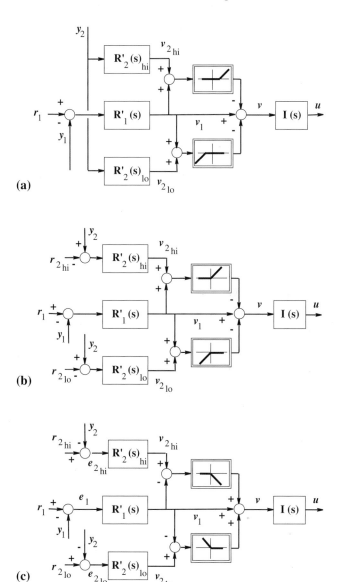

Fig. 3.13. Developing structure form **D** out of form **B**

- Fig. 3.13(b):
  - The breakpoints of the deadspan elements are shifted upstream across the R_2 controllers into a standard "+ measured value - setpoint" subtraction element delivering $-e_{2_{hi}}$ into $R_{2_{hi}}$.

- The nonlinear characteristic then has to be modified into

$$w = 0 \text{ if } v_1 + v_{2_{hi}} \leq 0 \quad \text{and} \quad w = v_1 + v_{2_{hi}} \text{ if } v_1 + v_{2_{hi}} > 0 \quad (3.36)$$

that is, a standard diode characteristic with unity slope.
And correspondingly for the "low" constraint, but there with the nonlinear characteristic

$$w = v_1 + v_{2_{lo}} \text{ if } v_1 + v_{2_{lo}} < 0 \quad \text{and} \quad w = 0 \text{ if } v_1 + v_{2_{lo}} \geq 0 \quad (3.37)$$

- Fig. 3.13(c): Finally the signs are adjusted to obtain standard regulator blocks for the R_2 controllers, and the nonlinear characteristics are modified accordingly.

Form E

The next step is to avoid the differentiations in the R_1'- and R_2'-blocks of Fig. 3.13. This is done by moving the integral action upstream into each R_i, *i.e.* to use a standard PI structure such as in Fig. 2.14. As the individual controllers are operating intermittently, an awf is required on each one. The open question is where to pick up the signals for forming the awf errors e_{a_i}, then the rest of the structure would be straightforward.

Hanus' Idea

This is resolved by slightly generalizing the construction rule for e_a used so far, which was to use the signals just upstream and just downstream of the saturation element. As proposed by Hanus, we shall replace this by using the output of the linear controller ($u_{i_{lin}}$) and the final u as it is applied to the plant. Note that this may contain several successive modifications of u_1 by saturations, nonlinear additions, *etc.*
Applying this generalized rule to R_1 and the previous rule to both R_2 controllers yields the structure shown in Fig. 3.14(top).

The final modification is to shift the signals to awf paths of the R_2 controllers as shown in Fig. 3.14(bottom), producing form **E**.
The structures in Fig. 3.14 are *equivalent*, as

$$\begin{aligned}
\text{from Fig.3.14(top)} \quad & e_{a_1} = -u_1 + u \\
\text{and} \quad & e_{a_{2_{hi}}} = -w + v \\
\text{where} \quad & w = u_{2_{hi}} - u_1 \\
\text{and} \quad & u = v + u_1 \quad \rightarrow \quad v = u - u_1 \\
\textit{i.e.} \quad & e_{a_{2_{hi}}} = -u_{2_{hi}} + u_1 + u - u_1 = -u_{2_{hi}} + u \quad (3.38)
\end{aligned}$$

$$\text{from Fig.3.14(bottom)} \quad e_{a_{2_{hi}}} = -u_{2_{hi}} + u \quad (3.39)$$

which is the same.

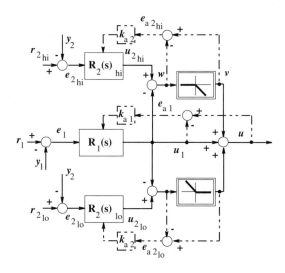

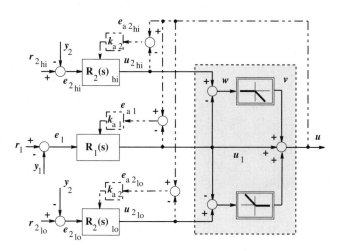

Fig. 3.14. Using integral action upstream of the nonlinear addition function and awf (top)   and Controller form **E**   (bottom).
The internal structure of the R_1 and R_2-blocks is form **D** from Sect. 2.4.

Form **E** now consists of three standard PI(aw) controllers with individual finite awf feedback gains k_{a_i}, and with a common tracking input u from the output of the "nonlinear additive" block, which is shaded in Fig. 3.14, bottom right.

This structure is more flexible and adaptable to different design needs than e.g. form **B**, because either R_1 or R_2 may also be designed without integral action.

Its performance will be investigated in more detail in Sect. 3.5.

3.4.2 The Selector Concept

Minimum and Maximum Selector blocks have been used for many decades in control engineering, and are the root of a second family of structures for output constraint control.

Form F

The "nonlinear additive" block of form **E** Fig. 3.14(bottom right) is the only non-standard function block in current process control systems. This shall be rectified next.

The *nonlinear characteristic* from w to v in the shaded block may be interpreted as

$$v = Min\,\{0,\ w\}$$
$$= Min\,\{0,\ u_{2_{hi}} - u_1\} \tag{3.40}$$

and as

$$u = u_1 + v = u_1 + Min\,\{\,0,\ u_{2_{hi}} - u_1\} \tag{3.41}$$

where u_1 may be added instead to both elements within the Minimum Selection, finally

$$u = Min\,\{\,u_1,\ u_{2_{hi}}\} \tag{3.42}$$

Similarly for the low side:

$$v_{lo} = Max\,\{\,0,\ w_{lo}\} = Max\,\{\,0,\ u_{2_{lo}} - u_1\}$$
$$\text{and}\quad u = u_1 + Max\,\{\,0,\ u_{2_{lo}} - u_1\}$$

i.e. finally

$$u = Max\,\{\,u_1,\ u_{2_{lo}}\} \tag{3.43}$$

Now if the "well-posedness" condition holds:

$$u_{2_{lo}} < u_{2_{hi}} \quad \forall \quad t \geq 0 \tag{3.44}$$

then the Min- and Max-selector blocks may be put in arbitrary sequence in the signal flow graph, such as putting the Max-selector first, and the Min-selector next:

$$u = Min\{ u_{2_{hi}}, Max\{ u_{2_{lo}}, u_1\}\} \tag{3.45}$$

or equivalently

$$u = Max\{ u_{2_{lo}}, Min\{ u_{2_{hi}}, u_1\}\} \tag{3.46}$$

Such Min- and Max-selector hardware blocks or software functions have been standard elements of industrial process control equipment since the 1960s.

This produces the controller form **F**, Fig. 3.15, which is *equivalent* to form **E**, Fig. 3.14, as we have just shown. We shall develop this into a "generic form" of output constraint controllers later.

If the condition of Eq. 3.44 should not hold, then the result depends on the sequence of the selector blocks. [2]

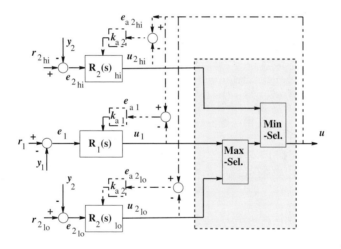

Fig. 3.15. Controller form **F**.
Note that the sequence of Min- and Max-selectors may be inverted

Exercise
- Investigate the effect of this sequence, if the problem is not "well-posed".
- and also with form **E** in this case.

---

[2] this property will be actively exploited later.

Form G

So far, the selection has been performed on the outputs of the linear controller parts. It has also been proposed to perform the selection on the individual control errors instead, *i.e.*

$$e = Min\{ e_{2_{hi}}, Max\{ e_{2_{lo}}, e_1\}\} \tag{3.47}$$

and then using a common control algorithm R for $e \to u$.

The advantage is that the integral action contained in R is always "in the loop", and thus will not wind up, and no awf is needed. The disadvantage is clearly that no reasonable closed-loop performance can be expected from a common R, as G_1 and G_2 are usually quite different. So this form will not be pursued any further, also because better alternatives are available.

Form H

This form tries to combine the advantage of **G** with having separate R_1 and R_2 controllers for individual tuning. This leads to placing the integral action in the common R, and having the R_1 and R_2 as generalized PD controllers, form **H**, in Fig. 3.16. It has been in use since the 1970s for boiler load control with multiple output constraints from critical temperatures, and temperature gradients, *i.e.* thermal stresses [18].

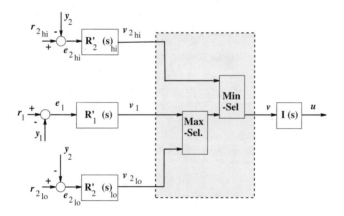

Fig. 3.16. Controller form **H**.
Again note that the sequence of Min- and Max-selectors may be inverted

And for continuous-time controllers:

$$u = \frac{1}{s}v$$
$$v_1 = k_{0_1}\left(r_1 - y_1\right) - sk_{1_1}y_1 - sk_{2_1}y_2 - \cdots$$
$$v_{2_{hi}} = k_{0_2}\left(r_{2_{hi}} - y_2\right) - sk_{1_2}y_2 \tag{3.48}$$

It may be convenient to insert an appropriate time scaling factor T_0. For discrete-time implementations it is recommended to use:

$$u = \frac{1}{1 - z^{-1}}v$$
$$v_1 = k_{0_1}T_s\left(r_1 - y_1\right) - \left(1 - z^{-1}\right)k_{1_1}y_1 - \left(1 - z^{-1}\right)k_{2_1}y_2 - \cdots$$
$$v_{2_{hi}} = k_{0_2}T_s\left(r_{2_{hi}} - y_2\right) - \left(1 - z^{-1}\right)k_{1_2}y_2 \tag{3.49}$$

rather than use as derivatives $\left(1 - z^{-1}\right)/T_s$, to improve scaling (especially useful for short T_s).

The strength of form **H** is the very simple and transparent structure. Unfortunately, there are numerous weaknesses: integral action must be implemented in all paths, whether it is required or not, and very different time scales in G_1, G_2 may generate scaling difficulties. The main problem, however, is its sensitivity to measurement noise, which must always be considered in real plants. The sensitivity is due to the differentiation in the R_1, R_2 together with the high-gain awf, as we shall demonstrate in Chapter 5.

Form J

This form is generated from **H** in its discrete-time form by splitting the one step delay of the integral action into three parallel ones and moving them upstream around the Max-Min-selector group; Fig. 3.17.

If these delay elements are all initialized to the same value, then by inspection form **J** is *equivalent* to form **H**:

$$u_{2_{hi}} = v_{2_{hi}} + uz^{-1}$$
$$\text{and} \quad u_1 = v_1 + uz^{-1}$$
$$\text{and} \quad u = Min\left\{u_{2_{hi}},\ u_1\right\}$$

$$i.e. \quad u = Min\left\{v_{2_{hi}} + uz^{-1},\ v_1 + uz^{-1}\right\}$$
$$\text{or} \quad u - uz^{-1} = Min\left\{v_{2_{hi}},\ v_1\right\}$$
$$\text{finally} \quad u\left(1 - z^{-1}\right) := v \tag{3.50}$$

$\square$

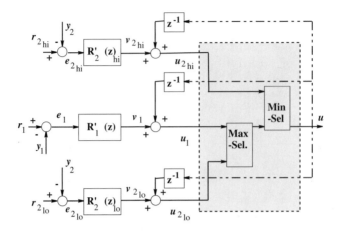

Fig. 3.17. Controller form **J**

Special Case of Form J: Input Saturation

Consider now the special case

$$G_2(z) = 1z^{-1} \quad i.e. \quad y_2(z) := u(z)z^{-1} \tag{3.51}$$

In other words, y_2 is the control input u, shifted by one sampling delay. This is generated by the output via the DA converter, and the input through the AD converter at the next sample.
Set also

$$R'_2(z) = 1 + 0\left(1 - z^{-1}\right) \quad i.e.: R_2 \text{ is a unity gain integral controller} \tag{3.52}$$

Then the two one-delay feedback paths in Fig. 3.18 cancel, and the "hi" input to the Min-selector block is reduced to $r_{2_{hi}}$, *i.e.*

$$r_{2_{hi}} := u_{hi} \tag{3.53}$$

In other words, this special form in Fig. 3.18 reverts to the input saturation structure of Fig. 2.12(c).

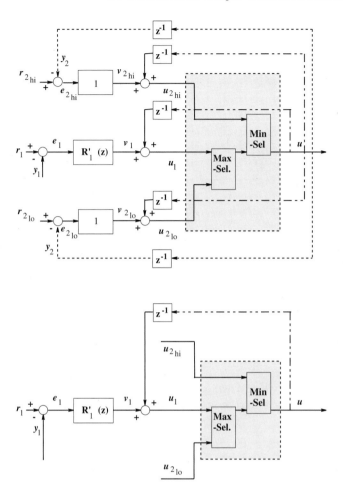

Fig. 3.18. Special case of form **J** (top), which reverts to an input saturation form (bottom)

Form K

The next modification is to shift the integral action fully upstream of the selector blocks, and add individual awf paths; Fig. 3.19.

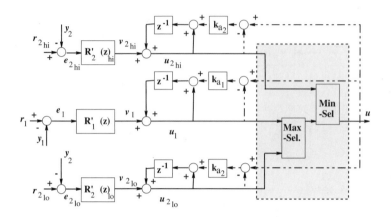

Fig. 3.19. Form K

Then, if and only if the awf gains are such that deadbeat response results in the awf loops, *i.e.* if and only if in Fig. 3.19

$$k_{a_1} = k_{a_2} = 1.0 \tag{3.54}$$

holds, then form **K** is *equivalent* to form **J**, and thereby to form **H**; see [34].

Note that now the serial structure of type $R'I(aw)$ in Fig. 3.19 may be transferred into a parallel one, $PI(aw)$, as shown in Chapter 2. This avoids the differentiation in the R' blocks.

Form L

This goes back to what Buckley [21] described for pneumatic control equipment; see Fig. 1.5.
Fig. 3.20(a) shows the block diagram for this arrangement. The first-order unity gain lag element with time constant T_f models the variable throttle and bellow arrangement in the positive feedback path, which generates the "reset" action.

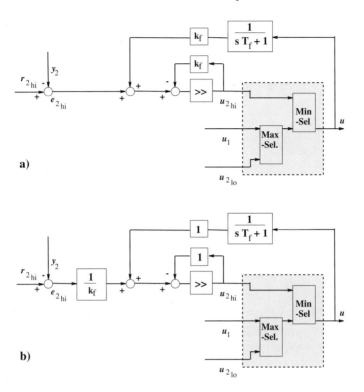

Fig. 3.20. Developing form **L**

This is developed further in Fig. 3.20(b) by shifting the feedback gain k_f into the input path as $1/k_f$. Then the unity negative feedback around the high-gain element can be replaced by a unity gain element, and thus drops out of the block diagram.

Form M

This leads to Fig. 3.21, where more general transfer functions E and F are inserted in the input and feedback path.
For *equivalence* of **M** to **L** in Fig. 3.20(b)

$$E = 1/k_f = k_p \quad \text{and} \quad F(z) = \frac{z^{-1}}{\frac{1-z^{-1}}{T_s}T_f + z^{-1}} \tag{3.55}$$

For *equivalence* of **M** to form **J** (and thus to forms **H** and **K**)

$$E = k_p\left[(1 - z^{-1}) + \frac{1}{T_i}\right] \quad \text{and} \quad F(z) = z^{-1} \tag{3.56}$$

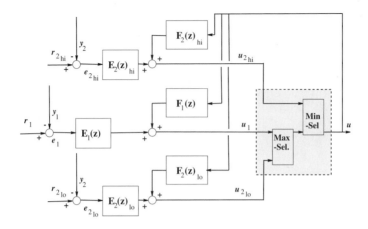

Fig. 3.21. Form M

Exercise

Develop the equivalent continuous structures.

Hint: look for RC networks around operational amplifiers; Fig.1.6.

Summary

From what we have shown, form **M** is a (mini-)generic structure for **H, J,** and **K**, but only with deadbeat awf gain, and the self-conditioning form **L**; see also Sect. 2.4.4. However, it is less general than what we shall present in Sect. 3.5.

We have also shown that the input constraint control problem may well be considered as a special form of the output constraint control problem, with $G_2 = z^{-1}$.

3.4.3 The Cascade-Limiter Concept

This concept (see Fig. 3.6(b) is very popular in the drives field, and is the root for a third family of output constraint control structures.

From Fig. 3.6(b) this may be seen as an input constraint control problem, with awf to the master controller R_1. From this, it is often concluded that the cascade limiter concept covers *all* output constraint control problems as well, and that thus any further development of structures is superfluous.

We shall show that this is not the case. In fact, the cascade-limiter forms are best suited only for one class of plants, a large one in fact, but do not cover other practically important cases. Overlooking this will lead to poor control performance.

Note that the benchmark plant defined in Sect. 3.3 belongs to this particular class, where the cascade-limiter concept is applicable. However the design of the linear controllers of Sect. 3.3 must be revised.

Assumptions on the Plant Model and the Control Structure

The basic idea of cascade control is to use an additional measured output y_2 for an inner feedback loop (often denoted as "slave"), where y_2 contains trend information on the future change of y_1 in the outer loop ("master"), and thus improve closed-loop performance of y_1.

More formally, define a class of plant models with the following properties:

1. Let the plant model be in control canonical form.
2. Let $y_1 = c_1 x_1$, *i.e.* no zero in $G_1(s) = y_1(s)/u(s)$ is admitted. Otherwise $y_1(t)$ would already contain such "trend information"
3. Let $y_2 = c_2 x_2$, *i.e.* from item 1: $y_2 = (c_2/c_1)(dy_1/dt)$, meaning that loosely speaking the "trend information" about the output y_1 should not be mixed up with information from other parts in the plant.

This holds for the two-integrators-chain specified as the benchmark plant in Sect. 3.3:

$$y_1(s) = \frac{1}{s^2 T_1 T_2} u(s) \quad \text{and} \quad y_2 = \frac{1}{sT_2} u(s) := sT_1 \, y_1(s) \qquad (3.57)$$

The feedback control structure is shown in Fig. 3.22.

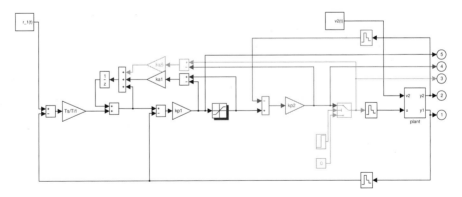

Fig. 3.22. Form N.
A second awf path has been added to R_1 to prevent windup if the run-up switch is in its "down" position

It is a cascade arrangement with control of y_2 by the inner loop with a P controller (gain k_{2_1}) and an outer loop for y_1 with a PI controller in a traditional arrangement of a P controller with gain k_{1_1} and a reset action on its reference by an integrator with gain k_{0_1}/T_0, driving $e_1 = r_1 - y_1 \to 0$.

It is also a state feedback structure for the plant (x_1, x_2) augmented by the controller integrator (x_0).

The controller parameters $k_{p_2}, k_{p_1}, T_{i_1}$ may be determined by the closed-loop pole assignment procedure via the state feedback gains $k_{0_1}, k_{1_1}, k_{2_1}$ to $(s + \Omega_1)^3 = 0$:

$$bk_{0_1} = \Omega^3 T_0 T_1 T_2 ; \quad a_1 + bk_{1_1} = 3\Omega^2 T_1 T_2 ; \quad a_2 + bk_{2_1} = 3\Omega T_2 \quad (3.58)$$

and then to the controller parameters

$$k_{p_2} := c_2 k_{2_1} ; \quad k_{p_1} k_{p_2} := c_1 k_{1_1} ; \quad \frac{1}{T_{i_1}} k_{p_1} k_{p_2} := c_1 \frac{k_{0_1}}{T_0} \quad (3.59)$$

and the standard awf (with gain $k_{a_S} = 1/k_{p_1}$) is used.

Note that this extends directly to plants of higher order, if the cascade loops are nested accordingly for each state variable.

The inner loop with P control will show a persistent control error $\bar{e}_2 \neq 0$, if a persistent load $\bar{v} \neq 0$ is applied in parallel to u; see Fig. 3.23.

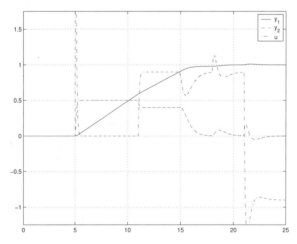

Fig. 3.23. Benchmark response for form **N**, the PI P cascade

An Additional Requirement and Modified Control Structure

However, the output constraint control task generally requires $e_2 \to 0$, *i.e.* the open loop transfer function $H_2 = R_2 G_2$ of the y_2 loop has to have the necessary limit property

$$\lim_{s \to 0} H_2(s) \sim \frac{1}{s} \tag{3.60}$$

Now the general case of $a_1 \neq 0$ in

$$G_2(s) = \frac{s\frac{c_2}{c_1}b}{s^2 + a_2 s + a_1} \quad \text{requires} \quad R_2 = \frac{s^2 k_p + s k_i + k_{ii}}{s^2} \tag{3.61}$$

i.e. a cascaded integral action as well. This situation occurs fairly often in practice (see Arthur's case), and requires to extend the basic controller blocks offered by industrial process control systems. It shall be investigated later (see Chapter 5).

But if $a_1 = 0$ (as in the benchmark plant), then the standard PI algorithm is sufficient for R_2.

$$G_2(s) = \frac{\frac{c_2}{c_1}b}{sT_2 + a_2} \quad \text{requiring} \quad u_2(s) = \frac{k_{0_2}}{sT_0}(r_2(s) - y_2(s)) + k_{1_2}(0 - y_2(s)) \tag{3.62}$$

A two-degrees-of-freedom structure is used here (which is often available in industrial process control systems), as this makes the pole assignment procedure much easier. Then for $(s + \Omega_2)^2 = 0$:

$$\frac{c_2}{c_1}bk_{0_2} = \Omega_2^2 T_0 T_2 \quad \text{and} \quad a_2 + \frac{c_2}{c_1}bk_{1_2} = 2D_2\Omega_2 T_2 \tag{3.63}$$

The "master" controller R_1 may be of P type only, as the main load disturbance has been suppressed by the integral action in R_2, and as $a_1 = 0$, the control error $e_1 \to 0$ for constant r_1; see Fig. 3.24. This structure is often used in mechatronic positioning loops.

The parameters are again determined by pole assignment. The closed-loop characteristic equation now is with

$$y_1(s) = \frac{c_{11}b}{(sT_2 + a_2)sT_1}u(s)$$

for the benchmark plant

$$0 = s^3 T_0 T_1 T_2 + (a_2 + \frac{c_2}{c_1}bk_{1_2})s^2 T_0 T_1 + (\frac{c_2}{c_1}bk_{0_2})sT_1 + c_{11}bk_{1_1}k_{0_2} \tag{3.64}$$

and with $(s + \Omega_1)^3 = 0$

$$a_2 + \frac{c_2}{c_1}bk_{1_2} = 3\Omega_1 T_2 \ ; \quad \frac{c_2}{c_1}bk_{0_2} = 3\Omega_1^2 T_2 T_1 \ ; \quad c_{11}bk_{1_1}k_{0_2} = \Omega_1^3 T_2 T_1 T_0 \tag{3.65}$$

i.e.

$$k_{1_1} = \frac{1}{3}\Omega_1 T_1 \tag{3.66}$$

If now the poles for the y_1 loop are assigned by Eq. 3.65, all three parameters are fixed, in particular the parameters of R_2. In other words, k_{0_2}, k_{1_2} are now given, and $\Omega_2, 2\zeta_2$ follow from Eq. 3.63, *i.e.*:

from k_{0_2} $3\Omega_1^2 T_2 T_1 \overset{!}{=} \Omega_2^2 T_2 T_1$ i.e. $\Omega_2 = \sqrt{3}\Omega_1$

and from k_{1_2} $3\Omega_1 T_2 \overset{!}{=} 2D_2\Omega_2 T_2$ i.e. $2\zeta_2 = \sqrt{3}$ (3.67)

In other words, considering the inner loop in isolated operation, its bandwidth is higher by a factor of $\sqrt{3}$ and the damping ratio $2\zeta_2$ is slightly lower than the damping ratio of the full system, but tolerably so.

Inserting the saturation for $r_{2_{hi}}$ and $r_{2_{lo}}$ leads to form **O**, Fig. 3.24. No awf is required for R_1, which makes the system particularly simple and transparent.

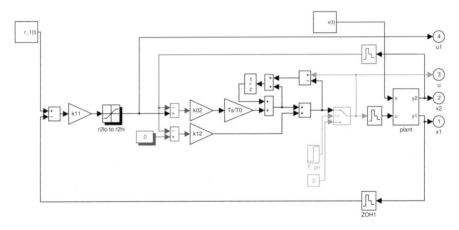

Fig. 3.24. Form **O**.
A deadbeat awf has been added to R_2 to avoid windup prior to enabling run-up

The response to the test sequence shown in Fig. 3.25 is well behaved. Should a PI type be required for R_1 due to additional elements in the specification, then obviously an awf has to be added to R_1.

Exercise
Investigate this further:

- What additional elements are needed to require a PI(aw) algorithm for R_1? Extend the benchmark accordingly.
  Hint: append an additional disturbance step input to the plant integrator for x_1; see Eq. 3.20
- Design the linear controller parameters for the PI PI cascade, see Eq. 3.33.
- Then, how do the closed-loop dynamic parameters Ω_2 and $2\zeta_2$ for isolated operation along the constraint relate to Ω_1 and $2\zeta_1$ from above?
- Check the transient response with the extended benchmark.

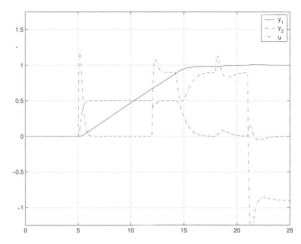

Fig. 3.25. Benchmark response for form **O**, the P PI cascade

Equivalence to Selector Control

Starting from Fig. 3.24, the saturation element is replaced by an equivalent
Max-Min-selector block, and R_2 is converted into serial form, see Fig. 3.26(a),
where[3]

$$r_2 = Min\{u_1,\ r_{2_{hi}}\} \tag{3.68}$$

and

$$u(s) = \frac{k_{0_2}}{sT_0}(r_2 - \tilde{y}_2) \quad \text{with} \quad \tilde{y}_2 = y_2\left(1 + sT_0\frac{k_{1_2}}{k_{0_2}}\right) \tag{3.69}$$

In the next step, as shown in Fig. 3.26(b), the Max-Min-selector block is
moved downstream in front of the integrator k_{0_2}/sT_0.
Subtracting $\tilde{y}_2$ everywhere in Eq. 3.68

$$r_2 - \tilde{y}_2 = Min\{\ u_1 - \tilde{y}_2,\ r_{2_{hi}} - \tilde{y}_2\ \} \quad i.e. \quad e_2 = Min\{\ e_{2_1},\ e_{2_{hi}}\ \} \tag{3.70}$$

and multiplying with k_{0_2} everywhere

$$k_{0_2}e_2 = Min\{\ k_{0_2}e_{2_1},\ k_{0_2}e_{2_{hi}}\ \} \quad i.e. \quad v = Min\{\ v_{2_1},\ v_{2_{hi}}\ \} \quad \square \tag{3.71}$$

as in Fig. 3.26(b). So Figs. 3.26 (a) and (b) are *equivalent*.

Now (b) corresponds to form **H**, and thus for the discrete-time versions
may be manipulated further by way of form **J** to form **K**, (only valid for deadbeat awf gain entries), finally to form **L**, all these structures being *equivalent*.

---

[3] For brevity, the argument is presented for the hi-constraint only.

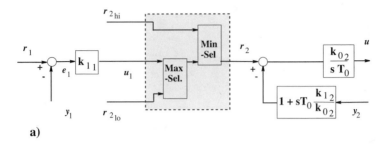

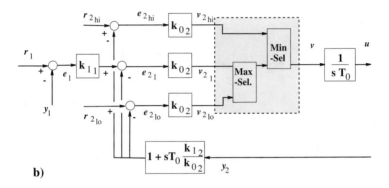

Fig. 3.26. Conversion of the cascade-limiter-type controller to a selector-type

Note that all three controllers for y_2 are the same. So there is no flexibility for tuning the y_1 and y_2 loops individually.

Exercise
Above, we have tuned for the y_1 loop, and thus fixed the y_2 dynamics. Investigate the other alternative of tuning first R_2 for the y_2 loop in isolated operation running along the constraint, and then tuning R_1.

The basic assumption for this subclass of cascade limiter control systems was

$$y_2(t) := T_1 \frac{d}{dt} y_1(t); \quad i.e. \text{ for the benchmark} \quad y_2(s) = sT_1 y_1(s) \qquad (3.72)$$

If this holds then the P PI cascade for y_1 can be manipulated further into the PI P cascade used previously. But again, the tuning can only be done for either the y_1- or the y_2-loop.

Exercise
Investigate this, show the equivalence to selector control, and check your results by the responses to the benchmark.

3.5 The Generic Structure

3.5.1 A First Version, for Direct Implementation

The Generic Structure for output constraint control using "control conditioning" feedback shall have the following properties:

- All structures forms **B** to **P** which perform adequately in the benchmark test can be generated from it by inserting appropriate transfer functions.
- It allows to do without integral action in any of the control loops (which is a key property for mechatronic systems).
- If an integral action is present, then the generic awf form of Chapter 2 shall be used.
- It can be directly implemented on current process control systems, where Min- and Max-selector and PI(aw) function blocks are standard library elements.

Form **F** is closest to these requirements. It leads directly to the first version of the Generic Structure, form **Q** in Fig. 3.27

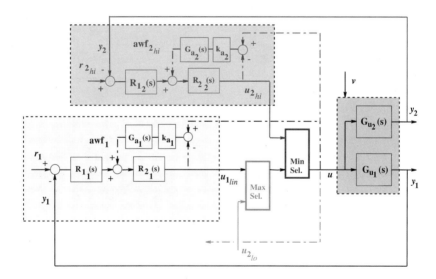

Fig. 3.27. The Generic Structure using standard pcs function blocks, suited for direct implementation

Its transient response is checked in the following simulations. The PI(aw) controllers are of the form **D** of Chapter 2.

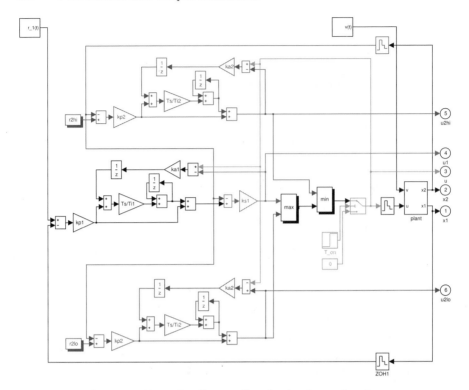

Fig. 3.28. Form **Q**, all controllers having integral action

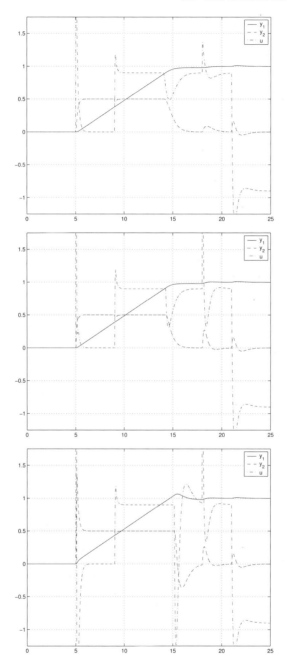

Fig. 3.29. Transient response with controller form **Q** with:
- deadbeat values k_{a_i}, $i = 1, 2$ of the awf gain (top),
- compensating values $k_{a_i}^*$, $i = 1, 2$ (center),
- and weak values $k_{a_i} = 0.2 \cdot k_{a_i}^*$, $i = 1, 2$ (bottom)

Next, the form variant with integral action in the main loop only is checked; see Figure 3.30:

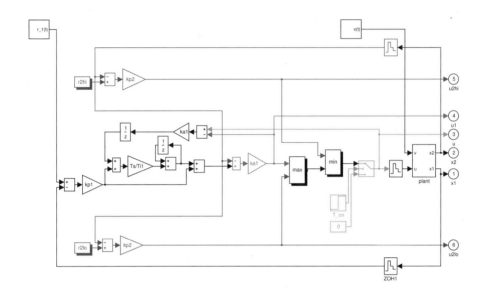

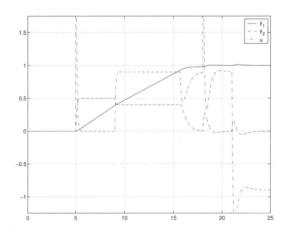

Fig. 3.30. Form $\mathbf{Q}$, R_1 with integral action, and compensating awf gain $k_{a_1}^*$, and both R_2 without integral action

and now the form with integral action only in the R_2 controllers; see Figure 3.31:

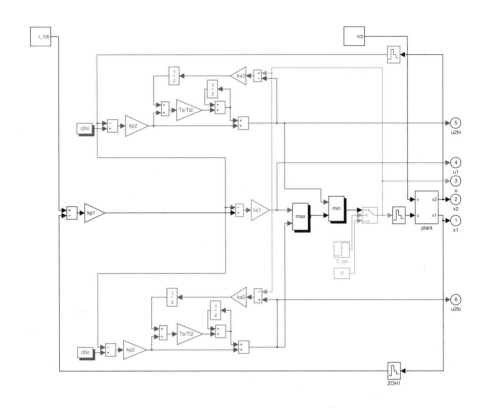

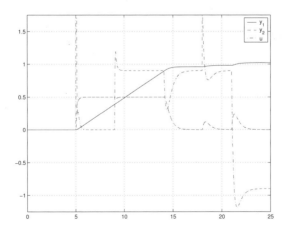

Fig. 3.31. Form **Q**, R_2 with integral action, compensating awf gain $k_{a_2}^*$, and R_1 without integral action

and finally the form with no integral action at all; see Figure 3.32:

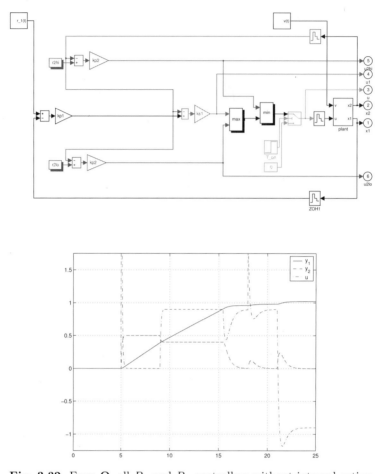

Fig. 3.32. Form **Q**, all R_1 and R_2 controllers without integral action

The responses are as expected, and well behaved.

It has been argued [55] that selectors "break" the main control loop, as they transfer control in a switching manner, and that this property may not be intuitively acceptable to users in safety-critical applications.

In such cases note that the selector blocks in form **Q** may be equivalently replaced by "nonlinear-adding" blocks, such as shown in form **E** above. This does not "break" the loop, but modifies the control signal $u(t)$ of the main loop via an addition block. Then, this resoning may be more easily accepted, but it does not change the substance.

3.5.2 A Second Version, for Stability Analysis

The Generic Structure with selectors is not well suited for nonlinear stability analysis. The canonical form to be used then requires the nonlinear element to have one input and not three as in form **Q**.
However this may be obtained using some weak assumptions:

- Use the equivalent nonlinear adding elements as in form **E** instead of the selectors.
- Assume that $R_{2_{lo}}$ and $R_{2_{hi}}$ have the same structure and the same coefficients, and differ only in their setpoint values $r_{2_{lo}}$ and $r_{2_{hi}}$.
- The setpoint values $r_{2_{lo}}$ and $r_{2_{hi}}$ are time-invariant,
  and are selected such that a "well-posed" design problem results:

$$r_{2_{lo}} < \bar{y}_2 < r_{2_{hi}} \quad \forall \quad \bar{y}_2 \qquad (3.73)$$

Then the $R_{2_{lo}}$ and $R_{2_{hi}}$ may be fused into one block R_2, and the two nonlinear characteristics into one of the deadspan type, yielding the second Generic Structure, form **R** in Fig. 3.33.

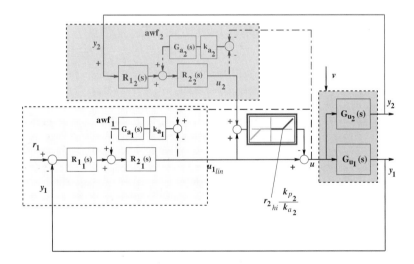

Fig. 3.33. The Generic Structure using a deadspan nonlinear element, adapted for stability testing

The break points $u_{2_{lo}}$ and $u_{2_{hi}}$ are determined as follows:

Consider first the case where R_2 does not contain an integral action. Then, by inspection of Figs. 3.30 and 3.32:

$$u_{2_{lo}} = r_{2_{lo}} \cdot k_{p_2} \quad \text{and} \quad u_{2_{hi}} = r_{2_{hi}} \cdot k_{p_2} \qquad (3.74)$$

Consider now the case, where R_2 contains an integral action, as shown in Fig. 3.33. Assume further that there is a non-vanishing awf gain, $k_{a_2} > 0$, as otherwise the R_2 would obviously wind up beyond all bounds in steady state operation of the main loop. If then the R_2 loops are needed in a transient, u_2 will surely not be selected in due time, see *e.g.* Fig. 3.29(bottom). Assume also that R_2 is of the form **D** of Chapter 2.

Consider now the phase of the benchmark test sequence when the controller has been switched on and the setpoint raised to the startup value, but the plant is still at standstill due to the shutdown switch being in its low position, *i.e.*

$$\bar{u} = 0 \quad \text{and} \quad \bar{y}_2 = 0 \tag{3.75}$$

For steady state of the awf loop in $R_{2_{hi}}$, the input to its integral action must be zero, *i.e.*

$$\bar{e}_2 k_{p_2} + \bar{e}_{a_2} k_{a_2} = 0 \quad \text{where} \quad \bar{e}_2 = r_{2_{hi}} - \bar{y}_2 := +r_{2_{hi}}$$
$$\text{and} \quad \bar{e}_{a_2} = -u_{2_{hi}} + \bar{u} := -u_{2_{hi}}$$
$$\text{that is} \quad u_{2_{hi}} = r_{2_{hi}} \frac{k_{p_2}}{k_{a_2}} \quad \text{and} \quad u_{2_{lo}} = r_{2_{lo}} \frac{k_{p_2}}{k_{a_2}} \quad \text{for } k_{a_2} > 0 \tag{3.76}$$

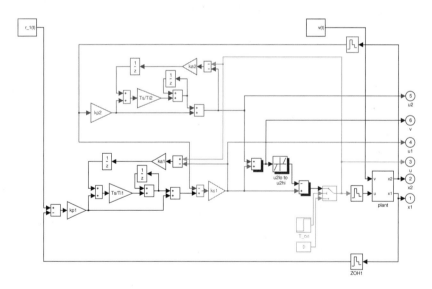

Fig. 3.34. Form **R**.
R_1 and R_2 controllers with integral action, and awf gains $k_{a_1}, k_{a_2} > 0$

In Fig. 3.35 this is checked by simulations on the control structure shown in Fig. 3.34. The transient responses are the same as in Fig. 3.29.

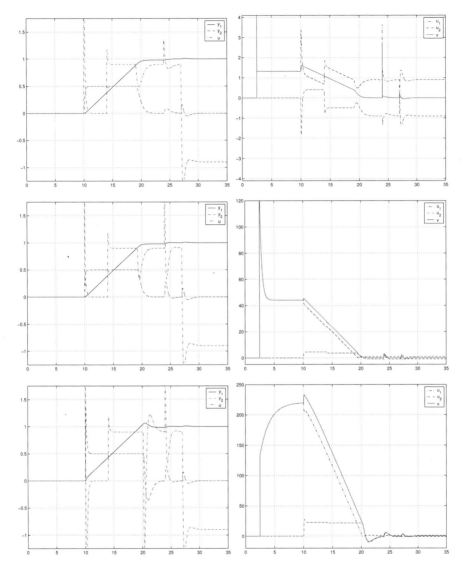

Fig. 3.35. Transient response with Generic Structure form **R** to the benchmark test sequence, with $\Omega_1 = 3$ and $\Omega_2/\Omega_1 = 3$,
- and deadbeat values of the awf gains k_{a_i}, $i = 1, 2$ (top row),
- compensation values $k_{a_i}^*$, $i = 1, 2$ (center row),
- and weak values $k_{a_i} = 0.2 k_{a_i}^*$, $i = 1, 2$ (bottom row)
The right column shows the inputs u_i and v to the deadspan nonlinear block

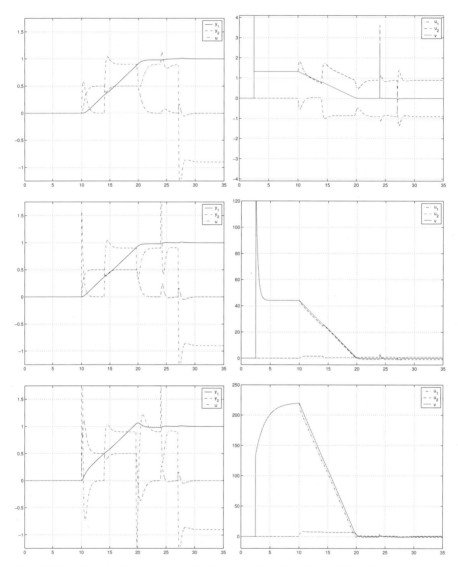

Fig. 3.36. Transient response with Generic Structure form **R** to the benchmark test sequence with $\Omega_2/\Omega_1 = 1.0$, and all other parameters and the plot arrangement being the same as in Fig. 3.35

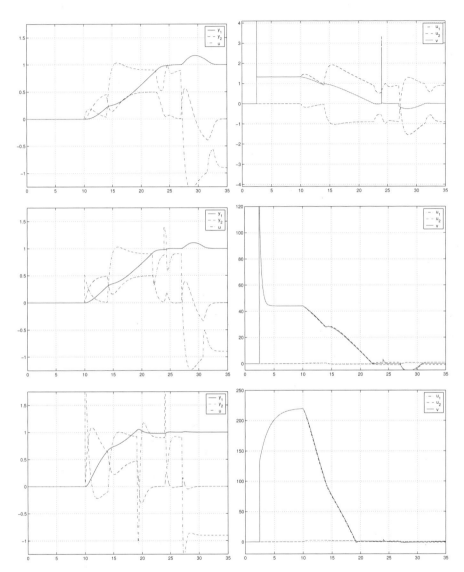

Fig. 3.37. Transient response with Generic Structure form **R** to the benchmark test sequence with $\Omega_2/\Omega_1 = 1/3$, and all other parameters and the plot arrangement being the same as in Fig. 3.35

3.6 Stability Analysis

3.6.1 The Cascade-Limiter Form

The cascade limiter version is quite similar to the structures from Chapter 2, and stability can be analyzed along the same lines. This shall not be done here, and is left to the reader as an exercise.

3.6.2 The Nonlinear-Additive and Selectors Forms

The analysis of the selectors forms is not that straightforward, because the nonlinear block at the plant input (consisting of the Maximum and Minimum Selector arrangement) has *three* time-varying inputs, instead of *one* required by the canonical structure; see Fig. 2.24.

However, such a canonical form can be derived if some weak assumptions are made, using the results from the previous section; Fig. 3.38.

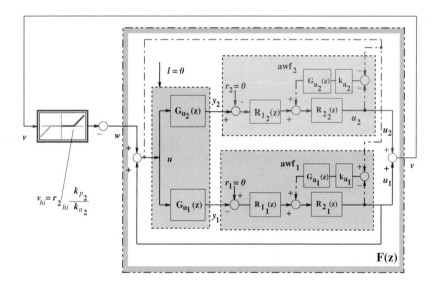

Fig. 3.38. Redrawing form **R** into its canonical form

Rearranging the linear elements in the Generic Structure of Fig. 3.33 leads to the *canonical structure* in Fig. 3.38, with $F(z)$ for the linear subsystem, the deadspan element from Fig. 3.33 for the nonlinear subsystem, and the sign inversion for negative feedback.

In the next subsections we shall look into the nonlinear subsystem first, into the linear subsystem next, and then discuss some general stability properties.

The Nonlinear Subsystem

The nonlinear characteristic in Fig. 3.38 is of the sector type. It is non-energy-storing (no hysteresis) and time-invariant.

Its *lower sector* slope a is zero

$$a = 0 \qquad i.e. \qquad -\frac{1}{a} \to -\infty \qquad (3.77)$$

and relates to the linear operation of the main control loop (for y_1).

Thus the prerequisites for the Popov test or for the off-axis circle test are met. And we may expect least conservative results from the test.

Loosely speaking, by inspection of Fig. 3.38, the *upper sector* slope b is determined by how far the extremal input v_{max}, which occurs during the transient to be considered in the stability test, exceeds the break point value v_{hi}, or alternatively v_{min} versus v_{lo}. For brevity of notation we shall write for indices *max* and *hi* only, but imply the "low" case as well.

Let $\bar{v}$ be the final equilibrium value of $v(t)$ for the transient considered in the test. Denote

$$\Delta v_{hi} := v_{hi} - \bar{v} \qquad \text{and} \qquad \Delta v_{max} := v_{max} - \bar{v} \qquad (3.78)$$

Then for b :

$$b = \frac{\Delta v_{max} - \Delta v_{hi}}{\Delta v_{max}} \qquad i.e. \qquad -\frac{1}{b} = -\frac{\Delta v_{max}}{\Delta v_{max} - \Delta v_{hi}} \qquad (3.79)$$

$$-\frac{1}{b} = -1 - \frac{\Delta v_{hi}}{\Delta v_{max} - \Delta v_{hi}} = -1 - \frac{1}{\frac{\Delta v_{max}}{\Delta v_{hi}} - 1} = -1 - \Delta_v \qquad (3.80)$$

which defines Δ_v. And

$$\Delta v_{max} \to \infty \qquad \text{or} \qquad \Delta v_{hi} \to 0 \qquad \text{yields} \qquad -\frac{1}{b} \to -1.0 \qquad (3.81)$$

The first condition $\Delta v_{max} \to \infty$ translates into initial deviations from steady state tending to infinity. The second condition $\Delta v_{hi} \to 0$ translates into having an equilibrium on the constraint border and not inside it, *i.e.* what we consider an "ill-posed" design case. Both conditions will not be met under reasonable design conditions, *i.e.*

$$-\frac{1}{b} < -1.0 \qquad (3.82)$$

A more specific estimate of b shall be discussed later.

The Linear Subsystem

The transfer function for the linear subsystem is derived next. From Fig. 3.38:

$$F(s) = \frac{v}{w} = \frac{u_1}{w} + \frac{u_2}{w}$$

and $\quad \dfrac{u_1}{u} = -G_1 R_{1_1} \dfrac{R_{2_1}}{(1 + k_a G_a R_2)_1} + \dfrac{(k_a G_a R_2)_1}{(1 + k_a G_a R_2)_1} \quad := \quad H_1$

and $\quad \dfrac{u_2}{u} = +G_2 R_{1_2} \dfrac{R_{2_2}}{(1 + k_a G_a R_2)_2} - \dfrac{(k_a G_a R_2)_2}{(1 + k_a G_a R_2)_2} \quad := \quad H_2$

and $\quad \dfrac{u}{w} = \dfrac{1}{1 - H_1}$

that is $\quad \dfrac{v}{w} = (H_1 + H_2)\dfrac{u}{w} = \dfrac{1}{1 - H_1}(H_1 + H_2)$ $\qquad$ (3.83)

Then

$$1 - H_1 = \frac{1}{(1 + k_a G_a R_2)_1}\left[(1 + k_a G_a R_2)_1 + (GR_1 R_2)_1 - (k_a G_a R_2)_1\right]$$

$$= \frac{(1 + GR_1 R_2)_1}{(1 + k_a G_a R_2)_1}$$

$$H_1 = \frac{1}{(1 + k_a G_a R_2)_1}\left[-(1 + GR_1 R_2)_1 + (1 + k_a G_a R_2)_1\right]$$

$$H_2 = \frac{1}{(1 + k_a G_a R_2)_2}\left[+(1 + GR_1 R_2)_2 - (1 + k_a G_a R_2)_2\right]$$

and finally

$$F = \frac{1}{1 - H_1}(H_1 + H_2)$$

$$= \frac{(1 + k_a G_a R_2)_1}{(1 + GR_1 R_2)_1}\left[-\frac{(1 + GR_1 R_2)_1}{(1 + k_a G_a R_2)_1} + 1 + \frac{(1 + GR_1 R_2)_2}{(1 + k_a G_a R_2)_2} - 1\right]$$

yielding the main result

$$F + 1 = \left(\frac{1 + k_a G_a R_2}{1 + GR_1 R_2}\right)_1\left(\frac{1 + GR_1 R_2}{1 + k_a G_a R_2}\right)_2 \qquad (3.84)$$

or also

$$F + 1 = \frac{F_1 + 1}{F_2 + 1} \qquad \text{with} \quad F_i + 1 = \left(\frac{1 + k_a G_a R_{2_i}}{1 + G R_{1_i} R_{2_i}} \right)_{i=1,2} \tag{3.85}$$

This compact result shows a nice symmetry. Note that it is not restricted to the low-order systems we are investigating here. This will be followed up in the second part of this book.

General Stability Properties

The nonlinear stability test requires $F + 1$ to be both proper and asymptotically stable. This shall be checked next.

- For preparation, the result Eq. 3.84 may be rewritten using the following notation
    - D_i for the closed-loop characteristic polynomial of control loop i
    - d_i for its open-loop characteristic polynomial
    - d_{p_i} for the open-loop characteristic polynomial of the plant
    - d_{a_i} for the parts of the controllers, that are inside the awf loop (the integral action), *i.e.* $d_i = d_{p_i} \cdot d_{a_i}$
    - D_{a_i} for the closed-loop characteristic polynomial for the awf loop.

    Then

$$F + 1 = \frac{d_1}{D_1} \frac{D_{a_1}}{d_{a_1}} \frac{D_2}{d_2} \frac{d_{a_2}}{D_{a_2}} \tag{3.86}$$

For controllers with integral action, d_{a_i} contains the corresponding pole at the origin, which will also be contained in the $d_i = d_{p_i} d_{a_i}$ $(i = 1, 2)$. Therefore, these poles and zeros in Eq. 3.86 will *cancel* without creating observability and controllability problems. And thus

$$F + 1 = \frac{d_{p_1}}{D_1} \frac{D_2}{d_{p_2}} \frac{D_{a_1}}{D_{a_2}} \tag{3.87}$$

- Inspecting the degrees *deg* of the polynomials yields

$$deg(D_1) = deg(d_{p_1}) + 1$$
$$\text{and} \quad deg(D_2) = deg(d_{p_2}) + 1$$
$$\text{and} \quad deg D_{a_1} = deg D_{a_2} = 1$$
$$\text{and also assume} \quad deg(d_{p_1}) > deg(d_{p_2}) \tag{3.88}$$

then the transfer function will be proper, but it is not strictly proper.

In other words, the first requirement is met.

We now discuss the second requirement, i.e. the stability of $F + 1$:

- Both D_1 and D_2 are asymptotically stable by design of the respective loops in their linear operating range.
- The same holds for the contributions of the awf loops D_{a_1} and D_{a_1}.
  In other words the stability properties of $F + 1$ depend on the open loop parts.
- If the transfer function G_1 of the plant should contain any non-asymptotically stable poles, they appear by way of d_{p_1} in $F + 1$ as zeros, *i.e.* they do not adversely affect the stability of $F + 1$.
- But any unstable poles in G_2 appear by way of d_{p_2} in the denominator of $F + 1$.
  In other words G_2 *must* be asymptotically stable, whereas G_1 need not be.

There is an important special case, where this condition may be relaxed:

- Consider G_1 and G_2 having physically common parts, which generate the *same* non-asymptotically stable poles in d_{p_1} and d_{p_2}, then they cancel without generating observability and controllability problems, and $F + 1$ will still be asymptotically stable.

Note that the benchmark plant used so far is such a special case!

A Practical Estimate of $1/b$

As used above, the value of $1/b$ determines the position of the circle with respect to the origin as a function of Δv_{max}:

$$-\frac{1}{b} = -1 - \frac{\Delta v_{2_{hi}}}{\Delta v_{max} - \Delta v_{2_{hi}}} \quad \text{for} \quad \Delta v_{max} > \Delta v_{2_{hi}} \qquad (3.89)$$

In many practical cases, the graphic test would admit only a Δv_{max} with a finite upper bound, if asymptotic stability is to be shown. For such cases consider the transients produced in the benchmark, Fig. 3.35. There, Δv_{max} is attained just before startup, when the controller structure and its awf loops are in operation and have reached their equilibrium with the setpoint moved at $r_1 = 0.98$, but the plant is still being held at standstill, *i.e.* $y_1 = y_2 = 0$. Then in Fig. 3.34

$$\overline{\Delta v} = \overline{\Delta u_1} + \overline{\Delta u_2}$$

$$\text{with} \quad \overline{\Delta u_2} = 0 \quad \text{because} \quad \overline{y}_2 = 0$$

$$\text{and in the awf loop of } R_1 \quad 0 = k_{p_1}\overline{e}_1 + k_{a_1}\frac{1}{k_{s_1}}\overline{e}_{a_1}$$

$$\text{with} \quad \overline{e}_{a_1} = \overline{\Delta u} - \overline{\Delta u_1} = 0 - \overline{\Delta u_1}$$

$$\overline{\Delta u_1} = r_1 \frac{k_{p_1}}{k_{a_1}\frac{1}{k_{s_1}}} = r_1 \frac{k_{p_1}k_{s_1}}{k_{a_1}} = \overline{\Delta v} \qquad (3.90)$$

Using this as an estimate for $1/b$ yields (with the notation $k_{1_2} := k_{p_2}$ and $k_{1_1} := k_{p_1} k_{s_1}$):

$$-\frac{1}{b} \approx -1 - \frac{\Delta v_{2hi}}{\Delta v - \Delta v_{2hi}} = -1 - \frac{r_{2hi} \frac{k_{1_2}}{k_{a_2}}}{r_1 \frac{k_{1_1}}{k_{a_1}} - r_{2hi} \frac{k_{1_2}}{k_{a_2}}}$$

$$< -1 - \frac{r_{2hi} \frac{k_{1_2}}{k_{a_2}}}{r_1 \frac{k_{1_1}}{k_{a_1}}} = -1 - \frac{r_{2hi}}{r_1} \frac{k_{1_2}}{k_{1_1}} \frac{k_{a_1}}{k_{a_2}} = -1 - \Delta v_a \qquad (3.91)$$

which defines Δv_a.

Discuss how the position of the vertical straight line is changed by modifying the design parameters in Δv_a as a small **exercise**.

3.7 Stability Properties of the Test Case

These will be investigated in two steps. The first aims at the shape of the Nyquist contour for the stability test and how it is influenced by the design parameters. For this the small delay D is omitted, and the continuous-time representation is used. In the second step, the original system is considered.

The Simplified System

The open- and closed-loop polynomials for the linear subsystem are

$$d_1 = s^3 T_{i_1} T_1 T_2$$
$$D_1 = s^3 T_{i_1} T_1 T_2 + s^2 T_{i_1} T_1 k_{s_1} + s T_{i_1} k_{s_1} k_{p_1} + k_{p_1} k_{s_1}$$
$$\text{and} \quad d_2 = s^2 T_{i_2} T_2$$
$$D_2 = s^2 T_{i_2} T_2 + s T_{i_2} k_{p_2} + k_{p_2}$$
$$\text{and for the awf loops} \quad d_{a_1} = s T_{i_1}; \quad D_{a_1} = s T_{i_1} + k_{a_1}$$
$$d_{a_2} = s T_{i_2}; \quad D_{a_2} = s T_{i_2} + k_{a_2} \qquad (3.92)$$

Inserting into Eq. 3.87 yields

$$F + 1 = \frac{s^3}{s^3 + \cdots + \frac{k_{p_1} k_{s_1}}{T_{i_1} T_1 T_2}} \frac{s + \frac{k_{a_1}}{T_{i_1}}}{s} \frac{s^2 + \cdots + \frac{k_{p_2}}{T_{i_2} T_2}}{s^2} \frac{s}{s + \frac{k_{a_2}}{T_{i_2}}}$$

$$= \frac{s^2 \left(s + \frac{k_{a_1}}{T_{i_1}}\right)}{(s + \Omega_1)^3} \frac{(s + \Omega_2)^2}{s \left(s + \frac{k_{a_2}}{T_{i_2}}\right)} = \frac{s (s + \Omega_2)^2 \left(s + \frac{k_{a_1}}{T_{i_1}}\right)}{(s + \Omega_1)^3 \left(s + \frac{k_{a_2}}{T_{i_2}}\right)}$$

$$= \frac{s}{s + \Omega_1} \left(\frac{s + \Omega_2}{s + \Omega_2}\right)^2 \frac{s + \gamma_1 \Omega_1}{s + \gamma_2 \Omega_2} \qquad (3.93)$$

where γ_j denotes

$$\gamma_j = \frac{k_{a_j}}{k_{a_j}^*} \quad \text{for} \quad j = 1, 2 \tag{3.94}$$

Then Eq. 3.93 may be rewritten as

$$F + 1 = \frac{(s/\Omega_1)}{(s/\Omega_1) + 1} \left[\frac{(s/\Omega_2) + 1}{(s/\Omega_1) + 1} \right]^2 \left[\frac{(s/\gamma_1\Omega_1) + 1}{(s/\gamma_2\Omega_2) + 1} \right] \frac{\Omega_2}{\Omega_1} \frac{\gamma_1}{\gamma_2} \tag{3.95}$$

The position of the straight line at $-1/b$ to the left of -1 on the negative real axis is determined by Eq. 3.91 and using Eq. 3.94:

$$-\frac{1}{b} + 1 < -\frac{r_{2_{hi}}}{r_1} \frac{2 \, \Omega_2 \, T_2}{3 \, \Omega_1^2 \, T_1 T_2} \frac{3\gamma_1}{2\gamma_2} = -\frac{r_{2_{hi}}}{r_1} \frac{1}{\Omega_1 T_1} \frac{\Omega_2}{\Omega_1} \frac{\gamma_1}{\gamma_2} \tag{3.96}$$

Discussion

- Note that the factors Ω_2/Ω_1 and γ_1/γ_2 appear in both Eq. 3.95 and 3.96. This can be seen as a radial expansion of the complex plane, which does not change the relative configuration in the graphical stability test.
- The Nyquist contour is independent of the Ω_1 value, as long as the relative closed-loop bandwidth Ω_2/Ω_1 and γ_1/γ_2 are not changed.
- The Nyquist contour of the first factor in Eq. 3.96 starts at the origin along the positive imaginary axis, and then evolves into the positive real half plane.
- The second factor is a lead-lag element, which deforms this Nyquist contour in the direction favorable for nonlinear stability, if $\Omega_2 \geq \Omega_1$, which intuitively is a very reasonable design choice.
- And the vertical straight line position moves from the left to $-1 + j0$ proportional to the fraction of the setpoints and to $1/(\Omega_1 T_1)$.

In other words, the closed-loop response of this system is not globally asymptotically stable, but will be asymptotically stable for up to very large initial initial conditions.

Particular cases
Consider

(a) deadbeat awf gains in both R_1, R_2, *i.e.*:

$$\frac{k_{a_j}}{T_{i_j}} = \frac{1}{T_s} \quad \text{and} \quad \gamma_j = \frac{1}{\Omega_j T_s}; \quad i.e. \quad \gamma_j\Omega_j = \frac{1}{T_s} \quad \text{for} \quad j = 1, 2 \tag{3.97}$$

Then from Eqs. 3.95 and 3.96

$$F + 1 = \frac{(s/\Omega_1)}{(s/\Omega_1) + 1} \left[\frac{(s/\Omega_2) + 1}{(s/\Omega_1) + 1} \right]^2 \frac{(s/\gamma_1\Omega_1) + 1}{(s/\gamma_2\Omega_2) + 1}$$

$$-\frac{1}{b} + 1 = -\frac{r_{2_{hi}}}{r_1} \frac{1}{\Omega_1 T_1} \tag{3.98}$$

(b) compensating awf gains in both R_1, R_2, *i.e.*:

$$\gamma_j = 1 \quad \text{for} \quad j = 1, 2 \tag{3.99}$$

then from Eqs.3.95 and 3.96

$$F + 1 = \frac{(s/\Omega_1)}{(s/\Omega_1) + 1} \frac{(s/\Omega_2) + 1}{(s/\Omega_1) + 1} \frac{\Omega_2}{\Omega_1}$$

$$-\frac{1}{b} + 1 = -\frac{r_{2hi}}{r_1} \frac{1}{\Omega_1 T_1} \frac{\Omega_2}{\Omega_1} \tag{3.100}$$

(c) and finally if $\Omega_2 = \Omega_1$ in cases (a) and (b), then

$$F + 1 = \frac{(s/\Omega_1)}{(s/\Omega_1) + 1}$$

$$-\frac{1}{b} + 1 = -\frac{r_{2hi}}{r_1} \frac{1}{\Omega_1 T_1} \tag{3.101}$$

Exercise
Discuss the shapes of $F + 1$ using Bode plot techniques and compare with the numerical results of Fig. 3.39.

The Original System

As in Chapter 2, the delay D and sampling time T_s are short compared to the closed-loop time constants $1/\Omega_1$, $1/\Omega_2$. So including them now will affect the Nyquist contours only in a frequency range well above the relevant one for the graphic stability test.

Fig. 3.39 shows the Nyquist contours from Matlab/Simulink for the Generic Structure (form **R**) with $T_s = 0.010$, and for the plant with delay $D = 0.025$, *i.e.* corresponding to the simulations Figs. 3.35, 3.36, and 3.37, and the corresponding vertical straight lines from Eq. 3.91 for the on-axis circle test.

This indicates asymptotic stability for almost all initial conditions, *i.e.* $v_{max} \to \infty$, for such design selections that

$$\text{ratio} \qquad \frac{\Omega_2}{\Omega_1} \geq 1$$

$$\text{and} \quad \text{ratio} \quad \left(\frac{k_a}{k_a^*}\right)_i \geq 1 \quad \text{for} \quad i = 1, 2 \tag{3.102}$$

Comparing this with the simulation results, this corresponds to good exploitation of the range available within the output constraints, and no overshoot due to windup.

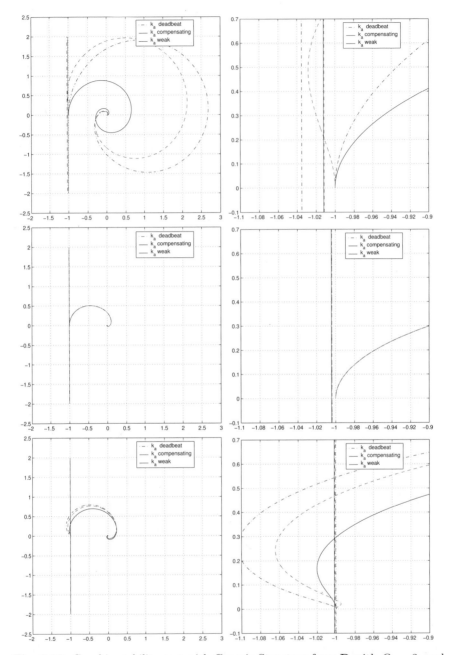

Fig. 3.39. Graphic stability test with Generic Structure form **R** with $\Omega_1 = 3$; and different awf gain values k_{a_i}, $i = 1, 2$;
for $\Omega_2/\Omega_1 = 3$ (top row), $\Omega_2/\Omega_1 = 1$ (center row), $\Omega_2/\Omega_1 = 1/3$ (bottom row). The right column shows an expansion of the area of interest for the test

Thus, both the constraint feedback loop gains (for y_2) and the awf gains k_{a_i} need not be made very large for stability reasons, but can be lowered to a comparable level to those of the main controller R_1. From an application point of view this is very convenient, considering such properties as sampling rate, robustness and high-frequency noise suppression.

Note that Eq. 3.102 has such a simple form that it is tempting to use it as a general design rule.
However it is only valid for this particular benchmark case, and should not be used elsewhere without circumspection. We shall come back to this item in the second part of this book.

3.8 Relations to Minimum Time Control

This short discussion will be along the same lines as in Sect. 2.7 for the input constrained case, and it will focus on the benchmark case considered so far.

By the transversality condition of the Maximum Principle for the minimum-time transient, the control $u(t)$ has to run along its saturation values, as long as state variables do not meet their constraint hypersurfaces (the 'regular' case). But if one state variables does run into its constraint hypersurface, then it has to run *on* it (with zero deviation from it), which then requires a $u(t)$ in-between the saturation values (the 'singular' case).

The benchmark case here deviates from this, because $u(t)$ has no saturation.[4] We also do not consider the small delay D. So the minimum-time trajectory for the run-up (see Sect. 3.3 d)) would consist of three successive phases:

A: move the state variable x_2 in minimum-time from the initial equilibrium at zero to $r_{2_{hi}}$ (by applying through u a Dirac-$\delta(t)$ of appropriate weight in the continuous-time case or a pulse of width T_s and appropriate height in the discrete-time case);

B: then controlling u such that $x_2 = r_{2_{hi}}$ irrespective of any disturbances, until $x_1 = r_1$ is attained in the continuous-time case, and for the discrete-time case $x_1 = r_1 - \epsilon_1$, where ϵ_1 is sufficiently large to accommodate the deadbeat response of phase C),

C: and finally move x_2 in minimum-time to zero, again by applying a suitable $\delta(t)$ or a pulse as above.

For the continuous case, the minimum transition time Δt_{min} then is

$$\Delta t_{min} = \frac{r_1}{r_{2_{hi}}} \qquad (3.103)$$

---

[4] We shall come to the normal case in Chapter 4.

This has to be compared with the performance of the 'output constraint control by feedback' scheme used above. The properties of the minimum-time transient described above would obviously require infinite gain feedback in the continuous-time case, and deadbeat response in the discrete-time case.

But for practical reasons we have resorted to finite gain feedback, *i.e.* finite closed-loop bandwidths Ω_1, Ω_2. We have also used integral action in all controllers in order to drive steady state errors to zero, and awf of course. We shall use compensating awf gains in the sequel.

For the three phases of the trajectory described above, the following results:

A: The run-up of $y_2 \leftarrow x_2$ from zero to its setpoint $r_{2_{hi}}$ is the same situation as discussed in Sect. 2.5, Analysis for form **D**: the response of $y_2(t)$ is of first order with time constant $\tau_2 = 1/\Omega_2$. And the response for the main output

$$\frac{y_1(s)}{u(s)} = \frac{1}{1 + s\tau_2} \frac{1}{sT_1}$$

is therefore a ramp delayed by τ_2.

B: The integral action is inserted to drive the steady state error to zero, *i.e.* $y_2 \rightarrow r_{2_{hi}}$ in the response for phase A. Thus there will be no additional delay in this phase, if:
- no disturbances are introduced,
- $r_{2_{hi}}$ is not time varying,
- and y_2 is the input to a final open integrator with y_1 as its output (see Sect. 3.3(a)).

C: This is the initial condition response of the linear control loop for y_1 as discussed in Sect. 2.5, and will contribute another $\tau \approx 2/\Omega_1$ until the final steady state is attained.

The overall transition time Δt then is

$$\Delta t \approx \frac{1}{\Omega_2} + \frac{r_1}{r_{2_{hi}}} + 2\frac{1}{\Omega_1} = \Delta t_{min}\left[1 + \frac{1}{\Omega_2 \Delta t_{min}} + 2\frac{1}{\Omega_1 \Delta t_{min}}\right] \qquad (3.104)$$

Exercise

- Give an estimate of the addition to Δt, if a step disturbance is introduced in phase B, see also the test sequence Sect. 3.3, item (d), as function of Ω_2.
- Investigate the transient in phase C and the contribution $2/\Omega_1$ given above.

3.9 Case study (continued): Temperature control in a chemical batch reactor

The example continues the case study from Chapter 2, but is closer to reality and to the original project. The model is more complex and there are output constraints in addition to the input saturation.

3.9.1 The Process and the Main Control Task

Fig. 3.40 is a simplified sketch of the batch reactor.

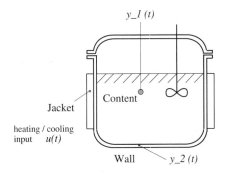

Fig. 3.40. Sketch of the plant

- The *main control* task is to keep the fluid temperature at a given setpoint. Overshoots are to be avoided, because over-temperatures may generate unwanted secondary products from the reactions. Under-temperatures will prolong the batch production time, as reaction rate goes down. Thus high performance temperature control is required.
- Available *instrumentation and operation subsystems*: the fluid content is continuously stirred. Its temperature ϑ_c is measured by a temperature sensor (y_1) in the fluid. There is a second temperature sensor (y_2) for the wall temperature ϑ_w.
  The jacket is a source or sink for thermal energy flowing to or from the wall to the content fluid. There is a third sensor for the jacket temperature ϑ_j (not used here). The response of the jacket subsystem is approximated by one short first-order lag with unity gain and time constant τ_u.
  The responses of the two temperature sensors shall also be approximated by two short first-order lags with time constants τ_y.
- Again, a *chemical reaction* shall take place in the content fluid, which starts at a given temperature threshold value ϑ_s below the nominal content temperature ϑ_r.

The *heat flow* from or to this reaction I_R^* shall be modeled by

$$I_R^* = -k_R(\vartheta_c - \vartheta_s) \quad \text{for} \quad \vartheta_c \geq \vartheta_s$$
$$I_R^* = \qquad 0 \qquad \text{for} \quad \vartheta_c < \vartheta_s$$

where

$$k_R \quad \begin{array}{l} < 0 \text{ for exothermal reactions (heat source)} \\ = 0 \text{ for inert reactions} \\ > 0 \text{ for endothermal reactions (heat sink)} \end{array}$$

The actual value of k_R is not known, but upper and lower bounds are specified.

- The *test sequence* consists of
  - starting from standstill (all variables at zero)
  - step up to the contents temperature setpoint r_1 to nominal, $r_1 = 1.0$.
  - After reaching the steady state, apply a small setpoint step to $r_1 = 1.05$
  - then introduce a small heat flow disturbance step $I_{d_1}^* > 0$ to the content
  - finally apply another step $I_{d_2}^* < 0$.
  This is to model a warmer and then a colder reactant inflow.

  These last three test steps will show the linear loop performance.

- No *measurement disturbances* shall be considered here.

- Three cases of *constraints* shall be considered,

  case A: input constraint from actuator saturation
  i.e. limits on u; $[-1.0 \cdots +1.0]$

  case B: output constraint on the
  "wall to fluid temperature difference",
  i.e. limits to heat flow

  case C: output constraint on
  "wall temperature above nominal fluid temperature",
  i.e. limits to max. fluid temperature at the wall

3.9.2 The Plant Model

In this model, two (well-stirred) main storage domains for thermal energy are considered, namely one for the fluid content and one for the wall,[5] resulting in two coupled ordinary first-order differential equations.

---

[5] Note that using the jacket subsystem as a second well-stirred storage domain and reducing the wall storage to the small value would lead to the same equation system.

Considering the range $\vartheta > \vartheta_s$ around the steady state at $r_1 = 1.0$ and introducing 'per unit' variables as in the case study of Sect. 2.8

$$\tau_1 \frac{d}{dt} x_1 = +(x_2 - x_1) - a(x_1 - x_s) \qquad + v_1 + v_2$$
$$= -(1 + a)x_1 + x_2 + ax_s \qquad + v_1 + v_2 \quad \text{for the fluid}$$
$$\tau_2 \frac{d}{dt} x_2 = -(x_2 - x_1) \qquad\qquad + u$$
$$= + x_1 - x_2 \qquad\qquad\quad + u \qquad\qquad \text{for the wall}$$
$$y_1 = x_1$$
$$y_2 = x_2 \tag{3.105}$$

This leads to the transfer functions

$$G_{u_1} = \frac{y_1}{u} = \frac{1}{s^2 \tau_1 \tau_2 + s[\tau_2(1+a) + \tau_1] + a} \tag{3.106}$$

$$\text{and} \quad G_{u_2} = \frac{y_2}{u} = \frac{s\tau_1 + (1+a)}{s^2 \tau_1 \tau_2 + s[\tau_2(1+a) + \tau_1] + a} \tag{3.107}$$

$$G_{u_{(y_2 - y_1)}} = \frac{y_2}{u} - \frac{y_1}{u} = \frac{s\tau_1 + a}{s^2 \tau_1 \tau_2 + s[\tau_2(1+a) + \tau_1] + a} \tag{3.108}$$

Again, the "fast" dynamics shall be represented by a series delay element e^{-sD} with $D := \tau_u + \tau_y$.

3.9.3 Parameter Values

$$
\begin{aligned}
\tau_1 &= & 900 \text{ s} \\
x_s &= & 0.90 & \qquad \text{reaction threshold temperature (in p.u.)} \\
a &= & -0.50 \cdots + 0.50 & \quad \text{reaction enthalpy gain (in p.u.)} \\
\tau_2 &= & 120 \text{ s} \\
D &= & 10 \text{ s} & \qquad \text{sum of actuator and sensor time constants} \\
u &= & -1.0 \cdots + 1.0 & \quad \text{actuator saturations} \\
x_2 - x_1 &= & -0.25 \cdots + 0.25 & \quad \text{bounds on temperature difference (in p.u.)} \\
x_2 &= & 0.0 \cdots + 1.15 & \quad \text{bounds on wall temperature (in p.u.)}
\end{aligned}
$$

3.9.4 The Controller

To avoid any overshoot in the linear range, a two-degrees-of-freedom linear PI algorithm with additional feedback of the wall temperature signal y_2 is suggested for the y_1 loop:

$$w_1(s) = k_{01}(r_1 - y_1) - s\tau_0 k_{11} y_1 - s\tau_0 k_{21} y_2 \quad \text{and} \quad u_1 = \frac{1}{s\tau_0} w_1 \tag{3.109}$$

and similarly for the y_2 override loop

$$w_{2_{up}}(s) = k_{02}(r_{2_{up}} - y_2) - s\tau_0 k_{12} y_2 \qquad \text{and} \quad u_{2_{up}} = \frac{1}{s\tau_0} w_{2_{up}} \tag{3.110}$$

3.9.5 The Design Program

It is suggested to proceed in four phases. Here are some additional hints:

1. Linear design:

   Design the controller settings by pole assignment as function of pole location Ω_1.

   What options are available to incorporate a?

   Select T_s.

   Build a suitable Simulink/Matab file pair with a discrete-time implementation of the controller.

   Check its performance for the test sequence starting from steady state at $\bar{y}_2 = 1.0$, *i.e.* without the run-up phase.

2. Case A: saturation on $u(t)$

   Start with the stability analysis for the range of a

   Determine first the steady state conditions $\bar{u}$ relative to the saturation values.

   What sizes of setpoint step deviations δr_1 and step disturbances $\delta \bar{v}$ are admissible for a well-posed stability problem?

   Draw the Nyquist contour for the reduced order and continuous loop and the actual control loop.

   Check the transient response for the full test sequence.

3. Case B: output constraint on the heat flow, *i.e.* on $y_2 - y_1$

   Proceed as for case A.

   Investigate a control structure *modified* as follows: to speed up the contents temperature run-up, enable the output constraints only if y_1 is larger than a design parameter $0 \le y_{1_{on}} \le 1.0$. Discuss the effect of variation of $y_{1_{on}}$ on stability and response.

4. Case C: output constraint on the wall temperature y_2

   Proceed as for case B.

5. Finale: combine constraints of cases A, B, and C.

3.9.6 Typical Results

The following Figs. 3.41–3.43 may help you to check your solutions.

Simulation results are displayed for $a = 0$ only, with $T_s = 5.0$, $\Omega_1 T_1 = 10$, $\Omega_2 T_2 = 2.5$.

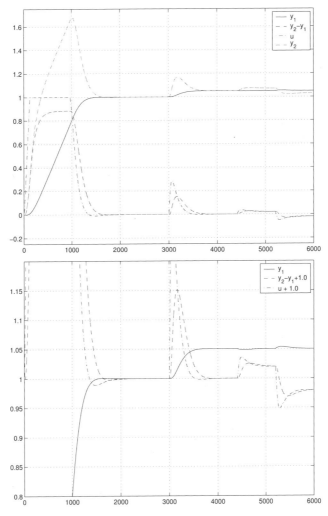

Fig. 3.41. Case A: response to the test sequence with input constraints only
The lower plot zooms in on the small signal range, *i.e.* on the linear responses

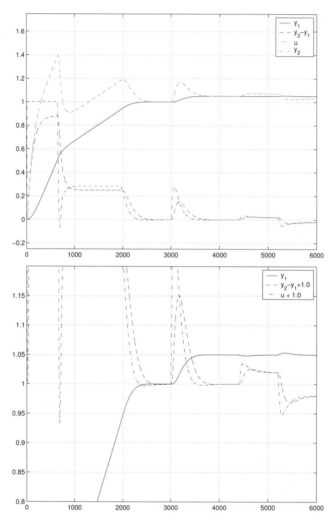

Fig. 3.42. Case B: response with output constraint on the temperature difference (heat flow), enabled for $y_1 > y_{1_{on}} = 0.5$

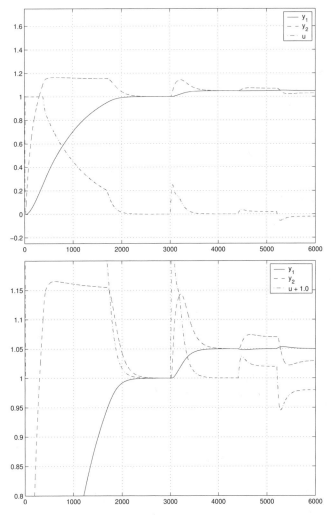

Fig. 3.43. Case C: response with output constraint on the wall temperature y_2

3.10 Summary

The basic PI control loop with one output constraint (but no input constraint) has been investigated.

Again, "well-posedness" is assumed, $i.e.$ the final steady state of the loop $\bar{y}_1 = r_1$ produces an output variable $\bar{y}_2$ inside the constraint limits $r_{2_{lo}} < r_{2_{hi}}$. In other words, the constraints are encountered transiently, but not permanently.

Structures

Starting from three quite different basic forms that are often used in practice, we have discussed numerous alternatives. Equivalence has been investigated. This led to a Generic Structure (Form $\mathbf{Q}$) in Sect. 3.4, which uses the generic awf structure form from Sect. 2.4.1, in conjunction with either nonlinear additive correction of u or Max-Min-selection.

It has also been shown that the cascade limiter version is more restricted in application than the other two basic forms.

Transient response

A benchmark has been proposed to test the performance of such structures by simulation. It uses for the plant the two-open-integrator chain, abstracted from positioning control, and with speed limitation as the output constraint. The Generic Structure has performed well from a practical design point of view, both for awf gain values being "high" (deadbeat) and "moderate" (compensating).

The run-up trajectory approaches the minimum-time one for increasing bandwidth of the main (Ω_1) control loop and the constraint (Ω_2) control loop. By observation, $u(t)$ is sufficiently smooth ('bumpless'). No chattering or other high-frequency phenomena were observed.

Stability properties

Using the standard Circle Criterion, the stability test has been prepared in general form for transfer functions $G_i, R_i, \quad i = 1, 2$; and for the characteristic polynomials for the open and closed loops (d_i and $D_i, D_{a_i}, \quad i = 1, 2$).

The result is an extension of the result for the generic awf form. Due to its symmetry properties, it is easy to memorize, and straightforward to use in practical design work.

It has been applied to the benchmark, where it nicely supports the results from the simulation, and provides much valuable insight into the effects of plant and design parameter variations.

In short, such output constraint control systems are straightforward to design, and to analyze, and they seem to perform quite well. This may well be the main reason for their popularity with industrial design engineers.

However, many important areas and topics have not been investigated so far, such as detailed below:

Regarding *structure*, the alternatives considered in Sect. 3.4 have a common root: the output constraining feedback acts directly on the control variable u, *i.e.* by "control conditioning". This is mainly motivated by the good performance achieved in Chapter 2. Another family of structures may be generated by letting the output constraining feedback act on the reference input r_1 of the main loop, *i.e.* by applying the "realizable reference" or "reference conditioning" stratagems. We shall address this topic in Chapter 7.

Considering the *transient response* aspect, using this particular benchmark may seem rather special, and the results on it may not be overextended. One element has already been mentioned: a more general transfer function for G_2 would be

$$G_2(s) = \frac{y_2(s)}{u(s)} = \frac{sT_1}{s^2 T_1 T_2 + sT_1 a_1 + a_0}$$

where $a_0 \neq 0$. Using a PI algorithm for R_2

$$R_2(s) = \frac{u(s)}{r_2 - y_2(s)} = \frac{sT_1 k_{1_2} + k_{0_2}}{sT_1}$$

yields

$$d_2(s) = s^2 T_1 T_2 + sT_1 a_1 + a_0 \quad \text{and} \quad D_2(s) = s^2 T_1 T_2 + sT_1 (a_1 + k_{1_2}) + (a_0 + k_{0_2})$$

and a steady state control error $\bar{e}_2$ for the constant setpoint $r_{2_{hi}} > 0$

$$\bar{e}_2 = \frac{a_0}{a_0 + k_{0_2}} r_{2_{hi}} < \frac{a_0}{k_{0_2}} r_{2_{hi}} > 0$$

To drive $\bar{e}_2 \to 0$, the control algorithm must be augmented to

$$R_2(s) \to \frac{sT_1 k_{1_2} + k_{0_2}}{sT_1} \frac{1}{sT_1} \tag{3.111}$$

How this is integrated in the override structure will be shown in Chapter 5.

Another topic is $G_i(s)$, $i = 1, 2$ being of second order. This is to be discussed further in the second part of this book. But we shall not go into the minimum-time problem for these cases.

The *stability test* has been presented already in a form that covers much more general situations than the benchmark. This shall be used extensively in the second part of the book.

From an applications point of view, however, the *main restriction* is that either one input or one output constraint is admitted, but not both. This is the most pressing need, and shall be addressed in the next chapter.

4

PI Control with Input and Output Constraints

From a practical point of view the most restricting assumption in Chapter 3 is that the control variable is required to be unbounded. In almost all applications, however, one also has to expect that $u(t)$ may transiently saturate in at least one phase of the transient for cases with an output constraint feedback. Again, we shall start by stating the general control problem, define a benchmark which can be handled sufficiently well within the scope of standard PI control techniques, and discuss possible controller structures and their transient response in parallel. Then the performance is investigated in relation to the minimum-time case, and stability properties are discussed. A case study is given as an exercise.

4.1 Problem Statement

4.1.1 The Plant Model

Fig. 4.1 shows the system to be controlled.

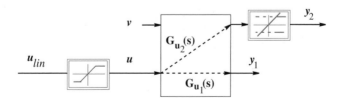

Fig. 4.1. Problem statement: the plant model

It is generated from the plant model of Fig. 3.5, with the same properties, having one pair of output constraints as (soft) operational limits on y_2:

$r_{2_{hi}}$, $r_{2_{lo}}$, by adding one pair of input constraints as (hard) control saturations on $u(t)$: u_{hi}, u_{lo}.

The main assumption again is that linear control of the main variable $y_1(t)$ by $u(t)$ is feasible for small but finite deviations from that steady state, without interference from both input and output constraints.

Otherwise this is considered an "ill-posed" control problem. It must be rectified by appropriate modifications in the plant layout and design, such that the constraints are not interfering with small deviations control, and by this to produce a "well posed" problem.

4.1.2 The Basic Control Idea

The basic idea is to combine the proven concepts of Chapters 2 and 3.

The first version, in Fig. 4.2, is based on the selector form from Chapter 3.

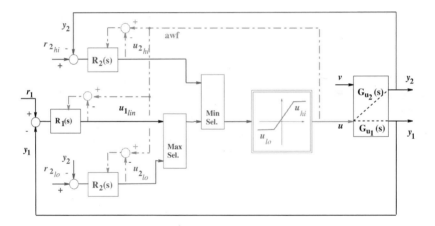

Fig. 4.2. Control concept using the elements of selection and saturation sequentially

Two main assumptions are made:
- the paradigm of control conditioning is used,
- and $u(t)$ is shaped appropriately by selection first
  and saturation second, in *"sequence"*.

Both have to be questioned, and shall be investigated later. At this instant, we intuitively argue that the first one has been simple and well-performing so far; for the second one, we observe that the input constraints are final limits to the working range of u, and may not be overrun by any output constraint action. So the saturation block should be placed downstream of the selection block, and not *vice versa*.

Remarks

- The awf reference for tracking of individual controllers must be picked up downstream of all nonlinear operations (Hanus's rule).
- The SAT block may be replaced by the equivalent Selector block, and then both Selector pairs by the equivalent Nonlinear-Additive blocks.
- As shown in Fig. 3.18, the saturation to u_{hi}, u_{lo} may be replaced by a feedback implementation with setpoints u_{hi}, u_{lo}. This can then be generalized into a second output constraint feedback (to be used later as an advanced technique).

The second basic version uses the cascade-limiter concept, shown in Fig. 4.3.

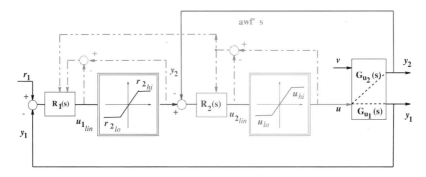

Fig. 4.3. Control concept combining the Cascade-Limiter and Control-Saturation elements

Note that if the master controller R_1 has integral action, then it needs a second awf to allow tracking of the final u-saturation as well; see the double-dot-dashed line in Fig. 4.3.

4.2 Benchmark System

We are using the same benchmark as in Chapter 3, with the addition of the control saturation at

$$u_{lo} = -1.0; \quad u_{hi} = +1.0 \tag{4.1}$$

and the test sequence in the following versions

v0: with no step load z_1 applied during run-up $z_1 = 0$, but then $z_2 = -0.90$ (shall require I(aw) action in R_1)

v1: with step load z_1 applied during run-up $z_1 = +0.90$ and $z_2 = -0.90$, as in Chapter 3, (shall require I(aw) action in the both R_2 as well)

v2: with step load $z_1 = -0.90$ and $z_2 = +0.90$
v3: same as v2, but $z_1 = -0.90$ is applied 1 s later.

The motivation for this will appear in the following sections.

4.3 Structures and Transient Responses

We build on the extended evaluation of various possible structures in Chapters 2 and 3 and focus on the final ones obtained there. This shortens the discussion considerably. Also, we consider only forms using the control conditioning paradigm.

4.3.1 Form A: Sequential Max-Min-Selection

The control structure is shown in Fig. 4.4. Here, the saturation element has been replaced by the equivalent Max-Min-selector combination, while respecting the sequence of nonlinear operations as discussed above.

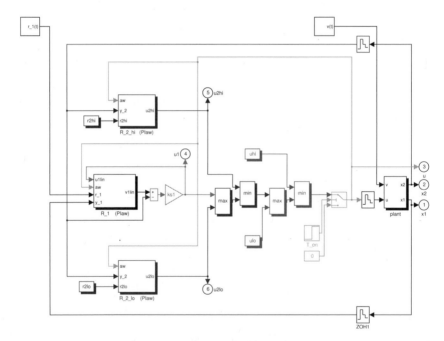

Fig. 4.4. Form **A**: sequential selection

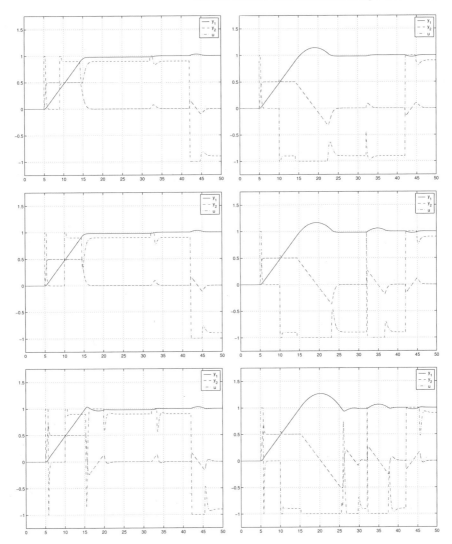

Fig. 4.5. Responses of form **A** to the benchmark v1 (left) and v2 (right), and with awf gains: deadbeat (top);   compensating $(k_{a_i}^*)$ (center);   low $(0.25k_{a_i}^*)$ (bottom)

Results

The transient responses in Fig. 4.5 show

- The constraints on u and y_2 are respected as long as the awf gains k_{a_2} are above or at least equal to the "compensating" values $k_a \geq k_a^*$. Below this there will be windup, *i.e.* the constraints on y_2 can no longer be respected.
- The response of y_1 is well behaved for benchmark v1 (left column). There is no significant deterioration compared with the non-saturated case of

Fig. 3.35. They are not sensitive to the awf gain k_{a_1}, as long as they are above or equal to the "compensation" value.

- However, the responses of y_1 for benchmark v2 (right column) are not acceptable: there is a large overshoot of $y_1(t)$ at the end of the run-up phase, for all awf gain values.
- But the constraints are respected for awf gains k_{a_2}, again if they are selected to be above or equal to the "compensation" value.

This misbehavior is due to the shift in the steady state of $u(t)$ due to z_1, which is now close to u_{lo}. That is, the available control span $\bar{u} - u_{lo}$ for bringing the plant to the steady state $\bar{y}_1 = r_1$ is much smaller here than in benchmark v1. This property is crucial for applications of antiwindup and override systems to plants of dominant order higher than one, and will be discussed in more detail later.

To get a better feeling for it, consider the following situation. You are driving your car at the statutory speed limit towards a traffic light. You start braking at a distance which, based on your driving experience, is sufficient for normal road conditions, a horizontal road and normal car mass.

If now the road is steeply downhill and you have loaded your car to the brim, and you start braking at the same distance, and apply the same braking force, then the deceleration will be much weaker, the speed will reduce more slowly (see Fig. 4.5 (right)), and your car will come to standstill beyond the traffic light, even if you apply the full braking force.

If, however, the road is steeply uphill and all other conditions are the same, then the capability for deceleration (the control span) is much higher and the problem disappears.

This example also suggests a possible remedy: you reduce your approach speed as the road gets steeper, such that you expect to be able to stop before the traffic light, with some safety margin. This strategy has been implemented in Fig. 4.6 (left column) on the benchmark v2 by stepping down the initial reference $r_{2_{hi_0}}$ (statutory speed) to a lower value $r_{2_{hi_1}}$ (safe speed), as soon as the control variable $u(t)$ has stabilized to the new z_1 and the reduced control span $\bar{u} - u_{lo}$ is established and thus may be detected. In the simulation, this is achieved 2 s after z_1 has been applied. And then $r_{2_{hi_1}}$ must be tuned to produce the required safety margin, in the simulation to 0.25 of the initial value. The responses are now acceptable. Again, the awf gains should be selected above or equal to the "compensating" ones.

Unfortunately, this simple strategy may fail, see Fig. 4.6 (right column) for the benchmark v3, where you are 1 s closer to the traffic light when the road gets steep. So your speed is too high for the remaining distance to the traffic light to avoid overshoot, but you apply full braking earlier than in Fig. 4.5 (right column), and thus the overshoot is at least reduced.

More robust strategies will be discussed later.

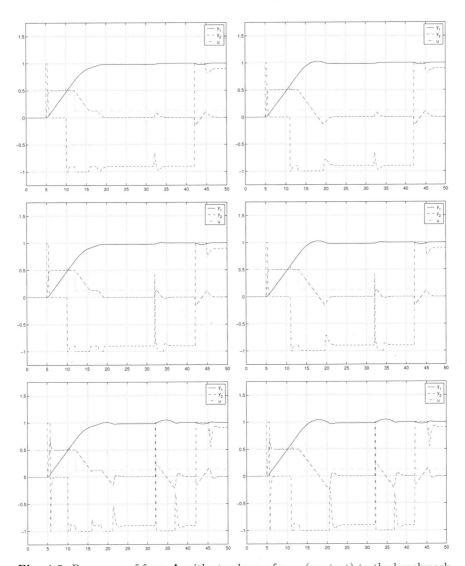

Fig. 4.6. Responses of form **A** with stepdown of $r_{2_{hi}}$ (see text) to the benchmark v2 (left) and v3 (right);

awf gains: deadbeat (top);   compensating $(k_{a_i}^*)$ (center);   low $(0.25k_{a_i}^*)$ (bottom)

4.3.2 Form B: Parallel Selection

Most process control systems offer function blocks for Min- and Max-selection of more than two inputs. This suggests replacing the sequential selection based on two inputs everywhere in Fig. 4.4 by such blocks, where all the Min- and Max-selections are compressed in one such block; see Fig. 4.7. This reduces the number of function blocks, and may increase transparency.

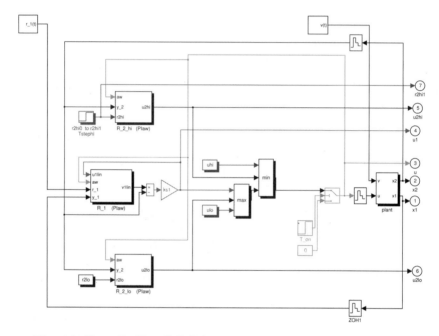

Fig. 4.7. Form **B**: "Parallel" Selection, with optional step-down of $r_{2_{hi}}$

One may say, that this is a "parallel" selection of the maximum and minimum $u(t)$, which suggests the name used for this form **B**. Note that the result of the "parallel" Max-selection block is still sequentially submitted to the "parallel" Min-selection block.

The transient responses in Fig. 4.8 are for the benchmark v2, with constant $r_{2_{hi}}$ (left column), and with $r_{2_{hi}}$-step down (right column).

The main result is that the constraints are not respected everywhere, in contrast to form **A**. Arrows have been placed to indicate where this occurs. For the steps $0 \rightarrow z_1$ and $r_{2_{hi}} \rightarrow r_{2_{hi_1}}$, the controller $R_{2_{hi}}$ produces transient control signal excursions $u_{2_{lo}}(t)$ far below u_{lo}. And this is not clipped before putting $u_{2_{lo}}$ to $u(t)$, because the Max-Selection, which should perform this clipping, is not at the appropriate place, but has migrated upstream.

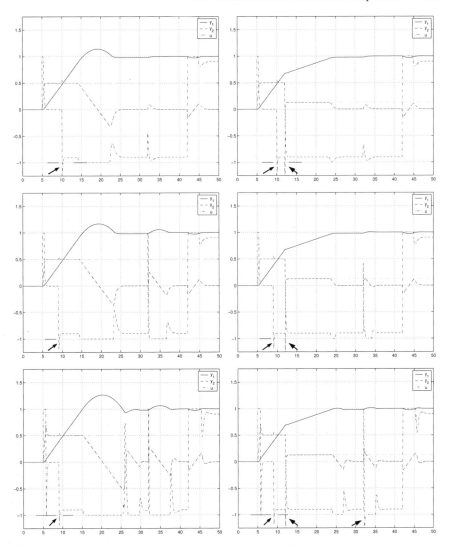

Fig. 4.8. Responses of form **B** to the benchmark v2 with stepdown of $r_{2_{hi}}$ suppressed (left column) and activated (right column);
awf gains: deadbeat (top);   compensating $(k_{a_i}^*)$ (center);   low $(0.25k_{a_i}^*)$ (bottom)

Going back to driving your car towards the traffic light, the controller outputs a control $u(t)$ such that the speed would be reduced nearly instantaneously to the new (safe) value, *i.e.* nearly infinite deceleration, which is clearly not feasible.

An idea to remedy this would be to change the sequence of the selections, *i.e.* selecting for the minimum first and then for the maximum. This would clearly fix this misbehavior, as now the clipping to u_{lo} is performed. But the problem would now appear during acceleration.

Exercise
Investigate this by adapting the benchmark appropriately and by simulations.

In other words, such "parallel selection" may produce a simpler structure, but it is considered *functionally non-sufficient*, and thus unfit for practical use. It will not be considered any further here.

4.3.3 Form C: "Lowest Wins"

Implying standard programming techniques, very probably the multi-input selector blocks in Form **B** internally will be implemented by sequential selection, e.g. for three inputs

$$out = Min\left\{in_1, Min\left\{in_2, in_3\right\}\right\}$$
$$out = Max\left\{in_1, Max\left\{in_2, in_3\right\}\right\} \tag{4.2}$$

This parses into four two input selector blocks. By arranging them appropriately in the signal flow graph, this may be interpreted directly as applying a *pre-selection* of type Max to all constraining $u_{j_{lo}}$ inputs.

$$v_{lo} = Max\left\{u_{2_{lo}}, u_{lo}\right\}$$

and another *pre-selection* of type Min to all constraining $u_{j_{hi}}$ signals

$$v_{hi} = Min\left\{u_{2_{hi}}, u_{hi}\right\}$$

and then feeding the results v_{lo}, v_{hi} of the pre-selections to the standard Max-Min-selector block with three inputs (the third input being u_1), as has been used exclusively so far. So the pre-selections may be seen as "parallel" operations. They produce the most constraining $u_{j_{hi}}; u_{j_{lo}}$. If only the Min part is considered, this motivates the designation "lowest wins" (which is also used by practitioners).

Thus, form **C** is equivalent to form **B**. Therefore, it is also considered "unfit for use", and not investigated any further.

4.3.4 Form D: Sequential Nonlinear Additive

To obtain this, replace each Max-Min-selector block in form **A** directly by the equivalent nonlinear-additive block which contains a corresponding deadspan. This is nothing new functionally, but provides a more suitable structure for a stability analysis using sector criteria.

4.3.5 Form E: Cascade Limiter

The cascade-limiter concept of Fig. 4.3 is implemented in Fig. 4.9.

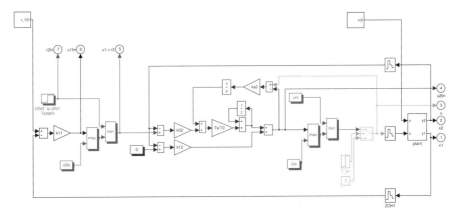

Fig. 4.9. Form **E**: Cascade-limiter structure with control saturations, with optional stepdown of $r_{2_{hi}}$

Remarks

- An integral action in R_2 is sufficient to drive the control errors e_1 and e_2 respectively to zero for the disturbances r and z specified in the benchmark. Thus R_2 will be of PI(aw) type.
- As no integral action is required for R_1 (valid within the scope of this benchmark), no multiple awf paths are needed (see Fig. 4.3), and thus need not be designed. Such more general cases will be addressed later.
- R_2 is designed as a two-degrees-of-freedom structure, where the setpoint to the proportional action is separated and set to a constant, here specifically to zero. This has the practical advantage that inputs to the control signal $u(t)$ are filtered in the integral action path and thus tend to be smoother. Also, computing the controller parameters is less involved in the dominant pole assignment method.

The transient responses in Fig. 4.10 for benchmark v2 show the following:

- Concerning the constraints, they do not exhibit such functional deficiencies as forms **B** and **C**.
- At first glance the performance of y_1 is comparable to what is produced by form **A**,
- both for constant $r_{2_{hi}}$ and with the stepdown option activated.

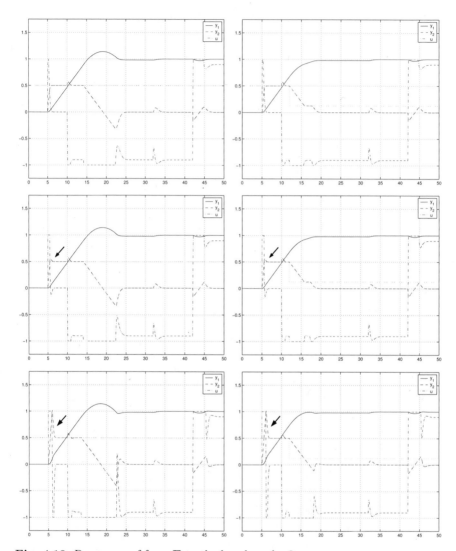

Fig. 4.10. Responses of form **E** to the benchmark v2
with stepdown of $r_{2_{hi}}$ suppressed (left column) and activated (right column);
awf gains: deadbeat (top);   compensating $(k_{a_i}^*)$ (center);   low $(0.25k_{a_i}^*)$ (bottom)

- Closer inspection reveals that there is now (in contrast to form **A**) an overshoot on y_2 during the initial phase of the run-up, if the awf gain is set to the compensating value. It disappears, however, for deadbeat awf gain. This is a slight drawback for applications.
- The performance of the full structure (the cascade of two PI(aw) controllers) remains to be seen.

To *summarize*: two structures, forms **A** and **E**, have emerged. They seem to perform adequately. However, the overshoot on y_1 caused by the control saturation must be investigated further. This is needed to provide a basis for designing more robust solutions.

4.4 Performance Analysis

4.4.1 Introduction

Again, the minimum-time transients shall be used as performance reference. We are particularly interested in the increments of settling time for the structures discussed before and the underlying differences of function.

To provide an intuitive understanding, we shall stay close to the benchmark situation used before:

- the plant shall be the two-open-integrator chain
- with time constants scaled to one, $T_1 = 1.0$ and $T_2 = 1.0$
- the delay is suppressed, $D = 0$
- and the sampling rate shall be reduced such that its effect is negligible, $T_s = 0.001$
- both constraints are normalized to one, *i.e.*

$$-1.0 \le u(t) \le +1.0 \quad \text{and} \quad -1.0 \le y_2(t) \le +1.0 \qquad (4.3)$$

- the focus is on the run-up phase from standstill $y_1 = 0$, $y_2 = 0$ to the final equilibrium $y_1 = r_1 = 1.0$, $y_2 = 0$
- with no load swing, $z = 0$
- a coordinate shift is performed to move the final steady state to zero, $x_1 = y_1 - r_1$, $x_2 = y_2$.

The phase plane method is a powerful analysis tool for such second-order systems: trajectories are depicted on the (x_1, x_2) plane, with the final steady state at the origin. They are to be supplemented by the usual time-response plots.

The first step will be to construct the minimum-time trajectories in the phase plane, for "almost all" initial conditions. Then we shall investigate the responses for the control structure form **A**, with P controllers first and then with integral actions.

Repeating this for the cascade-limiter and saturation Form **E** is left to the reader as an **exercise**.

4.4.2 Applying Open Loop Control Sequences $u(t)$

The standard procedure to construct the minimum-time trajectories in the phase plane is to let time run backwards, *i.e.* to start from the final equilibrium at the origin. We apply either $u_{lo} = -1.0$ or $u_{hi} = +1.0$ as the control signal. This produces a pair of minimum-time ejection trajectories of parabolic shape; see Fig. 4.11(top). Note the arrows indicating the direction of movement.

Let $\bar{u}$ be the value of the control variable needed at the final equilibrium and

$$\Delta u_{hi} = u_{hi} - \bar{u} > 0; \quad \Delta u_{lo} = u_{lo} - \bar{u} < 0 \tag{4.4}$$

The parabolic shape of the finals trajectory is (for u_{lo}):

$$x_{2_{lo}}(t_a) = -\frac{t_a}{T_2} \Delta u_{lo} \; > 0$$

$$\text{and} \quad x_{1_{lo}}(t_a) = \frac{1}{T_1 T_2} \Delta u_{lo} \frac{1}{2} t_a^2 \quad = \frac{1}{2} \frac{T_2}{T_1} \frac{1}{\Delta u_{lo}} x_{2_{lo}}^2(t_a) \quad < 0;$$

$$\text{i.e.} \quad x_{2_{lo}}(t_a) = a_{lo} \cdot x_{2_{lo}}^2(t_a) \quad \text{with} \quad a_{lo} := \frac{1}{2} \frac{T_2}{T_1} \frac{1}{\Delta u_{lo}} \quad < 0;$$

$$\text{or} \qquad x_{2_{lo}} = -\text{sign}(a_{lo}) \sqrt{\frac{x_{1_{lo}}}{a_{lo}}} \tag{4.5}$$

and correspondingly if u_{hi} is applied ($x_{2_{hi}} < 0$; $x_{1_{hi}} > 0$; $a_{hi} > 0$).
So the trajectory running from the origin to the upper left is generated by applying $u = u_{lo}$, and the opposite one running to the lower right by using $u = u_{hi}$.

Check Eq. 4.5 as an **exercise**.

Then starting at $-t_a$, the opposite control saturation signal is applied. In Fig. 4.11(center), two values for t_a have been entered, and the resulting continuation trajectories are depicted. They are of parabolic shape again, but of opposite curvature. If the reverse time continues up to $-t_e < -t_a$ then "almost all" end-points of the ejection trajectories in the phase plane may be reached with an associated pair of $t_a, t_e < +\infty$-values.

In Fig. 4.11(bottom) the constraints on the state variable x_2 have been inserted. Then the ejection trajectory for $u = u_{lo}$ will run upward until $x_2 = x_{2_{hi}}$ is encountered at time $-t_a^*$. For $-t < -t_a^*$ the $u(-t)$ must be such that $x_2(t)$ slides along the horizontal line at $x_{2_{hi}}$, up to a $-t_b \leq -t_a^*$ (in our case $u(t) = 0$ for $-t_b \leq -t \leq -t_a^*$).

Finally, the opposite control saturation $u = u_{hi}$ or $u = u_{lo}$ is applied, and the trajectory will be the corresponding parabola up to an arbitrary $-t_e \leq -t_b$. Thus, again "almost all" end-points in the state plane can be reached for $t_b, t_e < +\infty$.

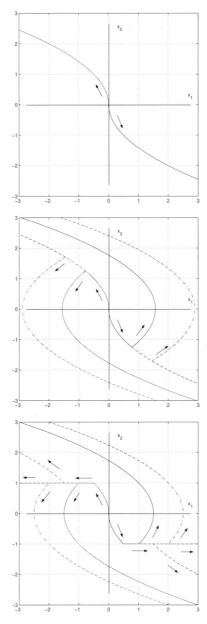

Fig. 4.11. Minimum-time trajectories on the phase plane with backward running time and with open-loop control sequences $u(t)$:

top: for $u = u_{hi}$ and $u = u_{lo}$ ("finals" trajectories);

center: for switching between $u = u_{hi}$ and $u = u_{lo}$ ("regular" problem);

bottom: with limitations on x_2 ("singular" problem)

The final move is now to let time run forward again. Then, the endpoint reached so far translates the initial point, and the final point translates the origin. And only the arrows in Fig. 4.11 must be reversed to obtain the phase plane portraits of the minimum-time trajectories. The ejection parts from Fig. 4.11(top) will translate in the final approach parts, the "finals" for short.

The phase portraits also suggest how u at an arbitrary time instant $t > 0$ can be generated from the values of x_1 and x_2 at that time instant, *i.e.* how a control function $u = f(x_1, x_2)$ can be set up. In other words we replace the feedforward control sequence $u(t)$ by a feedback control law $f(x_1, x_2)$. This is known as solving the synthesis problem.

In Fig. 4.11(top and center), all points x_1, x_2 below and to the left of the ejection parabolas have to produce $u = u_{hi} > 0$, whereas all points above and to the right have to produce $u = u_{lo} < 0$. On the ejection parabolas u changes to the opposite saturation. They are therefore called *switching curves*. Then

$$\text{if} \quad x_2 > 0$$
$$\text{if} \quad x_1 - a_{lo}x_2^2 \geq 0 \quad \text{then} \quad u = u_{lo} \quad \text{else} \quad u = u_{hi}$$
$$\text{if} \quad x_2 < 0$$
$$\text{if} \quad x_1 - a_{hi}x_2^2 > 0 \quad \text{then} \quad u = u_{lo} \quad \text{else} \quad u = u_{hi} \qquad (4.6)$$

Exercise
Generate the feedback control function for Fig. 4.11(bottom).

4.4.3 Generating $u(t)$ by Proportional Feedback and Selection

In the basic structures investigated above, the control $u(t)$ is generated by linear feedback using standard PI(aw) controllers and subsequent selection. This may be seen as piecewise linear control along the transient. For simplicity we shall look at the case of P control, Fig. 4.12, and shall insert the I(aw) actions later.

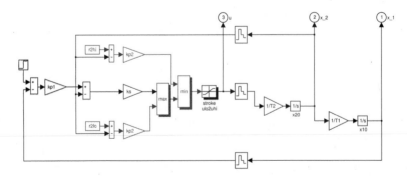

Fig. 4.12. The control system with proportional feedback and selection

We shall use the same approach as with the minimum-time control in Sect. 4.4.2, *i.e.* calculate u for each point of the phase plane.

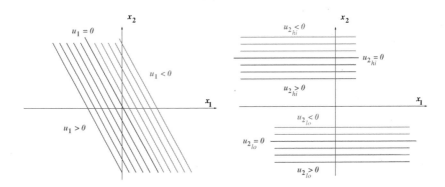

Fig. 4.13. Generating u by R_1 (left)   and by the R_2 controllers (right)

Consider first only the linear R_1 controller; Fig. 4.13(left)

$$u_1 = -k_{1_1} x_1 - k_{2_1} x_2$$

with $u_1 = const$ $\rightarrow$ $x_1 = -\dfrac{k_{2_1}}{k_{1_1}} x_2 - \dfrac{1}{k_{1_1}} u_1 := -\dfrac{2}{\Omega_1 T_1} x_2 - \dfrac{1}{(\Omega_1 T_1)^2} \dfrac{T_1}{T_2} u_1$

and for $u_1 = 0$: $x_{1_z} = -\dfrac{2}{\Omega_1 T_1} x_{2_z}$ (4.7)

where all points x_1, x_2 below and to the left of the $u_1 = 0$ line produce $u > 0$.

Consider next the linear R_2 controllers. For $R_{2_{hi}}$

$$u_{2_{hi}} = -k_{1_2} x_2 + k_{1_2} r_{2_{hi}}$$

with $u_{2_{hi}} = const \rightarrow x_2 = -\dfrac{1}{k_{1_2}} u_{2_{hi}} + r_{2_{hi}} := -\dfrac{1}{(\Omega_2 T_2)} u_{2_{hi}} + r_{2_{hi}}$ (4.8)

and correspondingly for $R_{2_{lo}}$. This produces horizontal lines with parameter $u_{2_{hi}}$, $u_{2_{lo}}$ on the phase plane; Fig. 4.13 (right).

The third step is to apply the selector block, as shown in Fig. 4.14 (left). Transition from $R_{2_{hi}}$ to R_1 will take place at

$$u_{2_{hi}} = u_1$$

i.e. $-k_{1_2} x_2 + k_{1_2} r_{2_{hi}} = -k_{1_1} x_1 - k_{2_1} x_2$

$$x_2 = \dfrac{1}{k_{1_2} - k_{2_1}} (k_{1_1} x_1 + k_{1_2} r_{2_{hi}})$$ (4.9)

and correspondingly for transition from $R_{2_{lo}}$ to R_1.

So $R_{2_{hi}}$ will be selected for points above and to the left of the dashed line in Fig. 4.14, R_1 for points in the corridor delimited by the dashed and the dash-dotted line, and $R_{2_{lo}}$ for points below and to the right of the dash-dotted line.

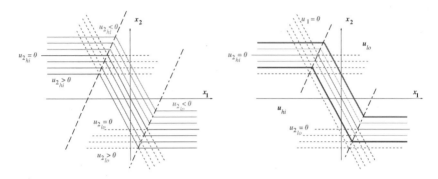

Fig. 4.14. Selecting u from u_1, $u_{2_{hi}}$, and $u_{2_{lo}}$ (left);   applying the saturation (right)

The last step is to apply the saturation on u, Fig. 4.14(right). Thus the phase plane is finally *partitioned* into
- a narrow corridor, where control is linear, along $u_{2_{hi}} = 0$,
    along $u_1 = 0$, and along $u_{2_{lo}} = 0$,
- a large area to the upper right, where $u = u_{lo}$,
- and a second large area to the lower left, where $u = u_{hi}$.

4.4.4 Comparison

We are now ready to compare the selection control law with the minimum-time control law from Sect. 4.4.1.

We consider an initial condition point at the left end of the $u_{2_{hi}} = 0$ horizontal line, Fig. 4.15. Then $x_1(t)$, $x_2(t)$ will move along this line to the right, until it meets the $u_1 = 0$ line, where transfer to R_1 control will occur. We now assume that $u(t)$ will be switched without any delay to its saturation value u_{lo}, as is the case for minimum-time systems. Then the trajectory will continue as the corresponding parabola.

As a special case, this may coincide with the "finals" parabola, Fig. 4.15(left), *i.e.* the Selector form would produce the minimum-time transient:

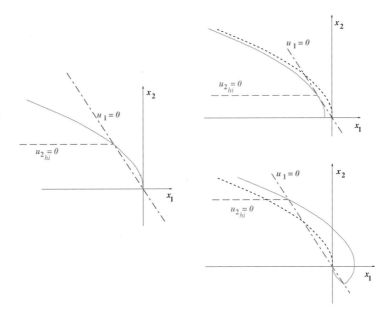

Fig. 4.15. Comparison of u above the phase plane by state feedback, selection and switching to u_{sat} and for the minimum-time system, for three values of $r_{2_{hi}}$

that is

for the "finals" parabola $x_1 = -|a_{lo}|x_2^2$

for the line $u_1 = 0$ $x_1 = -\dfrac{k_{2_1}}{k_{1_1}}x_2 = -\dfrac{2}{\Omega_1 T_1}x_2$

and $x_2 = r_{2_{hi}}$

i.e. $\Omega_1^* T_1 = \dfrac{2}{|a_{lo}|r_{2_{hi}}} = \dfrac{2}{\frac{1}{2}\frac{T_2}{T_1}\frac{1}{|\Delta u_{lo}|}r_{2_{hi}}}$

$= 4\dfrac{T_1}{T_2}\dfrac{|\Delta u_{lo}|}{r_{2_{hi}}}$

or $r_{2_{hi}}^* = \dfrac{4}{\Omega_1 T_2}|\Delta u_{lo}|$ (4.10)

and correspondingly for the $r_{2_{lo}}, a_{hi}$ case.

For a given set of parameters $T_1, T_2, \Delta u_{lo}, r_{2_{hi}}$ this allows you to calculate the value of Ω_1^* that would produce the minimum-time transient, or for a given $\Omega_1, T_1, T_2, \Delta u_{lo}$ the corresponding value for the speed constraint setpoint $r_{2_{hi}}^*$.

In other words, there is a clear tradeoff between performance of the linear R_1 loop (as determined by the value of Ω_1) and the "allowable" value of $r_{2_{hi}}$.

If now this value is reduced by a factor of two (while not changing Ω_1), then from the properties of the parabolic shape, the line $u_1 = 0$ is the local tangent to the parabola at the transfer point; Fig. 4.15 (top right). In other words, u_1 will only saturate at the very first instant, and not for the rest of the trajectory to the origin, *i.e.* the "finals" approach under R_1 control will be linear. The same effect can be obtained by reducing Ω_1 by a factor of $\sqrt{2}$, while not changing $r_{2_{hi}}$.

If, however. $r_{2_{hi}} > r_{2_{hi}}^*$ then a substantial overshoot of x_1 is unavoidable; Fig. 4.15 (bottom right). The trajectory is still nearly minimum-time in a mathematical sense, but violates the basic specification from the applications point of view.

We shall now drop the assumption that $u(t)$ switches without delay to the opposite saturation after x_2 crosses over the $u_1 = 0$ line. As R_1 is a proportional feedback with finite gains, *i.e.* the linear corridor in Fig. 4.14(right) is of finite width, $u(t) = u_1(t)$ will move more slowly to the saturation value, and thus run over the "finals" parabola. Consequently, there will be a noticeable overshoot.

In order to avoid this, we will either have to reduce Ω_1^* or $r_{2_{hi}}^*$ or both by some small amount. A conservative estimate for Ω_1^* (to Ω_1^{**}) may be derived as follows. The basic idea is to delay the transfer of $u(t)$ from u_{hi} to u_{lo} until the phase plane point $x_1(t), x_2(t)$ crosses over the parallel line for $u_1 = u_{lo}$, and there step from u_{hi} to u_{lo} without further delay, yielding

$$\text{from the ``finals'' parabola} \quad x_1 = -|a_{lo}|x_2^2$$

$$\text{from the } u_1 = u_{lo} \text{ line} \quad x_1 = -\frac{k_{2_1}}{k_{1_1}}x_2 + \frac{1}{k_{1_1}}|\Delta u_{lo}|$$

$$\text{with } x_2 = r_{2_{hi}} \text{ follows} \quad |a_{lo}|r_{2_{hi}} = \frac{2}{\Omega_1^{**}T_1} - \frac{1}{(\Omega_1^{**}T_1)^2}\frac{T_1}{T_2}\frac{|\Delta u_{lo}|}{r_{2_{hi}}}$$

$$\rightarrow |a_{lo}|r_{2_{hi}}(\Omega_1^{**}T_1)^2 - 2(\Omega_1^{**}T_1) + \frac{T_1}{T_2}\frac{|\Delta u_{lo}|}{r_{2_{hi}}} = 0$$

$$\text{i.e. } (\Omega_1^{**}T_1) = \frac{1}{|a_{lo}|r_{2_{hi}}}\left[1 \pm \sqrt{1 - \frac{|\Delta u_{lo}|}{|a_{lo}|r_{2_{hi}}^2}(|a_{lo}|r_{2_{hi}})^2\frac{T_1}{T_2}}\right]$$

$$\text{(where the } - \text{ of } \pm \text{ is irrelevant)}$$

$$= \frac{1}{|a_{lo}|r_{2_{hi}}}\left[1 + \sqrt{1 - |\Delta u_{lo}||a_{lo}|\frac{T_1}{T_2}}\right] = \frac{1}{|a_{lo}|r_{2_{hi}}}\left[1 + \sqrt{1 - \frac{1}{2}\frac{T_2}{T_1}\frac{T_1}{T_2}}\right]$$

$$\rightarrow (\Omega_1^{**}T_1) = \frac{\Omega_1^*T_1}{2}\left(1 + \sqrt{1/2}\right) = 0.854(\Omega^*T_1) \tag{4.11}$$

and correspondingly for the $r_{2_{lo}}, a_{hi}$ situation.
This is illustrated by the simulations in Fig. 4.16.

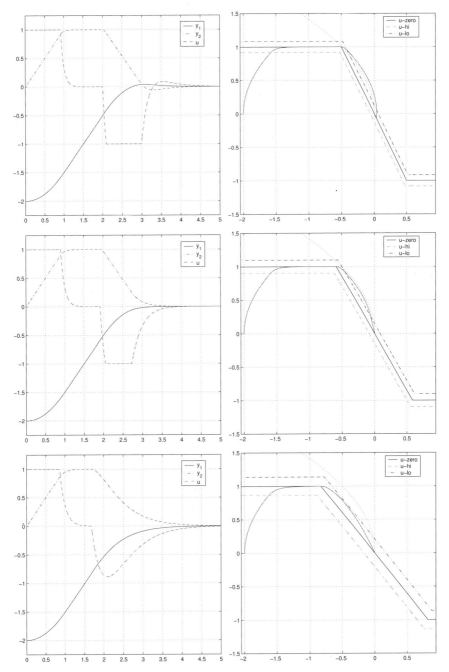

Fig. 4.16. Transient response along the time axis (left) and on the phase plane (right);   top: $\Omega_1 = \Omega_1^*$;   center: $\Omega_1 = \Omega_1^{**}$;   bottom: $\Omega_1 = \sqrt{1/2}\, \Omega_1^{**}$

Consider now a given set

$$T_1 = 1.0, \quad T_2 = 1.0, \quad u_{hi,lo} = \pm 1.0, \quad r_{2hi,lo} = \pm 1.0, \quad x_1(0) = -1.0, \quad x_2(0) = 0$$

for the system used in Fig. 4.16 and Ω_1/Ω_1^* as a parameter.
Then for $\Omega_1/\Omega_1^* < 1$ the transfer to R_1 will be before the $u_{2hi} = 0$ line crosses the "finals" parabola; consequently, the settling time will be larger than the minimum-time. And if $\Omega_1/\Omega_1^* > 1$ the transfer to R_1 will be after this crossover. Then x_1 will overshoot, which again leads to a larger settling time.

This is illustrated by Fig. 4.17, where the results of simulations on the system used in Fig. 4.16 are shown. The settling time for any system with a linear trailing phase is usually defined by observing from when the control error $e_1(t)$ stays in a given corridor, typically $|e_1| < 0.05$. Here, we use u instead ($|u - \bar{u}| < 0.05$), as this is a more sensitive criterion.

The minimum-time response for this system settles in 3.0 s to the origin.[1]

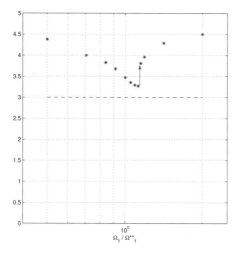

Fig. 4.17. Settling time to $|\Delta u| \leq 0.05$ for different Ω_1/Ω_1^{**}, compared with the minimum-time 3.0 s; with $\Omega_1^{**} = 3.414$

The discontinuity in the measured settling times indicated by the arrow appears because $u(t)$ now overshoots outside its corridor, and re-enters it distinctly later. So, for practical design purposes, one would recommend using

$$\Omega_1/\Omega_1^{**} := 1 \tag{4.12}$$

to provide a small safety margin against overshoot.

---

[1] You may want to check this as a short **exercise**.

The transient responses also indicate that the dominating part of the additional settling time is due to the linear transient from the "sliding equilibrium" at $r_{2_{hi}}$ to the final equilibrium at 0. This suggests increasing Ω_1. But then (from the basic tradeoff mentioned above) $r_{2_{hi}}$ must be decreased to avoid overshoot. And this will lengthen the settling time again.

4.4.5 Adding Integral Action with Antiwindup Feedback

The next move to a more practical setup is to add integral action to both R_1 and R_2 controllers in order to suppress step disturbances; Fig. 4.18.

The awf gains must be designed such that there are no overshoots either on $x_1(t)$ or on $x_2(t)$. From the previous simulations we have to look into two transfers, *i.e.*
- from the saturation on u to R_2
- and from R_2 (with $x_2 = r_{2_{hi}}$) to R_1

The first transfer has been investigated in depth in Chapter 2, from where the awf gain should be selected at least as the compensating value and up to the deadbeat value $2 \leq k_{a_2} \leq (T_{i_2}/T_s)$

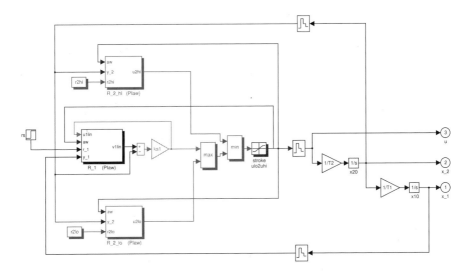

Fig. 4.18. The control system with PI(aw) feedback, selection and saturation

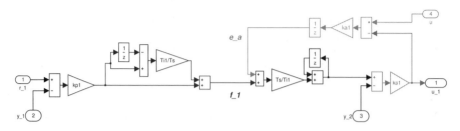

Fig. 4.19. Details of the R_1 controller with PI(aw) P structure

The second transfer can be investigated using the same approach. Consider the continuous equivalent form of Fig. 4.19.

$$u_1 = \frac{\frac{1}{sT_{i_1}}k_{s_1}}{1 + \frac{1}{sT_{i_1}}k_{s_1}k_{a_1}}f_1 - \frac{k_{s_1}}{1 + \frac{1}{sT_{i_1}}k_{s_1}k_{a_1}}y_2, \quad \text{using} \, u = const. = 0$$

$$= \frac{1}{k_{a_1}}\left[\frac{1}{1 + \frac{sT_{i_1}}{k_{s_1}k_{a_1}}}f_1 - \frac{sT_{i_1}}{1 + \frac{sT_{i_1}}{k_{s_1}k_{a_1}}}y_2\right]$$

As $y_2 = const.$:

$$u_1 = \frac{1}{k_{a_1}}\frac{1}{1 + \frac{sT_{i_1}}{k_{s_1}k_{a_1}}}f_1 = \frac{1}{k_{a_1}}\frac{1}{1 + s\tau_{a_1}}f_1; \quad \text{where} \ \tau_{a_1} = \frac{T_{i_1}}{k_{s_1}k_{a_1}} \quad (4.13)$$

and again from Fig. 4.19

$$
\begin{aligned}
f_1(s) &= k_{p_1}(1 + sT_{i_1})e_1(s); \quad \text{where} \quad e_1 = r_1 - x_1 \\
\text{i.e. with} \quad r_1 &= const = 0 : \\
f_1(t) &= k_{p_1}\left[-x_1(t) - T_{i_1}\frac{d}{dt}x_1(t)\right] \\
&\rightarrow k_{p_1}\left[-x_1(t) - T_{i_1}\frac{1}{T_1}r_{2_{hi}}\right]
\end{aligned}
$$

with zero crossover at $(x_1)_f$:

$$(x_1)_f = -T_{i_1}\frac{1}{T_1}r_{2_{hi}} \rightarrow (x_1)_f = -\frac{3}{\Omega_1 T_1}r_{2_{hi}} \quad (4.14)$$

where $x_1(t)$ (and thus $f_1(t)$ as well) ramps up with constant speed $\frac{1}{T_1}r_{2_{hi}}$. In other words, $u_1(t)$ will ramp up as well, but for $t \gg \tau_{a_1}$ it will be delayed by τ_{a_1}. That is, the zero crossover of $f_1(t)$ is also delayed by $\Delta t_{u_1} := \tau_{a_1}$. During this time interval, $x_1(t)$ travels horizontally in the phase plane by Δx_1

$$\Delta x_1 = +\tau_{a_1}\frac{1}{T_1}r_{2_{hi}} = \frac{T_{i_1}}{k_{s_1}k_{a_1}}\frac{1}{T_1}r_{2_{hi}} = \frac{T_{i_1}}{T_1}\frac{1}{k_{s_1}k_{a_1}}r_{2_{hi}} \quad (4.15)$$

and thus the delayed zero crossover will take place at

$$x_{1_{zc}} = \frac{T_{i_1}}{T_1}\left[-1 + \frac{1}{k_{s_1}k_{a_1}}\right]r_{2_{hi}} = -\frac{3}{\Omega_1 T_1}\left[1 - \frac{1}{k_{s_1}k_{a_1}}\right]r_{2_{hi}} \qquad (4.16)$$

Now we are ready to consider specific entries of the awf gains:

1. For the deadbeat case:

$$\frac{1}{k_{s_1}k_{a_1}} = \frac{\Omega_1 T_s}{2} \ll 1.0; \text{ that is} \qquad x_{1_{zc}} = -\frac{3}{\Omega_1 T_1}r_{2_{hi}} \quad (4.17)$$

2. For the compensating case

$$\frac{1}{k_{s_1}k_{a_1}} = \frac{1}{3}; \quad i.e. \quad x_{1_{zc}} = -\frac{3}{\Omega_1 T_1}\left[1 - \frac{1}{3}\right]r_{2_{hi}} = -\frac{2}{\Omega_1 T_1}r_{2_{hi}} \quad (4.18)$$

This is the same result as for the P-controller version from above, *i.e.* the same transient response can be expected.

3. For the "low-gain" case

$$\frac{1}{k_{s_1}k_{a_1}} = \frac{1}{\gamma}\frac{1}{3}; \quad \text{with } \gamma \text{ set to, say, } \frac{1}{4}; \text{ then}$$

$$x_{1_{zc}} = -\frac{3}{\Omega_1 T_1}(1 - 4/3)\,r_{2_{hi}} \qquad = +\frac{1}{\Omega_1 T_1}r_{2_{hi}} \quad (4.19)$$

Note the change of sign of $x_{1_{zc}}$: the transfer of u from one saturation to the opposite one is now delayed beyond $x_1 = r_1$.

Finally, note that the zero crossover of $u_1(t)$ is the point in time where control transfers from $R_{2_{hi}}$ to R_1.

Remarks

- Item 2 in the list above suggests setting again $\Omega_1 = \Omega_1^{**}$. Then we can expect to get a near minimum-time response again.
- For item 3, we have to expect strong overshooting of $x_1(t)$.
- For item 1 the transfer to R_1 will be earlier than necessary, the opposite saturation will not be run into so strongly as in item 2, and the settling time will be distinctly longer than near minimum-time.

These findings are checked by simulations in Fig. 4.20.

The last remark suggests modifying the design rule, but only for the deadbeat awf case, to $\Omega_1 = 1.5\Omega_1^{**}$. Then we would again expect a near minimum-time response.

Exercise
Check this by simulations and discuss the results.

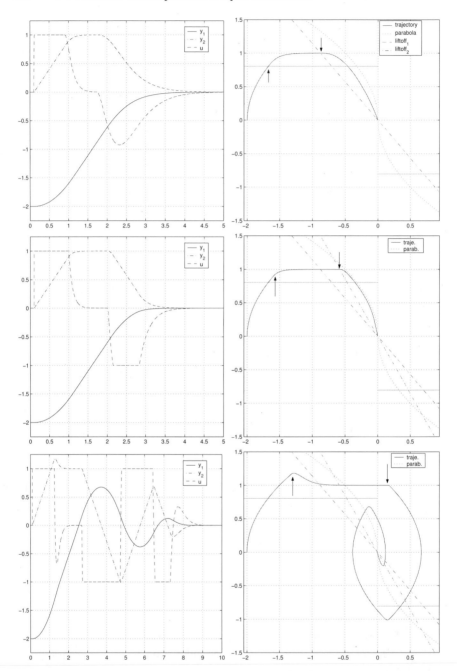

Fig. 4.20. Transient response along the time axis (left) and on the phase plane (right), with $\Omega_1 = \Omega_1^{**}$ from Fig. 4.16, with integral actions and with awf gains: deadbeat (top);   compensating, k_a^* (center);   low, $0.25k_a^*$ (bottom)

4.4.6 Back to the Benchmark

The benchmark in Sect. 4.2 provides a more realistic setup than the simple two-integrator-chain plant with initial condition response back to the origin, and we shall use it to check the results from above. It features an additional delay D, a larger T_s, and also different values for $T_1, T_2, r_{2_{hi,lo}}$. And the disturbance sequence specifies additional load swings during the run-up phase, which then motivated using PI(aw) controllers in the output constraint loops as well. Again, we focus on controllers with integral action, and consider only the case of compensating awf gains $k_{a_j} = k^*_{a_j}; j = 1, 2$.

Benchmark v0

Here

$$a_{lo} = \frac{1}{2} \frac{T_2}{T_1} \frac{1}{\Delta u_{lo}} := \frac{1}{10}$$

and from

$$a_{lo} r^2_{2_{hi}} \stackrel{!}{=} \frac{2}{\Omega^*_1 T_1} r_{2_{hi}} \quad \rightarrow \quad \Omega^*_1 = \frac{2}{a_{lo} r_{2_{hi}} T_1} := 8.0$$

Including the delay $D = 0.025$, the sampling time $T_s = 0.010$ and the finite width of the proportional band of R_1 leads to an estimate for Ω^{**}_1:

$$a_{lo} r^2_{2_{hi}} + \frac{D + T_s}{T_1} < \frac{2}{\Omega^{**}_1 T_1} \quad \rightarrow \quad \Omega^{**}_1 < \frac{2}{a_{lo} r_{2_{hi}} T_1 + (D + T_s)} \approx 7.0$$

From the simulations Fig. 4.21(top), setting $\Omega^{**}_1 = 7.01$ produces a very small overshoot (not shown), which disappears for $\Omega^{**}_1 = 6.86$.

In any case, the design value from the benchmark $\Omega_1 = 3.0$ is smaller than $\Omega^{**}_1/2$. So there is no overshoot to be expected. And the "finals" trajectory in the R_1 regime will avoid the opposite saturation on u, *i.e.* will be linear. These predictions are confirmed by the simulations in Fig. 4.21(bottom).

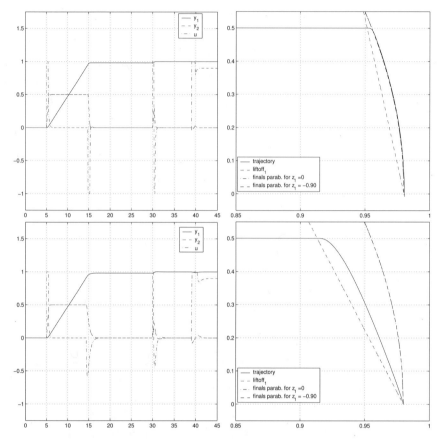

Fig. 4.21. Transient response for benchmark v0.
Top: with $\Omega_1 = \Omega_1^{**} = 6.86$; $\Omega_2 = \Omega_1$;   bottom: with $\Omega_1 = 3$; $\Omega_2 = 3\Omega_1$

Benchmark v2

As specified for version v2, we apply $z_1 = -0.90$ during run-up (instead of $+0.90$ as in Chapter 3). Then

$$ a_{lo} = \frac{1}{2}\frac{T_2}{T_1}\frac{1}{\Delta u_{lo}} = 1.0; \text{ and therefore } \Omega_1^* = \frac{2}{a_{lo}r_{2_{hi}}T_1} = 0.8 $$

As the current design value for Ω_1 is much larger than Ω_1^*, a large overshoot can be predicted. This is confirmed by the simulation in Fig. 4.22.

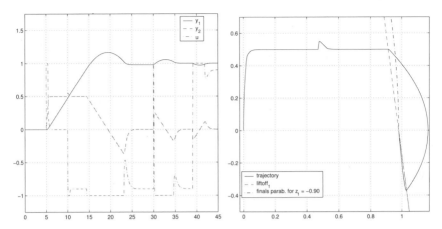

Fig. 4.22. Transient response for benchmark v2 with $\Omega_1 = 3$, $\Omega_2 = 3\Omega_1$; $r_{2_{hi}}$ is not modified

A way out, continued

In Sect. 4.3 we proposed an intuitive way out by stepping down the speed constraint setpoint $r_{2_{hi}}$ from its specification value 0.5 down to a reduced value $r_{2_{hi_1}}$, which was determined by trial and error.

Now we are able to pre-calculate at least a good estimate. While neglecting all delays

$$a_{lo}r^*_{2_{hi_1}} = \frac{2}{\Omega_1 T_1} \quad \rightarrow \quad r^*_{2_{hi_1}} = \frac{2}{15} = 0.133$$

This is indicated by the dashed arrow in Fig. 4.23.

And including the delay D and T_s, but neglecting the proportional band

$$a_{lo}r^*_{2_{hi_1}} < \frac{2}{\Omega_1 T_1} - \frac{D + T_s}{T_1} \quad \rightarrow \quad r^*_{2_{hi_1}} < \frac{2}{15} - \frac{0.035}{5} = 0.126$$

In the simulations for $r_{2_{hi_1}} = 0.126$ (full line arrow in Fig. 4.23) a residual small overshoot is found. It finally can be suppressed by slight re-tuning, to $r_{2_{hi_1}} = 0.120$ (not shown).

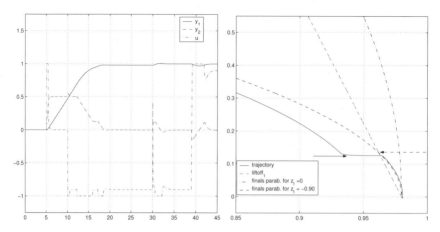

Fig. 4.23. Transient response for benchmark v2, with $\Omega_1 = 3$, $\Omega_2 = 3\Omega_1$, and with stepdown of $r_{2_{hi}}$ to pre-designed $r_{2_{hi_1}}$

Other alternatives to avoid overshoot will be explored in a more systematic way in the "Advanced Techniques" part.

4.5 Stability Analysis

Stability means that the response trajectories to almost all initial conditions $(1 \ll |x_1(0)|, |x_2(0)| < \infty)$ end up at the origin.

This shall be investigated by using two different approaches. The first one is the standard one. It considers the system as having two separate nonlinearities of the sector type and a linear subsystem with two inputs (the outputs of the two nonlinearities) and two outputs (the inputs to the nonlinearities). In other words, we shall have to apply the *multivariable circle criterion*.

The second approach will use the concept of reaching the original initial conditions in state space by letting the trajectory evolve backwards in time. So we start from the final equilibrium of the control system at $\overline{y}_1 = y_1(0) = r_1$; $\overline{y}_2 = y_2(0) = 0$, increase the initial conditions in successive steps, and investigate the stability properties of the resulting trajectory.

This will result in *partitions* of the phase plane.

In the following we shall consider again the double integrator chain as plant and cascaded P-controllers as R_1 and two poles assigned to $-\Omega_1$, and the R_2 as P-controllers, with one pole assigned to $-\Omega_2$ for each output constraint loop, where $u \in [u_{lo}, u_{hi}]$ as input constraint; see Section 4.4.1 *ff*.

4.5.1 The Multivariable Circle Criterion Approach

We start with the "nonlinear additive" form of the system, Fig. 4.24 (top), and redraw it into the canonical structure for the multivariable[2] circle criterion, Fig. 4.24 (bottom).

For this system to be globally asymptotically stable, all four transfer functions in the 2×2 matrix $Z(s)$ must be positive real, *i.e.* all the Nyquist contours must evolve exclusively in the right hand half plane.

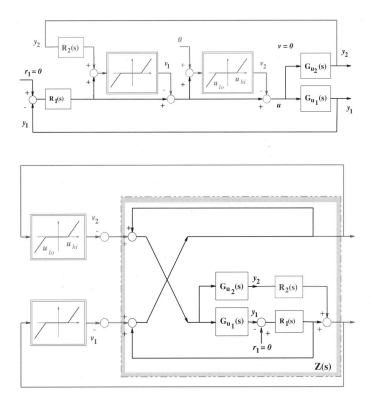

Fig. 4.24. The system with both input and output constraints
in its nonlinear additive form (top) and redrawn in the canonical form for the multivariable circle criterion (bottom)

Note that this is not a standard full two-by-two problem, because the inputs to the two nonlinear blocks are not independent, and not generated by separate sets of transfer functions or initial conditions.

---

[2] two-inputs-two-outputs

We shall look at four cases:

case 1

Let first $v_1(t) = 0$ and $v_2(t) = 0$ $\forall$ $t > 0$, *i.e.* small enough initial conditions, such that the transient will be linear. This leads to the trivial requirement that the R_1 loop has to be stable by design of R_1.

case 2

Consider next the case of transients where $v_1(t) = 0$ but $v_2 \neq 0$ at some length along the trajectory. This leads to the linear subsystem in Fig. 4.25, and to the requirement that the R_1-loop with saturation has to be stable; see Chapter 2.

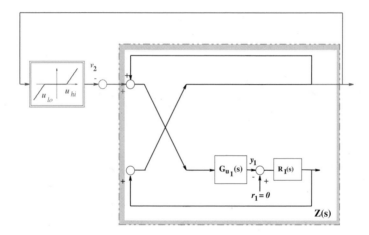

Fig. 4.25. The subsystem with both $v_1 = 0$ and $v_2 \neq 0$

case 3

Consider now the case of transients where $v_1(t) \neq 0$ at some length along the trajectory, but $v_2(t) = 0$. This leads to the linear subsystem in Fig. 4.26 and to the requirement that the R_1-loop with the R_2 overrides, but without the saturation being active, has to be stable; see Chapter 3.

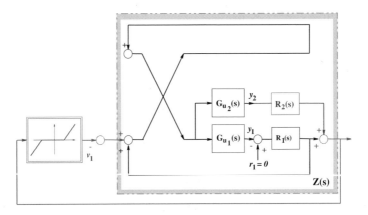

Fig. 4.26. The subsystem with both $v_1 \neq 0$ and $v_2 = 0$

case 4

Consider finally the case of transients where $v_1(t) \neq 0$, *i.e.* one of the R_2 controllers is selected first (and thus the output of R_1 is masked), and then it saturates at some length along the trajectory, *i.e.* $v_2(t) \neq 0$. Then the contribution of the feedbacks of $G_{u_1} R_1$ through the nonlinear block cancels with the one added downstream of the nonlinearity v_1.

This leads to the linear subsystem in Fig. 4.27, and to the requirement that the R_2-loop with saturation has to be stable, see Chapter 2.

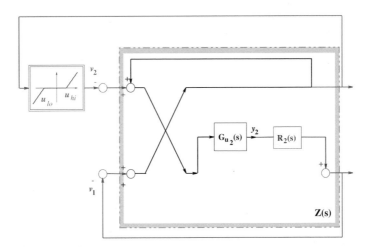

Fig. 4.27. The subsystem with saturation while R_2 is selected

4.5.2 An illustrative example

Applying this to the simple example given above (with state feedback gains k_{1_1}, k_{2_1}) yields for the four cases discussed above

case 1

$$0 = 1 + \frac{k_{2_1}}{sT_2} + \frac{k_{1_1}}{s^2T_1T_2} \rightarrow 0 = (s + \Omega_1)^2 \quad \text{where} \quad \Omega_1 > 0 \qquad (4.20)$$

which is covered by design of R_1.

case 2

We start with the on-axis circle test.
Applying the main stability result from Chapter 2 yields for the linear subsystem

$$F(s) + 1 = \frac{s^2}{(s + \Omega_1)^2} \qquad (4.21)$$

Its Nyquist contour is, with the normalized frequency $\alpha = \omega/\Omega_1$

$$F(j\alpha) + 1 = \frac{(j\alpha)^2}{(j\alpha + 1)^2} = \frac{-\alpha^2}{1 - \alpha^2 + j2\alpha} = \frac{-\alpha^2\left[(1 - \alpha^2) - j2\alpha\right]}{(1 + \alpha^2)^2} \qquad (4.22)$$

which is plotted in Fig. 4.28

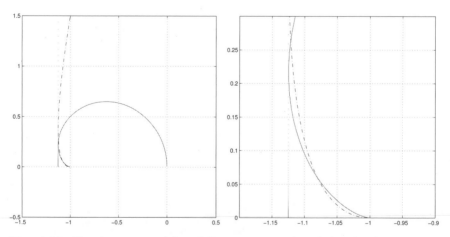

Fig. 4.28. Nyquist contour ($-$) and Popov contour ($\cdot - \cdot$) for the R_1-loop with saturation

and from Eq. 4.22 the real part $\Re(F(j\alpha) + 1)$ has its minimum at

$$\alpha = \frac{\sqrt{3}}{3} \quad \text{and} \quad \Re(F(j\alpha) + 1)_{min} = -\frac{1}{8} \qquad (4.23)$$

Again, from Chapter 2:

$$|\Re(F(j\alpha) + 1)_{min}| \le \frac{\Delta u}{\Delta u_{1_{max}} - \Delta u} \quad i.e. \quad \frac{\Delta u_{1_{max}}}{\Delta u} \le 9 \qquad (4.24)$$

Finally, for $|\Delta u_{1_{max}}/\Delta u|$ *outside* the range given by the on-axis circle criterion, asymptotic stability (*i.e.* asymptotic convergence to the origin) can be proved by the Popov test, as follows:

$$P(j\frac{\omega}{\Omega_1}) = P(j\alpha) = \Re(F(j\alpha) + 1) + j\alpha\Im(F(j\alpha) + 1)$$

$$= -\frac{\alpha^2(1 - \alpha^2)}{(1 + \alpha^2)^2} + j\,2\,\frac{\alpha^4}{(1 + \alpha^2)^2} \qquad (4.25)$$

the imaginary part of $P(j\alpha)$ is positive for all $\alpha > 0$, *i.e.*

$$1 \le |\frac{\Delta u_{1_{max}}}{\Delta u}| < |\infty| \qquad (4.26)$$

from which follows $\quad |x_1(0)|, |x_2(0)| < |\infty|$.

case 3

From chapter 3 the Nyquist contour is given by

$$F(s) + 1 = \frac{s^2}{(s + \Omega_1)^2}\frac{s + \Omega_2}{s} = \frac{s}{s + \Omega_1}\frac{s + \Omega_2}{s + \Omega_1} \qquad (4.27)$$

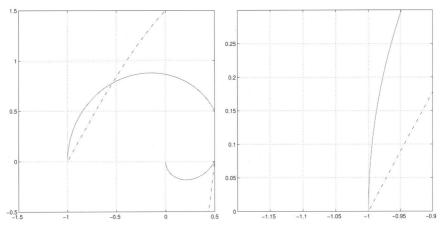

Fig. 4.29. Nyquist contour (−) and Popov contour (· − ·) for the R_1 loop with R_2 override and no saturation

Therefore, the response is stable to the origin, as long as the initial conditions are such that $u(t)$ does not transiently saturate.

case 4

From Chapter 2 the Nyquist contour is given by

$$F(s) + 1 = \frac{s}{s + \Omega_2} \qquad (4.28)$$

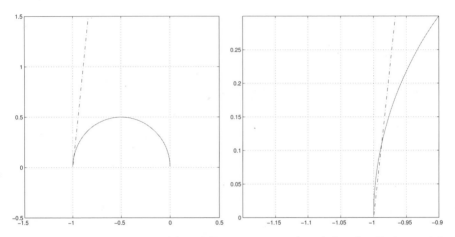

Fig. 4.30. Nyquist contour (–) and Popov contour ($\cdot - \cdot$) for the R_2 loop with saturation

Therefore the response is stable to the origin $e_2(t) \to 0$ for almost all initial conditions.

The conditions for showing asymptotic stability are least favorable in case 2. This point has to be investigated further. The main idea is to use the fact that the output constraints are shaping the trajectories such that the transfer to the R_1 loop with saturation will not take place at arbitrarily large initial conditions, but rather in a quite restricted region of the phase plane which contains the origin. And the size of this area is determined by the design parameter values of $r_{2_{hi}}$, $r_{2_{lo}}$.

4.5.3 The Phase Plane Partitions Approach

Step 1: Linear Control Zone of R_1

We start by applying initial conditions small enough such that $u_1(t)$ stays within the linear operating range, *i.e.* neither the input nor the output constraint are run into. Then all the trajectories starting from these initial conditions will decay asymptotically to the origin, by design of R_1. In other words, the final equilibrium is an asymptotically stable attractor for this compact partition of the phase plane around the origin.

Such linear trajectories will evolve in the phase plane in the polytope depicted in Fig. 4.31 (right column). They are bounded in the horizontal direction

$$\text{to the left by} \quad -k_{1_1}x_1 - k_{2_1}x_2 = u_{hi}$$
$$\text{and to the right by} \quad -k_{1_1}x_1 - k_{2_1}x_2 = u_{lo} \tag{4.29}$$

The top and bottom bounds of this polytope will be derived in step 2.

The main interest will be on trajectories starting at initial conditions $x_1(0), x_2(0)$ such that

$$-k_{1_1}x_1(0) - k_{2_1}x_2(0) = 0 \tag{4.30}$$

i.e. along the line u_z in the polytope, where

$$x_1(0) := -\frac{k_{p_1}}{k_{2_1}}x_2(0) \; ; \quad i.e. \quad x_2(0) := -\frac{2}{\Omega_1 T_1}x_1(0) \tag{4.31}$$

from the pole assignment.

Step 2: Saturating Control Zone of R_1

We now continue to augment the initial conditions along the line u_z. We will come up to a value $x_2(0)^*$, where the trajectory will be the finals parabola for $u = u_{lo} \; \forall \; t \geq 0$, and ending at the origin.

$$\text{For the } u_z \text{ line} \quad (x_1(0)^*)_\ell = -\frac{2}{\Omega_1 T_1}x_2(0)^*$$
$$\text{and for the parabola} \quad (x_1(0)^*)_p = a_{lo}(x_2(0)^*)^2$$

$$\text{where} \quad (x_1(0)^*)_\ell \overset{!}{=} (x_1(0)^*)_p$$

$$i.e. \quad x_2(0)^* = -\frac{1}{a_{lo}\frac{2}{\Omega_1 T_1}} \tag{4.32}$$

$$\text{Inserting also} \quad a_{lo} = \frac{1}{2}\frac{T_2}{T_1}\frac{1}{\Delta u_{lo}}$$

$$\text{finally yields} \quad \frac{x_2(0)^*}{\Delta u_{lo}} = -\frac{4}{\Omega_1 T_1}\frac{T_1}{T_2} \tag{4.33}$$

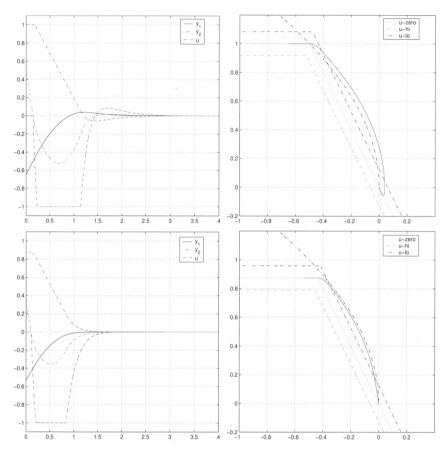

Fig. 4.31. Phase plane trajectories with $\Omega_1 = 4$; $\Omega_2 = 12$ for initial conditions:
(top)   such that $u(t)$ would produce the finals trajectory if not delayed;
(bottom)   for initial conditions reduced by $\Delta x_2(0)$ as estimated in Eq. 4.34

Note that this implies instantaneous transition of $u(t)$ from its previous value u_{hi} to u_{lo} when crossing the u_z line, and also instantaneous transition to the steady state value $\bar{u}$ when reaching the origin. However, in this case, both changeovers will be delayed, see Fig. 4.31 (top).

This is due to the finite width proportional band of R_1.[3] Thus, the effective $x_2(0)^*$ will be lower by $\Delta x_2(0)$ than the value calculated above.

For a first estimate of this $\Delta x_2(0)$, by inspection of Fig. 4.31 (top), the finals parabola may be estimated to start at the $u = u_{lo}$ line instead of at the u_z line. Therefore (replacing the parabola by its local tangent)

---

[3] And in more realistic setups to other elements, such as the delay D in the plant, the sampling delay T_s, etc.

$$\Delta x_2(0) = -\frac{\Omega_1 T_1}{2} \frac{\Delta u_{lo}}{\Omega_1^2 T_1 T_2} = -\frac{1}{2\Omega_1 T_2} \tag{4.34}$$

This has been introduced in Fig. 4.31 (bottom).

We can now determine the upper bound on the polytope in Step 1:

$$x_2(0) \le 0.5 \, x_2(0)^* \tag{4.35}$$

This is derived from the observation that the tangent in the phase plane at the initial condition point to the u_{lo} parabola must be the u_z line. Then $u(t)$ will not exceed u_{lo} while moving the plant to the origin. Note that this again implies instantaneous transition of $u(t)$ at the u_z line.
With the finite P band of the R_1 controller, then (see Fig. 4.32)

$$x_2(0) \le 0.5 \left(x_2(0)^* - \frac{1}{2\Omega_1 T_2} \right) \tag{4.36}$$

So this simple estimate is fairly conservative.

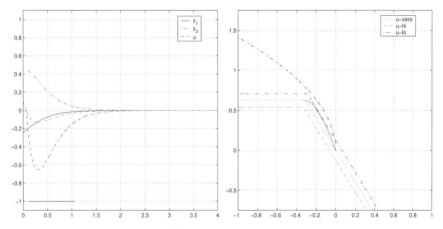

Fig. 4.32. Trajectory for initial conditions such that $u(t)$ stays within its linear range, with correction by Eq. 4.36

Step 3: Radius of Attraction for the Saturating R_1 Control

We increase initial conditions along the u_z line further, i.e. $x_2(0) > x_2(0)^*$. Then the phase plane trajectory will be a Δu_{lo} parabola, until the u_z line is met again in the fourth quadrant; Fig. 4.33

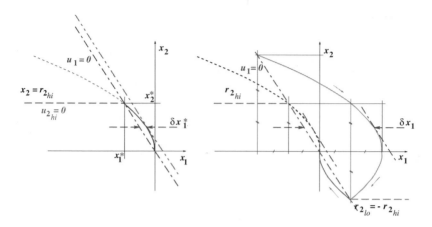

Fig. 4.33. Trajectories for $u(t) = u_{lo}$ and instantaneous transfer of u with initial conditions: (left) at $x_1(0)^*, x_2(0)^*$ on the finals parabola; (right) such that the u_{lo}-parabola meets $r_{2_{lo}} = -r_{2_{hi}}$ on the u_z line

The question now is whether the trajectories on the full line parabolas in Fig. 4.33 can be shown to be asymptotically stable to the origin.

The system to be analyzed is the R_1 loop with control saturations (input constraints) at Δu_{lo}, Δu_{hi}. Then, for the graphical stability tests from Chapter 2, we assume the saturation to be symmetrical, i.e. $|\Delta u_{lo}| = |\Delta u_{hi}| = \Delta u$, and we have to determine

- the Nyquist contour $F(j\omega) + 1$ of the linear subsystem, and
- the position of the straight line on the real axis at $-\frac{1}{b} = -\frac{\Delta u}{u_{1_{max}} - \Delta u}$,

  i.e. $u_{1_{max}}$.

The Nyquist contour for the on-axis circle test has been discussed previously:

$$\rightarrow \quad \frac{\Delta u_{1_{max}}}{\Delta u} \leq 9 \tag{4.37}$$

and for $|\Delta u_{1_{max}}/\Delta u|$ *outside* this range, asymptotic stability has been proved by the Popov test (see above).

The next step is to determine $|\Delta u_{1_{max}}/\Delta u|$ from the trajectory. Transitions of $u(t)$ shall be instantaneous at crossing the u_z line.

Consider first the situation in Fig. 4.33 (left). From above, parallel lines to the u_z line are lines of constant Δu_1. So Δu_1 will be at its maximum for the parallel line that is tangential to the finals parabola. From the basic properties of the parabola, this tangent will be at $x_{2_t} := (1/2)x_{2_{hi}}(0)^*$.

There the horizontal distance δx_1^* is

$$\delta x_1^* = 0.5 x_1^* - (0.5)^2 x_1^* = 0.25 x_1^*$$

where

$$x_1^* = a_{lo}(x_2^*)^2; \quad \text{and with} \quad x_2^* = \frac{1}{a_{lo}}\frac{2}{\Omega_1 T_1}$$

$$i.e. \quad x_1^* = \frac{1}{a_{lo}}\left(\frac{2}{\Omega_1 T_1}\right)^2 = 2\frac{T_1}{T_2}\Delta u_{lo}\left(\frac{2}{\Omega_1 T_1}\right)^2 = 8\frac{\Delta u_{lo}}{\Omega_1^2 T_1 T_2}$$

$$= 8\frac{\Delta u_{lo}}{k_{1_1}} \tag{4.38}$$

and thus $\quad \delta x_1^* = 2\dfrac{\Delta u_{lo}}{k_{1_1}} \tag{4.39}$

Furthermore, the P band of the R_1 controller is

$$\Delta u_{lo} = k_{1_1}\Delta x_{1_{lin}}$$

i.e.

$$\delta x_1^* = 2\,\Delta x_{1_{lin}} \quad \longrightarrow \quad \frac{\Delta u_{1_{max}}}{\Delta u} = 2 < 9 \tag{4.40}$$

i.e. the finals trajectory from $x_1(0) = x_1^*$; $\quad x_2(0) = x_2^*$ shown in Fig. 4.33 (left) is asymptotically stable to the origin by the on-axis circle criterion.

Consider now the situation in Fig. 4.33 (right), where the initial conditions $x_1(0)$, $x_2(0)$ have been chosen such that the full-line parabolic trajectory meets the u_{1_z} line at $x_2 = r_{2_{lo}} = -x_2^*$.

From the properties of the parabola in Fig. 4.33 (right) it follows that

$$x_1(0) = 2x_1^*; \quad x_2(0) = 2x_2^*$$

$$\text{and} \quad \delta x_1 = 2x_1^* + \delta x_1^* \quad = \quad 2.25x_1^* \quad = \quad 9\delta x_1^*$$

$$i.e. \quad \frac{|\Delta u_{1_{max}}|}{|\Delta u|} = 18 \ > \ 9 \ , \ \text{but} \ < \infty \tag{4.41}$$

In other words, asymptotic stability to the origin can no longer be shown by the on-axis circle criterion for this trajectory. But it can be shown using the (less conservative) Popov criterion.

Step 4: Linear Control Zone of R_2

We now increase the initial conditions along the $r_{2_{hi}}$, $r_{2_{lo}}$ lines in the phase plane. More precisely we consider $x_1(0)$, $x_2(0)$ for which the output u_2 of the R_2 controller, which is currently switched through to u by the selectors, is within the saturation limits.

In the phase plane this corresponds to introducing initial condition values $x_1(0)$, $x_2(0)$ within corridors along $r_{2_{hi}}$, $r_{2_{lo}}$, which are delimited by horizontal lines parallel to the u_{2_z} lines (denoted by $u_{2_{lo}}$ and $u_{2_{hi}}$).

We assume first that the setpoint values $r_{2_{hi}}, r_{2_{lo}}$ are sufficiently low such that, after the transition from the R_2- to the R_1-control, $u(t)$ does not saturate, *i.e.*

$$r_{2_{hi}} \leq 0.5x_2^* \quad \text{and correspondingly for} \quad r_{2_{lo}} \tag{4.42}$$

A typical such transient is shown in Fig. 4.34.

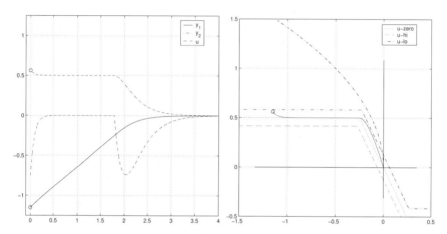

Fig. 4.34. Trajectory for initial conditions corresponding to the assumptions of the stability test of Chapter 3

Then $u(t)$ will saturate nowhere along the initial condition response. And only the selection nonlinearity will be active, *i.e.* the stability test of Chapter 3 applies. Thus for the system considered here, where

$$F(s) + 1 = \frac{s}{s + \Omega_1} \frac{s + \Omega_2}{s + \Omega_1} \tag{4.43}$$

see Fig. 4.29, the initial condition response will be asymptotically stable to the origin for almost all initial conditions within the horizontal corridors delimited above, *i.e.*

$$x_1(0) \quad \forall \quad |x_1(0)| < \infty$$

Consider now increasing the constraint setpoints *beyond* the values used above, keeping everything else constant:

$$0.5x_2^* < r_{2_{hi}} \leq 2x_2^* \quad \text{and correspondingly for} \quad r_{2_{lo}} \tag{4.44}$$

Then u will transiently saturate, but only in the R_1 regime; see Fig. 4.35

Exercise

Derive why this particular choice of $r_{2_{hi}}/x_2^*$ has been made in Fig. 4.35.

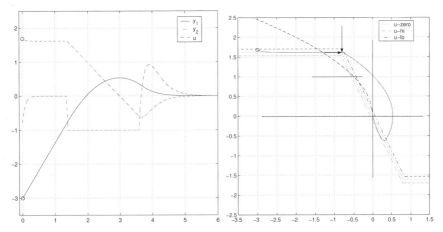

Fig. 4.35. Trajectory for initial conditions along the $r_{2_{hi}}$ line, at a level corresponding to Fig. 4.33 (right). Here, $r_{2_{hi}} = 0.5(1 + \sqrt{5})x_2^*$

To check the *stability* of such trajectories we argue that the circle criterion is based in fact on the Ljapunow stability concepts with a corresponding Ljapunow function V. This means that for all parts of a stable trajectory $dV/dt \leq 0$ holds. So we have to show that each part of the trajectory is stable by applying the appropriate circle criterion. Specifically:

- In the first part, the trajectory decays to $r_{2_{hi}}$ by control of $R_{2_{hi}}$. This part is non-saturating, and by applying the stability test of Chapter 3, this part converges, *i.e.* the state variables decay from their initial conditions down to the values x_1, x_2 indicated by the arrows in Fig. 4.35 (right).
- When $x_1(t)$, $x_2(t)$ are crossing the u_z line at x_{1_t}, x_{2_t}, the selection mechanism will transfer control to the R_1 loop.
- The values of the state variables there (x_{1_t}, x_{2_t}) are now the initial conditions for the trajectory controlled by R_1. The movement along the next part of the trajectory may either be linear, as u does not saturate, which is the case if $r_{2_{hi}}$ was set complying to $0 < r_{2_{hi}} \leq 0.5x_2^*$, or else nonlinear.
- The system to be considered then is the combination of R_1 and the u_1-saturation. As shown above, the trajectory to the origin will be asymptotically stable by the on-axis circle criterion if $r_{2_{hi}}$ was set according to $0.5x_2^* < r_{2_{hi}} \leq 0.5(1 + \sqrt{5})x_2^*$. [4]

In other words, the key element is the choice of $r_{2_{hi}}$ with respect to x_2^*. The stability test then is conducted for the two parts of the trajectory

    (a) leading up to the transfer condition in the selector and

---

[4] And for larger $r_{2_{hi}}$ the trajectory would still be asymptotically stable by the Popov test.

(b) leading from there on to the origin.

This "consecutive parts procedure" will be used again in the following steps.

Step 5: Saturating Control Zone of R_2

We now continue to expand $x_2(0)$ vertically; Figs. 4.36 to 4.38.

First look at the Min-selection between u_1 and $u_{2_{hi}}$. This partitions the phase plane by the straight line '$v_{2_{hi}}$' in Fig. 4.36, which is determined by

$$u_1 = -k_{1_1} x_1 - k_{2_1} x_2 \quad \text{and} \quad u_{2_{hi}} = -k_{1_2} (x_2 - r_{2_{hi}})$$

$$\text{where} \qquad -u_1 \;=\; -u_{2_{hi}}$$

$$\rightarrow \; x_2 (k_{1_2} - k_{2_1}) \;=\; k_{1_2} r_{2_{hi}} + k_{1_1} x_1$$

and thus
$$x_2 \;=\; \frac{1}{\frac{\Omega_2}{\Omega_1} - 2} \left[(\Omega_1 T_1) x_1 + \frac{\Omega_2}{\Omega_1} r_{2_{hi}} \right] \quad (4.45)$$

In other words, for all initial conditions in the upper left partition, the $R_{2_{hi}}$ controller will be selected initially. For initial conditions $x_1(0)$, $x_2(0)$ above the linear corridor around $r_{2_{hi}}$, $u = u_{lo}$, the corresponding u_{lo} trajectory 'open to the left' will be generated; see Fig. 4.36.

We now investigate where the u_{lo} trajectories lead to and use this to partition the phase plane.

Area A

If $x_1(0)$, $x_2(0)$ is above the linear corridor of $R_{2_{hi}}$ and below the u_{lo} parabola through the transfer point $u_{2_{hi_z}}, u_{1_z}$, then the trajectory will enter the linear corridor of $R_{2_{hi}}$.

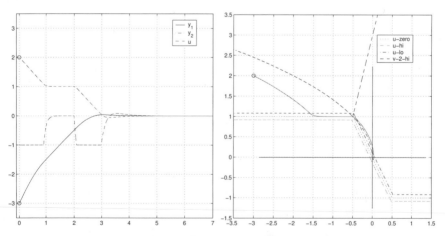

Fig. 4.36. Phase plane trajectory for initial condition $x_1(0) = -3.0$, and $x_2(0)$ in area A

And for the stability of this part of such a trajectory, it must be asymptotically stable for all initial conditions in this area of the phase plane, extending outward to near-infinite values. Then the system to be investigated reduces to the $R_{2_{hi}}$ loop with the input saturation $[u_{lo}, u_{hi}]$. Its stability properties are determined as in Chapter 2 by the Nyquist contour of

$$F_2 + 1 = \frac{s}{s + \Omega_2} \tag{4.46}$$

See Fig. 4.30. It indicates asymptotic stability for all $|x_2(0)| < |\infty|$. In other words, $x_2(t)$ will decay towards $r_{2_{hi}}$ for almost all initial conditions in this specific area.

*Area **B***

This is delimited by the u_{lo} trajectory through the transfer point $u_{2_{hi_z}}, u_{1_z}$ (read $_z$ as zero) to the lower and left side and by the u_{lo} trajectory through the transfer point $u_{2_{lo_z}}, u_{1_z}$ (associated to $R_{2_{lo}}$) to the upper right side, and again extending outward to near-infinite values, and inward to the $v_{2_{hi}}$-separation line; see Fig. 4.37.

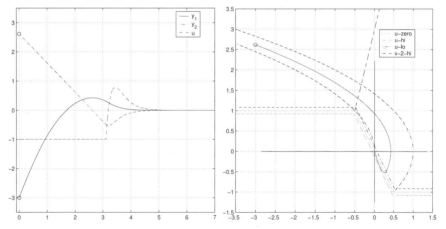

Fig. 4.37. Phase plane trajectory for initial condition $x_1(0) = -3.0$, and $x_2(0)$ in area ***B***

Trajectories from such initial conditions will first be formed by the $R_{2_{hi}}$ loop with saturation, up to this separation line. In this part, the same stability test as above applies.

After crossing the separation line, the selector transfers control to the R_1 loop with saturation and the u_{lo} trajectory will proceed to the linear R_1 corridor adjacent to the u_{1_z} line, and from there on to the final equilibrium at the origin. Therefore, the system to be investigated in the stability test is the R_1

loop with saturation, but with initial conditions $x_1(0)$, $x_2(0)$ delimited within the finite, (and comparatively small) area around the origin.

In other words the R_1 loop with saturations need not be asymptotically stable for almost all initial conditions, but only for those in this restricted area. Furthermore, the key design parameter for this area is the value attributed to $r_{2_{hi}}$ with respect to x_2^* for the particular set of plant parameters. As we have shown above, stability can be shown with a substantial margin for $0 < r_{2_{hi}}/x_2^* \leq 2$ with the Popov test. This nicely covers the delimited area around the origin, where the R_1 loop with saturation will be active.

Area *C*
The third area for $x_1(0)$, $x_2(0)$ in the phase plane is delimited to the lower left by the u_{lo} trajectory through the transfer point $u_{2_{lo_z}}, u_{1_z}$ (associated to $R_{2_{lo}}$), and again extending outward to the upper right side to near-infinite values.

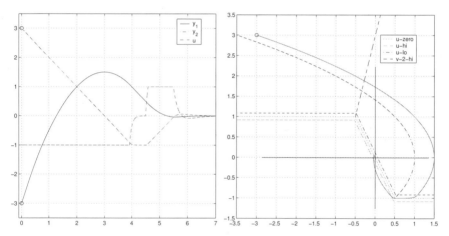

Fig. 4.38. Phase plane trajectory for initial condition $x_1(0) = -3.0$, and $x_2(0)$ in area *C*

There the u_{lo} trajectories will proceed to the $u_{2_{lo_z}}$ horizontal line, associated with the R_2 loop with saturations. In other words the system to be considered for stability of this part of the trajectory is the $R_{2_{lo}}$ loop with saturation, where stability to its own origin at $x_2 = r_{2_{lo}}$ must be shown for almost all initial conditions in this area.

Exercise
Repeat this for initial conditions below the line $u_{2_{hi_z}}$, u_{1_z}, $u_{2_{lo_z}}$ in the phase plane.

4.5.4 A Modified Approach

Another approach to stability analysis would be to consider the selection between the three controllers first. This partitions the phase plane into three areas, see Fig. 4.39, by the separation lines '$v'_{2_{hi}}$ and '$v'_{2_{lo}}$.

For initial conditions in area **A** from above, the first part of the trajectory converges to $r_{2_{hi}}$ within the partition to the left of '$v'_{2_{hi}}$, *i.e.* the $R_{2_{hi}}$ loop with saturation has to be tested for convergence of x_2 to $r_{2_{hi}}$ as before.

For initial conditions in area **C** from above, as shown in Fig. 4.39, the first part of the trajectory up to the '$v'_{2_{hi}}$ separation is generated by the $R_{2_{hi}}$ loop with saturation, the next part up to the '$v'_{2_{lo}}$ separation by the R_1 loop with saturation, the next part up to the $r_{2_{lo}}$ line and further to the transfer at the point $u_{2_{lo_z}}$, u_{1_z} by the $R_{2_{hi}}$ loop with saturation again and the next part (the final one in Fig. 4.39) again by the R_1 loop with saturation.

For the whole trajectory to converge, each part has to be generated as stable; that is, *each* of the three loops with saturation has to be stable for almost all initial conditions.

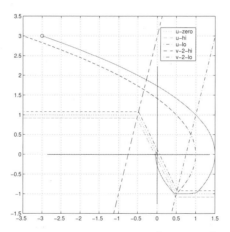

Fig. 4.39. Phase plane with separation lines '$v'_{2_{hi}}$ and '$v'_{2_{lo}}$ from the Max-Min-selection between the three controllers, and downstream saturation.
Trajectory for initial condition $x_1(0) = -3.0$, and $x_2(0)$ in area **C**, as in Fig. 4.38

This approach is much less involved than the first one; but, as we have seen, it puts more restrictive assumptions on the R_1-loop with saturations. This is a disadvantage for some applications, where this simpler approach cannot show stability, whereas the more-involved one succeeds.

4.5.5 A Fourth Approach to Design for Stability

The stability test would be further simplified by restricting the design values of $r_{2_{hi}}$ to $0 < r_{2_{hi}} \leq 0.5x_2^*$, and $r_{2_{lo}} = -r_{2_{hi}}$.

Then the trajectories consist of two parts, first the approach to the $r_{2_{hi}}$, $r_{2_{lo}}$ lines by the R_2-loop with saturation (and continuing along them in a 'sliding mode'), and then, after transfer to the R_1 loop, a linear final approach to the origin.

4.5.6 Summary and Generalization

Looking backwards from the final equilibrium of the R_1 loop at $x_1 = 0$, $x_2 = 0$ along the trajectories to the starting conditions, we have extended step by step the range of "admissible" initial conditions to almost all points on the phase plane, where "admissible" initial conditions means that the trajectories starting there can be shown to be asymptotically stable, *i.e.* to converge to the final equilibrium, and where "almost all" means: $\forall \; |x_1(0)|, |x_2(0)| < |\infty|$. This has been discussed for a specific example, *i.e.* the two-integrator chain with simple proportional controllers R_1, R_2 and zero load disturbances.

For this we have needed and used the following elements:

1. Both the $R_{2_{hi}}$ and $R_{2_{lo}}$ loops with the u saturations must be asymptotically stable to their respective equilibria at $e_{2_{hi}} \to 0$ and $e_{2_{lo}} \to 0$ for almost all initial $y_2(0)$.

2. The R_1 loop with the u saturations must be asymptotically stable to the origin, $e_1 \to 0$ for a bounded, compact area in the phase plane, the size of which is chiefly determined by the design choice of $r_{2_{hi}} = r_{2_{lo}}$ in relation to x_2^*. It need *not* be asymptotically stable for almost all initial conditions.

Discussion

- Note that the actual shape of the u_{lo}, u_{hi} trajectories (of parabolic shape in the phase plane for the case above) does not enter explicitly into the stability analysis (but of course it will in the performance analysis). This indicates that this stability test procedure may be applied to other plants of dominant second order.

- For practical purposes, the P feedbacks used so far must be augmented by I(aw) actions. We have shown before that if the awf gains k_{a_j} are set to their compensating values, then the response and the stability properties are the same as for the P feedbacks investigated so far.
  But if higher awf gains are introduced, then the transfer from the R_2 control to the R_1 control will occur *earlier* (because of less transient windup). For example, if deadbeat awf gains are selected, the transfer will occur on the line[5]

---

[5] this implies $T_s \to 0$ and $D = 0$.

$$x_2 = -\frac{\Omega_1 T_1}{3} x_1; \text{ instead of on } x_2 = -\frac{\Omega_1 T_1}{2} x_1 \text{ for compensating } k_{a_j}$$

$$(4.47)$$

This may be utilized for increasing robustness to variations of a_{lo}, a_{hi}.

- About selecting the $r_{2_{hi}}$ or $r_{2_{lo}}$ values, there are two different aspects.

  The first one is that these setpoints are equal to operational limits imposed on the secondary output variable $y_2(t)$. In other words, they are fixed by plant properties.

  The second one is from the control system dynamic performance, *i.e.* from the design rules derived above to obtain responses, which comply to performance specifications (such as no perceptible overshoot).

  If the first set of $r_{2_{hi}}$ or $r_{2_{lo}}$ values (from operations) is more constraining than the second one, then there is no conflict, as the first set will automatically produce responses complying to control specifications. If, however, the second set is the more constraining one, then implementing it would not exploit what the plant offers, and overall performance will suffer.

  One possible way out is to apply the more constraining set only in the final approach phase, as demonstrated in Fig. 4.6 (left column). Others will be discussed later in the "Advanced Techniques" part.

- The assumption in the fourth approach to stability testing (about not permitting any saturating u in the final approach under R_1 control) may seem too conservative. However, from a practical design point of view, it has the advantage of increasing robustness against overshoot caused by additional unknown loads and variations of T_1, T_2, *i.e.* to off-design values of a_{lo}, a_{hi}. Of course, one then has to accept some increase of settling time, as indicated in Fig. 4.17.

- Considering the general control problem with both input and output constraints shown in Fig. 4.2, the same elements from the list above apply for the stability analysis. But for the performance analysis, the change of shape of the u_{lo} and u_{hi} trajectories must be taken into account.

- Finally, the set of initial conditions considered so far must be carefully reviewed.

  From a theoretical point of view, looking at all or at least almost all points of the phase plane is the standard procedure for stability analysis.

  However, the initial conditions which may appear in the actual control loop are much more restricted. So, from an applications point of view, the bounds on the compact set of initial conditions to be investigated are also more restricted. That is, proving asymptotic stability for almost all initial conditions is "nice to have", but in general there is "no need to have".

  From this it follows that the physical bounds of initial conditions must be carefully checked for a specific application. This becomes a key issue, if the general procedure is **not** able to prove asymptotic stability for almost all initial conditions.

- In practice, actuator subsystems often have so-called slew and stroke constraints. This may either be seen as a special case of combined input and output constraints, or as a more general anti windup problem. We shall investigate this next.

Exercise

- Evaluate the physical bounds on the initial conditions for the benchmark specified in Sect. 4.3 and discuss the consequences.
- Apply the concept and procedures from above to the example in Sect. 4.1 to show why using the opening limitation applied there has such a beneficial effect on the run-up response.

Extended Exercise
The cascade-limiter form with I(aw) on both controllers has not been investigated with respect to both performance and stability so far. Do this as an extended exercise.

4.6 Case Study: Elevator Positioning Control

4.6.1 Plant Description and Data

Consider the typical elevator system shown in Fig. 4.40, where the cabin and a constant counterweight mass are suspended by a cable from a drive drum situated above the top floor. The drum is part of an electrical drive subsystem.

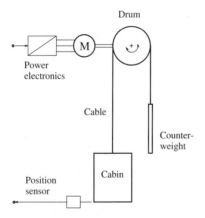

Fig. 4.40. The elevator (schematic)

Plant Data

Mass of cabin (empty)	m_K	500	kg
Mass of counterweight	m_W	500	kg
Payload mass	m_P	0 to 250	kg
Drum inertia	Θ	0	kg m^2
Elevator operating height	L	3 to 43	m
Reference value for masses	m_R	250	kg
Reference value for cable force	$F_R := m_R g$	2500	N
Reference value for speed	v_R	2.0	m/s
Reference value for position	s_R	40.0	m
Speed constraints	v_{hi}, v_{lo}	± 2.0	m/s
Cable force constraints			
from drive torque constraints	u_{hi}	$+2F_R$	N
	u_{lo}	$-1F_R$	N
Position sensor gain	k_{y_1}	$1.0/s_R$	1/m
and time constant	τ_{y_1}	0.010	s
Speed sensor gain	k_{y_2}	$1.0/v_R$	1/(m/s)
and time constant	τ_{y_2}	0.10	s
Drive subsystem gain	k_u	$F_R/1.0$	N
and time constant	τ_u	0.040	s
Cable mass	m_C	0	kg
and cable stiffness coefficient	c_C	∞	N m/m
Friction and drag forces are neglected			

4.6.2 Mathematical model

The equation of momentum for all (rigidly coupled) masses is

$$(m_K + m_P + m_W)\frac{d}{dt}v = F_u - (m_K + m_P)g + m_W g$$

i.e. $\quad \dfrac{(m_K + m_P + m_W)v_R}{F_R}\dfrac{d}{dt}\dfrac{v}{v_R} = \dfrac{F_u}{F_R} - \dfrac{(m_K + m_P)g}{m_R g} + \dfrac{m_W g}{m_R g}$

finally $\quad \tau_2 \dfrac{d}{dt}x_2 = u_l - z \qquad\qquad (4.48)$

with the notations

$$\tau_2 = \frac{(m_K + m_P + m_W)v_R}{F_R} = 0.8 \text{ to } 1.0 \text{ s};$$

and

$$x_2 = \frac{v}{v_R}; \quad u_l = \frac{F_u}{F_R} = -1.0 \text{ to } +2.0; \quad z = \frac{m_P}{m_R} = 0 \text{ to } 1.0;$$

and with the drive subsystem equation

$$\tau_u \frac{d}{dt} u_l = -u_l + k_u u \tag{4.49}$$

From cabin speed to cabin position

$$\frac{d}{dt} s = v; \quad \text{that is} \quad \frac{s_R}{v_R} \frac{d}{dt} \frac{s}{s_R} = \frac{v}{v_R} \quad \text{and with } x_1 := \frac{s}{s_R}$$

$$\rightarrow \quad \tau_1 \frac{d}{dt} x_1 = x_2 \tag{4.50}$$

where

$$\tau_1 := \frac{s_R}{v_R} = 20 \text{ s};$$

and with the sensor equations

$$\tau_{y_j} \frac{d}{dt} y_j = -y_j + k_{y_j} x_j; \quad j = 1, 2 \tag{4.51}$$

4.6.3 Suggestions for the Control System Design and Analysis

Case A

Consider the system as specified above. Then all necessary design and analysis elements for the position control system are available from Chapters 2 to 4. We suggest using as design parameters

$$T_s = 0.020 \text{ s}; \quad \Omega_1 = 1.0 \text{ rad/s}; \quad \Omega_2 = 3.0 \text{ rad/s}$$

but check and confirm that this is feasible.

Case B

Extend the system by adding a "jerk" limitation, i.e. the change of force and cabin acceleration is not instantaneous but has a speed of change limit, such that an increase of one unit of F_R takes 5 s.
Hint: focus on the speed control first and then extend to the position control.

Case C

Extend the system by letting the cable be elastic, such that for the cabin being at the lowest position, a resonance frequency of 10 rad/s is observed.
Hint: use additional feedback for active damping.

4.7 Summary

After investigating control loops with input constraints (a pair of saturation limits) in Chapter 2, and one pair of output constraints in Chapter 3, the focus is now on control loops combining both constraints.

Regarding the design of the *controller structure*, the intuitive approach is to combine the structure elements from the two previous chapters. The additional new element used here is to implement the constraints by sequential selection on the control variable u. The output constraint feedback action on u is selected for first, followed by the input constraints action, as those are hard actuator limits and cannot be moved outward by the output constraint feedback action. So this makes use of particular masking properties. Thus the output and input constraints act on the control variable, *i.e.* control conditioning is used (and not reference conditioning).

The *transient response* for the benchmark system is well behaved, but unfortunately not for all cases: the saturation of the control variable may be so restrictive that, figuratively speaking, the kinetic energy at the transition to the finals trajectory is too high for being "braked off" in the remaining interval of the control error e_1, and an excessive overshoot is unavoidable. So one possible countermeasure is to reduce the approach speed by lowering the speed constraints appropriately.

The *stability* properties are more involved, as there are now two separate nonlinear elements, although both are of the deadspan type. There are several methods available. A sequential approach using the one-nonlinearity form has been demonstrated.

Performance is now a less trivial matter, due to the overshoot tendency. Here, the analysis in the phase plane is a powerful additional tool. In this framework, a standard element is looking at the trajectories with their particular shape (parabolas in the benchmark case). An additional element is the lines, along which control is transferred by the sequential selection block, and which partition the phase plane.

This opens an additional approach to *stability analysis* by starting with small deviations (initial conditions) from the final equilibrium, testing stability with the sector criteria, and then increasing the initial conditions into adjacent partitions of the phase plane, again checking the stability, until this covers a sufficient part of the phase plane for the particular application. This may be seen as "moving backwards in time" along the trajectory from the final equilibrium to the actual initial conditions. Note that this approach to stability is closely related to Ljapunow's.

So far, the focus has been on plants of dominant first and second order with PI control and PI-P cascaded control, where the visualization in the phase plane is straightforward. But this poses the problem of higher order systems. Again, the intuitive approach would be to apply the same basic method to this new problem class, *i.e.* using the selection block for implementing the constraints as above, and replacing the standard linear PI(aw) algorithms for R_1 and the R_2 by more advanced ones, such as state feedbacks (with observers if needed), and augmented by I(aw) actions.

However, this shall be investigated later, as there are some open questions within the standard PID-control area that must be addressed first.

5

Further Topics on PI(aw) Control

So far, the basic situations of constrained control with PI controllers have been investigated. In practical applications additional questions arise. Three such topics are selected here for further discussion:

- *PI control with actuator slew and stroke constraints.*
  This situation will be present in almost all applications. It takes into account that moving an actuator will always require an inflow or outflow of mass or energy. Such flows are always bounded, leading to such slew constraints. The question is for which such constraint values the transient and stability properties will start to deviate significantly from the case of pure stroke constraints, as they have been assumed so far.
  This will provide an indication how much dynamic performance (and thus cost) of the actuator subsystem may be reduced without significantly affection closed-loop performance.

- *PI(aw) control with derivative action.*
  The standard structure is in fact PID control, and not just PI control as discussed so far. So the D action, which is available, has been set to zero. In cases where such additional D action would be required to stabilize the loop and produce acceptable performance, it has been replaced by a cascade structure of PI(aw)-P type. The reasons for this will become apparent in this section.

- *PI(aw) control with measurement noise.*
  So far, the measured signals $y_i(t)$ used for feedback have been considered to contain no measurement noise component. This is unrealistic, of course. Even if high-quality signal conditioning and filtering are applied, some low-level component will remain.
  Here, the focus shall be on the effect of a low-level high-frequency disturbance on the measured variables to PI(aw) control as developed above.

5.1 PI Control with Actuator Slew and Stroke Constraints

5.1.1 Motivation

So far, no actuator dynamics have been considered. In other words, all moves to new actuator position are instantaneous. There are no rate or speed constraints, *i.e.* no bounds on "slew".

However, real-world actuators generally have some storage (integrating) elements, such as servo cylinders in electrohydraulic actuators, or screw-and-nut arrangements in linear actuators with rotative electric drives, *etc.* And the flow to the servo cylinder or the rotative speed of the electric motor is always bounded by physical reasons. This introduces rate constraints or "slew saturations" in addition to any position limits or "stroke saturations".

High slew-rate actuators require high power input, such as to produce the high flow at given cylinder pressure, or high speed at the given load torque on the screw. And this will be more expensive. So there is a strong economic motive to use not only the smallest actuator, which has the lowest stroke constraints, but also the least power-consuming actuator, *i.e.* with the lowest slew constraints, such that the control loop still can meet its dynamic specifications.

On the other hand, slew saturations slow down the movement of u, which will deteriorate performance and stability properties. And this will be more pronounced if the slew saturations are lowered.

The dynamic effects of such additional slew saturations are visible in John's case; Fig. 2.2. Such slew saturations are even introduced expressly for constraining the pressure surge in the penstock of the hydroelectric power plant, and not only to reduce power supply costs.

So there is a strong need in the design phase to consider additional slew constraints on stroke-constrained actuators. Any such design procedure or stability test must be simple and straightforward to be practicable.

5.1.2 Actuator Modeling

Fig. 5.1 shows a simplified layout of a typical electrohydraulic actuator subsystem. In the range of linear operation, the rate of change $w = dh/dt$ of position h is assumed as proportional to oil inflow Q_{oil}. It is set proportional to the pilot valve opening v; see Eq. 5.1, which defines τ_2.

$$\frac{h_R}{w_R} \frac{d}{dt} \left(\frac{h}{h_R} \right) = \tau_2 \frac{d}{dt} \left(\frac{h}{h_R} \right) = \frac{Q_{oil}}{Q_{oil_R}} = \frac{v}{v_R} \qquad (5.1)$$

That is, variations of oil pressure from the pump, from the oil reservoir and in the servomotor due to changing load forces and acceleration of the closing weight are neglected.

Two nonlinearities are considered. The first one is the effect of the orifices shown in Fig. 5.1. They limit the oil flow to and from the servomotor, and thus the transient water pressure rise. This is a standard safety measure in hydropower plants. The limiting effect is modelled by input saturations on flow to/from the servomotor,

$$\frac{v_{dn}}{v_R} \leq \frac{v}{v_R} \leq \frac{v_{up}}{v_R} \tag{5.2}$$

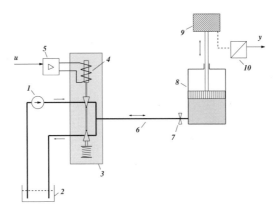

Fig. 5.1. The actuator subsystem
pump (1), oil reservoir (2), spring loaded servo valve (3), with drive solenoid (4) and power amplifier (5), pipe (6) with safety orifices (7) to and from the servomotor (8), closing weight (9) for fail safe shutdown, position transducer (10)

The second nonlinearity is caused by the mechanical limits imposed on the servomotor position, again consult Fig. 5.1. It is modelled by a stiff spring coming into action, if the position exceeds the limit values h_{lo}, h_{hi}. To attain equilibrium of forces, the servomotor pressure p_{SM} will rise to the supply pressure p_P, which then reduces the oil inflow to zero for standstill. The relation to use would be $Q_{oil} \sim v\sqrt{p_P - p_{SM}}$, and correspondingly at the low end. This effect is approximated here by an additive high-gain feedback on the servomotor integrator input, if the servomotor position would exceed the mechanical limit values at h_{lo} or h_{hi}. The servo valve dynamics are modeled by a first-order lag with unity gain and time constant τ_{sv}. Any further small time constants in the servo valve, its power amplifier, and the position transducer are neglected. Finally, the position shall be controlled by a proportional controller with gain k_s, see Fig. 5.2. Numerical values are:

$$h_{lo} = 0.0 \; ; \quad h_{hi} = 0.80; \quad \tau_{sv} = 0.10 \; s; \quad k_s = 9.0;$$

$$\tau_2 = \frac{h_R}{w_R} = 5.0 \; s; \quad \frac{v_{dn}}{v_R} = -0.2; \quad \frac{v_{up}}{v_R} = +0.2; \quad \rightarrow \quad 100 \; \% \text{ in } 25 \; s. \tag{5.3}$$

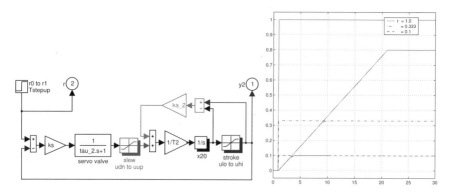

Fig. 5.2. The actuator control loop model, and typical setpoint step responses (small, medium, high)

5.1.3 Control Structures and Transient Response

This actuator subsystem shall now be embedded in a speed control loop, where it manipulates the driving torque. The plant is shown in Fig. 5.3, with numerical values $T_1 = 5.0$ s; $a = 0.025$.

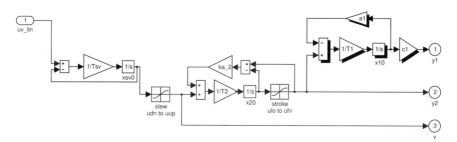

Fig. 5.3. The plant model for speed control with slew- and stroke-saturated actuator The P control loop for actuator positioning from Fig. 5.2 is omitted here

Again, several alternative controller structure forms are used in practice. Here, we shall build on what we have discussed so far; see Fig. 5.4:

-   An input constraint (saturation on u) is representing the working range of flow through the servo valve. This implies constant oil pressures and a linear valve characteristic.
-   An output constraint is used to capture the mechanical limitations on servomotor position y_2.

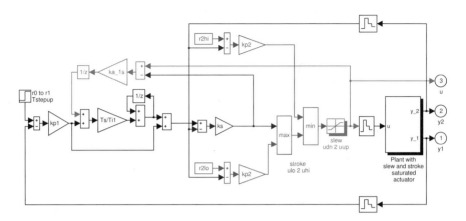

Fig. 5.4. Implementing the speed control with input saturation (for slew) and output saturation (for stroke)

Note that if a saturation block is introduced at the controller output, with saturation values u_{dn}, u_{up} at a small distance inside of the working range of flow $[u_{dn_p}, u_{up_p}]$, then the input saturation on the actuator will not become active, and can be deleted from the actuator model.

Similarly, if the setpoints $r_{2_{hi}}$, $r_{2_{lo}}$ are adjusted at a very small distance inside of the stroke working range, then the mechanical limitations are not met, and the high-gain feedback in the actuator model will not become active, and can again be deleted from the actuator model.

In other words, the actuator model is now linear, and the corresponding constraints are implemented in the controller structure.

The controller R_1 is a cascade structure of
- a P controller for the actuator positioning (as used above)
- a PI(aw) controller for turbine speed.

The three parameters are calculated by pole assignment, using $c_1 = 1.0$:

$$(s + \Omega_1)^3 = 0 \quad \text{with} \quad \Omega_1 = 0.6; \quad \text{that is} \quad \Omega_1 T_1 = 3.0$$

$$\text{yielding} \qquad k_s = 9.0 \text{ (as above)}; \quad k_{p_1} = 3.0; \quad \text{and} \quad T_{i_1} = 5.0 \quad (5.4)$$

This fits well with typical values for such hydroelectric units.

And for the override controller R_2:

$$s + \Omega_2 = 0 \quad \text{with} \quad \Omega_2 = 3\Omega_1 \quad \text{yielding} \quad k_{p_2} = 9 \qquad (5.5)$$

The transient response in Fig. 5.5 shows a large overshoot, which cannot be tolerated. It is the consequence of a large buildup of the actuator position, which cannot be reduced in time due to the downward slew saturation.

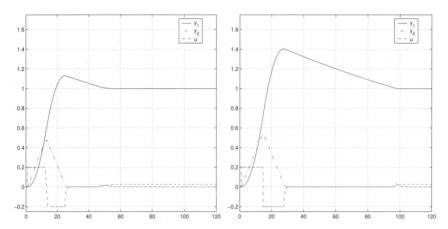

Fig. 5.5. Runup response of the speed control loop in Fig. 5.4,
with awf gains: deadbeat (left);   compensating (right)

A standard countermeasure is to introduce an "opening limit". This can be implemented quite simply in Fig. 5.4 by lowering $r_{2_{hi}}$ appropriately from the upper mechanical limit of 0.80. In this case, 0.20 is a suitable value (and a typical one for such applications); see Fig. 5.6. The overshoot is suppressed, and the run-up time is reduced from 110 s to 40 s.

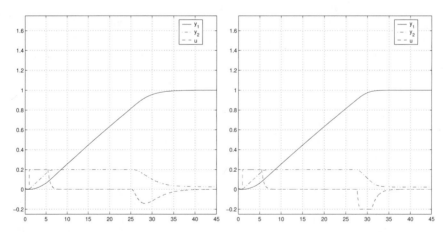

Fig. 5.6. Runup response of the speed control loop in Fig. 5.4 with the "opening limit" set to $r_{2_{hi}} = 0.20$, and with awf gains: deadbeat (left); compensating (right)

5.1.4 An Approximation

The Basic Idea

The aim is to reduce the effort for the stability test from above.
Consider a system that has only an output constraint feedback of $\tilde{y}_2$ and no input constraints, where $\tilde{y}_2$ is an appropriate linear combination of the stroke $y_2 = u$ and slew dy_2/dt, with parameter τ_a in the approximation.

$$\tilde{y}_2 := y_2 + \tau_a \frac{d}{dt} y_2 \tag{5.6}$$

For steady state in the constraint feedback loop:

$$dy_2/dt \to 0 \quad i.e. \quad \tilde{y}_2 \to y_2 \quad \text{and thus} \quad \tilde{y}_2 \to r_{2_{hi}} \quad (\text{or} \to r_{2_{lo}}) \tag{5.7}$$

At $t = T_0$ the transfer to R_1 shall occur. Let $u_1(t)$ move stepwise to its final steady state value $\bar{u} = 0$. Then $y_2(t)$ will decay to zero with time constant τ_a; that is, the downward slew rate on $y_2(t)$ will not exceed

$$\frac{d}{dt} y_2 \bigg|_{dn} = -\frac{r_{2_{hi}} - \bar{u}}{\tau_a} \quad \text{where} \quad r_{2_{hi}} = u_{hi} \quad \text{and} \quad u_{hi} - \bar{u} = \Delta u_{hi} \tag{5.8}$$

On the other hand, from the slew saturation in the actuator in the original system:

$$\frac{d}{dt} y_2 \bigg|_{dn} = \frac{v_{dn}}{T_2} \tag{5.9}$$

Let both left-hand sides be equal. Then the value of τ_a in the replacement system with the same slope of $y_2(t)$ is

$$-\frac{r_{2_{hi}} - \bar{u}}{\tau_a} = \frac{v_{dn}}{T_2} \quad i.e. \quad \tau_a = -T_2 \frac{\Delta u_{hi}}{v_{dn}} \tag{5.10}$$

and correspondingly for a step of u_1 from zero upward.

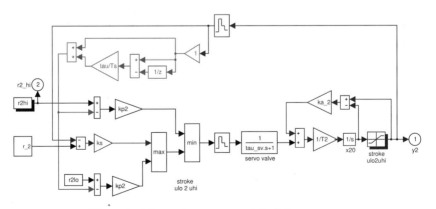

Fig. 5.7. The replacement model of the actuator

Transient Responses

This is checked by simulations; see Fig. 5.8.

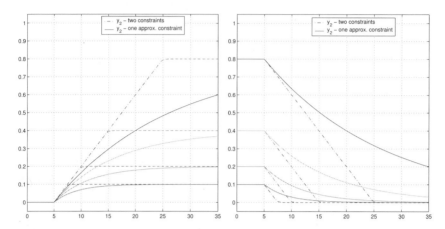

Fig. 5.8. Responses of the replacement system Fig. 5.7 for different values of $r_{2_{hi}}$ (full lines); compared with those of the original system (dotted lines), for u_1 steps $0 \to 1$ (left) and $1 \to 0$ (right)

Within the speed control loop (Fig. 5.4) the approximation is conservative, see Figure 5.9, as the movement of u is slower than in the original system. The opening limitation at left is set to $r_{2_{hi}} = 0.6$, such that it does not interfere in the original system, and the awf gain k_{a_1} is set to the compensating value.

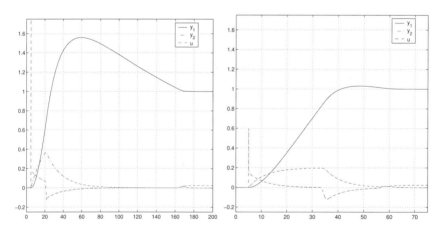

Fig. 5.9. Speed run-up responses with the replacement model of the actuator for $r_{2_{hi}} = 0.6$ (left) and $r_{2_{hi}} = 0.2$ (right)

5.1.5 Stability Analysis

The general result of Chapter 3 may be applied now.
For the Nyquist contour

$$F(s) + 1 = \frac{1 + k_{a_1} G_{a_1} R_{2_1}}{1 + G_1 R_1} \frac{1 + G_2 R_2}{1 + 0} \tag{5.11}$$

Note that the output constraint controller is of P type, and needs no awf, $i.e.$ $k_{a_2} = 0$. Furthermore set k_{a_1} to its compensating value. Then

$$\text{with} \quad 1 + G_2 R_2 = 1 + \frac{1}{sT_2}\left(1 + s\tau_a\right) k_{p_2}$$

$$\text{and} \quad F(s) + 1 = \frac{s^3}{(s + \Omega_1)^3} \frac{s + \Omega_1}{s} \frac{s\left(1 + \tau_a \frac{k_{p_2}}{T_2}\right) + \frac{k_{p_2}}{T_2}}{s}$$

$$\text{where} \quad \frac{k_{p_2}}{T_2} = \Omega_2$$

$$\rightarrow \quad F(s) + 1 = \frac{s}{s + \Omega_1} \frac{s\left(1 + \tau_a \Omega_2\right) + \Omega_2}{s + \Omega_1} \tag{5.12}$$

We now discuss the second factor in Eq. 5.12.
Let

$$\Omega_2 = \kappa \, \Omega_1 \tag{5.13}$$

which defines κ (here $\kappa = 3$).
Its denominator and numerator zeros cancel if

$$s\left(1 + \tau_a \Omega_2\right) + \Omega_2 = 0 \quad \text{and} \quad s + \Omega_1 = 0 \quad \text{are the same}$$

$$\text{that is} \quad \frac{\kappa \Omega_1}{1 + \tau_a \kappa \Omega_1} = \Omega_1$$

$$\text{yielding} \quad \tau_a = \frac{\kappa - 1}{\kappa} \frac{1}{\Omega_1} \tag{5.14}$$

$$\text{where from Eq. 5.10} \quad \tau_a = -T_2 \frac{r_{2_{hi}} - \bar{u}}{v_{dn}}$$

This allows one to determine directly the value for $r_{2_{hi}}^*$ for such cancelation. Using the numerical values from above:

$$r_{2_{hi}}^* = \bar{u} + (\kappa - 1)(-v_{dn})\frac{1}{\Omega_2 T_2} = 0 + (3 - 1)\frac{1}{5}\frac{1}{9} = \frac{2}{45} \approx 0.0444$$

Inserting this into Eq. 5.12 yields

$$F^* + 1 = \kappa \frac{s}{s + \Omega_1} \tag{5.15}$$

and a Nyquist contour with the real part $\Re(F^*(j\omega) + 1) > 0$ for $\omega > 0$, that is asymptotic stability for almost all initial conditions by the on-axis circle criterion. Note that this nicely covers the case with the opening limiter set to $r_{2_{hi}} = 0.20$ in the example.

Fig. 5.10 shows the Nyquist contours for opening limiter setpoints beyond $r_{2_{hi}}^*$. Up to $r_{2_{hi}} = 0.10$, the Nyquist contour of $F + 1$ still evolves to the right of the vertical line in 0, $j0$ for $\omega > 0$. So there is an ample safety margin. For $r_{2_{hi}} = 0.20$ and $r_{2_{hi}} = 0.80$ the Nyquist contour evolves to the left of this vertical line, but then stability can still be demonstrated for almost all initial conditions by using the Popov test. This covers the cases of $r_{2_{hi}}$ from the simulations.

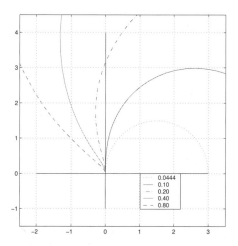

Fig. 5.10. Nyquist contour for the stability test with the replacement model of the actuator, for different values of $r_{2_{hi}}$ at and beyond $r_{2_{hi}}^*$, according to Eq. 5.14

A More General Result

So far, the goal of a much simpler stability test has been attained, but the main result in Eq. 5.12 is still based on the speed control case.
However, it may be interpreted in a more general way, by rewriting it as follows:

$$
\begin{aligned}
F + 1 &= \frac{s^2 \left(1 + \tau_a \Omega_2\right)}{\left(s + \Omega_1\right)^2} + \frac{s \Omega_2}{\left(s + \Omega_1\right)^2} \\
&= \alpha \frac{s^2}{\left(s + \Omega_1\right)^2} + \beta \frac{s}{\left(s + \Omega_1\right)^2} \\
&= \alpha \cdot \left(F_{slew} + 1\right) + \beta \cdot \left(F_{stroke} + 1\right)
\end{aligned}
\tag{5.16}
$$

In other words, the contour of $F + 1$ for the system with the replacement actuator model is a *linear combination* of the Nyquist contours for the systems with slew saturation only ($F_{slew}+1$) and with stroke saturation only ($F_{stroke}+$ 1), with weights α and β. Note that this statement does not depend on the specific speed control system any more. It generalizes Eq. 5.12 for all cases with plant dynamics $G(s)_{y_2 \to y_1}$ of dominant first order at least.

In the Nyquist plot the contour of $F + 1$ will evolve within the area delimited by those two contours, which are more directly determined.

Another consequence is that, in the frequency range relevant for the stability properties, the contour with additional slew constraints will always evolve to the left of the contour with no such slew constraints, *i.e.* the stability properties will deteriorate. However, we see from the numerical example in Fig. 5.10, that the effect is starting slowly with increasing slew constraints, *i.e.* increasing τ_a.

On the other hand, the case $\beta = 0$ would be the most critical situation for stability, see chapter 2. However this is not a realistic design situation. It would require that the stroke limits are set to infinity. But they are at least bounded to their finite physical limits. As we see from the example, the area of attraction from the stability test is not reduced significantly for realistic slew constraints.[1]

Note that Eq. 5.16 also holds for non-compensating awf gain k_{a_1}.

Exercise
- Check the last statement.
- Analyze the effect of τ_a on the contour F in more detail.
- Check Eq. 5.16 for plants with dominant order > 1.

5.1.6 An Implementation Alternative

The structure in Fig. 5.4 we have investigated so far is rather complex. It needs an internal signal from the actuator, *i.e.* the flow from the servo valve with its constraints for awf, meaning a rather expensive sensor, and also Max-Min-selectors. The aim here is to develop a simpler alternative, which should not perform significantly worse.
We shall get there by several block diagram modification steps; Fig. 5.11.

- Consider first form **(a)** in Fig. 5.11. The shaded block covers the servo valve stroke saturation and the integrating servo cylinder $1/sT_s$ with the mechanical end limitations acting back to the inflow by the very high-gain feedback through k_{a_s}.
  There is a standard awf with gain g_{a_1} from the servovalve opening saturation (which is *a priori* known) back to the integral action of R_1, but no awf from the inflow and from the end limitations.
  As long as the upper end limit $y_{2_{hi}}$ is not encountered, only the servo valve stroke limit v_{hi} is active, and the R_1 output would track to $r_2(t) = y_2(t) + v_{hi}/k_s$ for a deadbeat awf g_{a_1}.

---
[1] But performance will deteriorate.

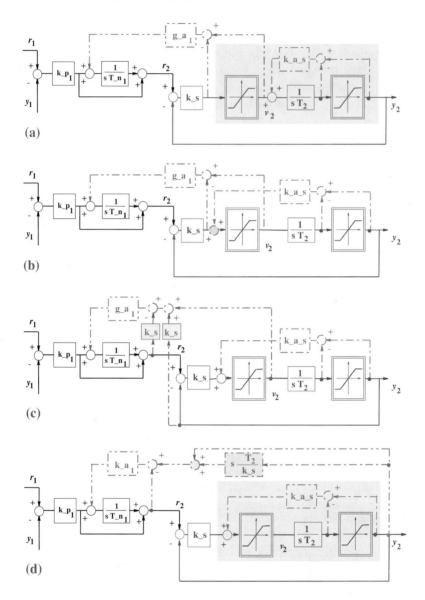

Fig. 5.11. Redrawing of the block diagram of the actuator subsystem

Now if the upper end limit $y_{2_{hi}}$ is reached, then y_2 will stop moving at $y_{2_{hi}}$. The integrator output will stabilize at a small value outside $y_{2_{hi}}$, due to the very high gain k_{a_s}. Then the R_1 output will stabilize at $r_2(t) = y_{2_{hi}} + v_{hi}/k_s$. In other words, there will be a windup to v_{hi}/k_s; and correspondingly for the downward moving and the low end limit.

- Form **(b)** shows how to suppress this windup at the end limits. It consists of moving the summing point of the end limit feedback upstream to the input of the servo valve saturation, but still downstream of the negative input to the awf error for R_1.

  As long as $y_2(t)$ is below its end limit $y_{2_{hi}}$, the system behaves as form **(a)**. If $y_{2_{hi}}$ is reached, then the feedback through k_{a_s} reduces the integrator input to zero. But in this form, it moves the positive input to the R_1 awf error down to zero as well. By this, r_2 will now track to $y_{2_{hi}}$. In other words, the windup at the end limits vanishes, as it does in the case of Fig. 5.4 with output constraints to $y_{2_{hi}}$, $y_{2_{lo}}$ and selectors. From a physical viewpoint this is a limitation of the inflow to the servo cylinder.

- Then form **(c)** is produced by shifting the negative input to the R_1 awf error (the output of the k_s block) upstream of the summing point.

- Finally, form **(d)** is obtained by two steps.

  The first one is to shift the positive input to the R_1 awf error (the output of the v_{hi}, v_{lo} saturation) downstream to the output y_2. This leads to a block sT_2 in this path.

  The second one uses from above that

$$g_{a_1} = \frac{k_{a_1}}{k_s}$$

with k_{a_1} as the standard awf gain in the generic controller structure. This leads to unity gains in the r_2- and y_2-paths from form **(c)**, and to the block sT_2/k_s.

Now this may be interpreted as a basic input conditioning structure from Chapter 2 for the speed control loop with R_1 and $G = G_{y_2 \to y_1}$, with two modifications.

The first one is that the non-dynamic input saturation from Chapter 2 is now replaced by the closed loop for actuator positioning with internal slew and stroke saturations. In other words, this is now a dynamic, but passive nonlinearity.

The second modification is that, to account for the dynamic response of the actuator loop, a derivative action sT_s/k_s has been placed in the awf path.

Regarding *implementation*, no additional signal need be measured besides the standard sensor for y_2. For R_1 this requires a standard PI(aw) controller, but with an input for an externally generated awf tracking signal.[2] And the derivative action in the awf path of y_2 can be implemented by a standard PD module.

Note that its contribution decreases with higher actuator position controller gain k_s. Note also that the differentiation in the awf-loop may require faster sampling than without it for the awf loop to be numerically stable. And it

---

[2] This may not be available in all industrial process control systems.

is a liability in the presence of measurement noise on y_2. So one may rather force it to zero and put up with the poorer performance.

These block diagram modifications are implemented in Simulink in Fig. 5.12. Note the additional factor γ inserted in the awf path. For $\gamma = 1$, this implements the design as derived above. With $\gamma = 0$, the derivative action sT_2/k_s is suppressed.

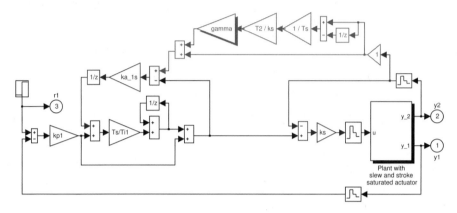

Fig. 5.12. Simulink implementation of the block diagram modifications. "γ" switches in/out the derivative action in the awf path

From the simulation, Fig. 5.13, the performance deterioration with $\gamma = 0$ turns out to be acceptable here, but this should always be checked.

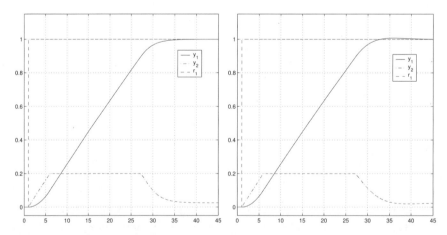

Fig. 5.13. Transient responses for the system in Fig. 5.12:
(left) with the design derivative action on y_2; (right) and if it is forced to zero

5.2 PI(aw) Control with Derivative Action

5.2.1 Introduction

Industrial controllers today are almost exclusively of the PID type. But, from experience, in almost all applications the D part is set to zero during commissioning, for various reasons. Some of them will become clearer in the following. One of the historical roots of the D action seems to be in the 1920's for autopilots keeping a ship on a reference heading by acting on the rudder angle. More generally speaking this is position control on a system of dominant second order. The control problem is similar to the benchmark of Chapter 3, but without speed constraints. There we have used for linear range control a P-PI or PI-P cascade structure instead of a PID type.

Intuitively, the D action provides a look-ahead correction to the $u(t)$ produced by the PI part, which increases with the finals approach speed of $y(t)$. In short, it reduces u earlier, and therefore reduces overshoot and improves closed-loop damping.

From experience, a very significant improvement can be attained for plants of dominant second order, such as positioning control. The effect is not so marked on plants of dominant first order, such as speed or temperature control, where the Ziegler–Nichols rules apply and provide design values for T_d in the D action.

We shall proceed as follows. First we shall put the qualitative statements on linear control from above on a more quantitative basis, using a specific case (again a benchmark) for illustration. Next, we insert a saturation on u, and demonstrate the effects. Then we implement a standard awf scheme from Chapter 2, and again demonstrate the effects on transient response. Finally we shall discuss improvements.

5.2.2 The Benchmark

The Plant

The plant $u \to y$ shall be modeled by its transfer function

$$G = \frac{y}{u} = e^{-sD}\frac{1}{a_2 + sT_2}\frac{1}{sT_1} \tag{5.17}$$

with $T_1 = 5.0$, $T_2 = 1.0$ and a small delay $D = 0.025$ to cover the non-modeled dynamics. The only difference to the plant from Chapter 3 is a_2: it was $a_2 = 0$ there, and is now $a_2 = 1.0$.

The physical background to this would, for instance, be a DC-servo positioning loop with no cascaded current or torque feedback control, and with an input saturation on the armature voltage, and no speed constraint.

The Controller Structure

The standard Ziegler–Nichols arrangement is used, with $e = r - y$

$$R = \frac{u}{e} = k_p \left(1 + \frac{1}{sT_i} + sT_d \frac{1}{1 + s\tau_d} \right) \tag{5.18}$$

As usual, the D action is first-order filtered with $\tau_d = T_d/5$, and the PI controller can be obtained by setting $T_d = 0$.

The Controller Design

For the *PID case* we shall use again the pole assignment procedure on the dominant dynamics model ($D = 0$ and $\tau_d = 0$)

$$0 = 1 + RG = 1 + \frac{s^2 k_p T_d T_i + s k_p T_i + k_p}{sT_i} \frac{1}{1 + sT_2} \frac{1}{sT_1}$$

$$0 = s^3 T_i T_2 T_1 + s^2 (T_i T_1 + k_p T_d T_i) + s k_p T_i + k_p$$

$$= s^3 + s^2 \frac{1}{T_2} \left(1 + k_p T_d \frac{1}{T_1} \right) + s k_p \frac{1}{T_2 T_1} + k_p \frac{1}{T_2 T_1 T_i}$$

$$\stackrel{!}{=} (s + \Omega)^3 \tag{5.19}$$

yielding for the controller parameters

$$k_p = 3(\Omega T_1)(\Omega T_2) ; \quad \frac{T_i}{T_1} = \frac{3}{\Omega T_1} ; \quad \frac{T_d}{T_2} = \frac{3(\Omega T_2) - 1}{(\Omega T_2)^2} \tag{5.20}$$

From the simulations shown later, we obtain a well-damped linear range closed-loop performance with $\Omega = 1.5$.

For the *PI case*, we set $T_d = 0$, from where

$$\Omega T_2 = \frac{1}{3} ;$$

$$\text{and thus} \quad k_p = 3 \frac{T_1}{T_2} (\Omega T_2)^2 = \frac{1}{3} \frac{T_1}{T_2}; \quad \frac{T_i}{T_2} = \frac{3}{\Omega T_2} = 9; \tag{5.21}$$

Compare the design results in the following table:

Controller type	Bandwidth Ω	Gain k_p	Reset T_i
PI	1/3	1.667	9
PID	1.5	33.75	2

By inserting the D action, the closed-loop bandwidth can be improved by a factor of 4.5, *i.e.* the gain k_p is increased by a factor of $4.5^2 \simeq 20$. This is a significant improvement in control performance indeed, and is a strong motivation for using the derivative action at all.

The sampling time for the discrete-time implementation is set to $T_s = 0.010\ s$, as in Chapter 3. The discrete controller is implemented by

$$R = \frac{u}{e} = k_p \left(1 + \frac{T_s}{T_i} \frac{1}{1 - z^{-1}} + \frac{1 - z^{-1}}{T_s} T_d \right) \tag{5.22}$$

Using a non-filtered derivative may seem unrealistic, but it simplifies the following arguments considerably. And the usual first-order filtered derivative will be investigated in an exercise later.

The Input Saturations

These are set to

$$u_{hi} = +1.0 \ ; \quad u_{lo} = -1.0 \tag{5.23}$$

The Test Sequence

This shall be similar to Chapter 2:

- We start at $t = 0$ with the the plant at standstill: $x_1 = 0$; $x_2 = 0$.
- At $t = 0.5$ the loop is closed, with $r = 0$.
- Then at $t = 1.0$ s, the setpoint r is stepped up to 0.95. There is no load disturbance ($z = 0$) during run-up.
- After completion of run-up and stabilization, a small setpoint step is applied; while $z = 0$, $\quad r = 0.95 \rightarrow 1.0$ [3]
- Finally a load step $\quad z = 0 \rightarrow +0.90$ is introduced.

The Anti Windup Feedback

The obvious move is to install the simple awf scheme from Chapter 2, with awf gain
- first at its deadbeat value, $\qquad\qquad k_a = T_i/T_s$
- and then at its compensating value, $\qquad k_a = 3.0$

---

[3] The step is larger for better visibility. You may check that this does not change the typical response.

5.2.3 Transient Response

Experimental Findings

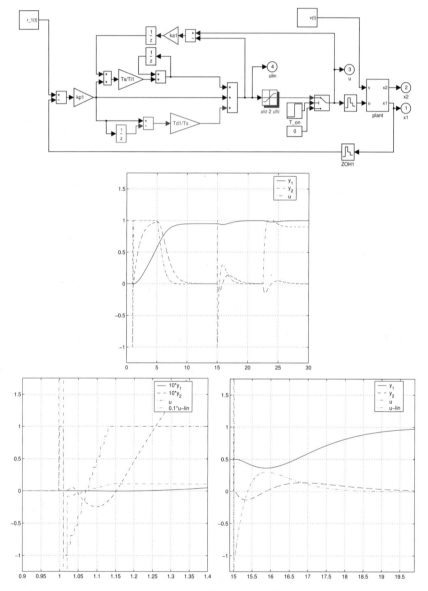

Fig. 5.14. The control loop with a "PI(aw) and D" controller and its responses with deadbeat awf gain, $k_a = T_i/T_s$.

In the bottom right graph $y_1' = 10\,(y_1 - 1.0) + 1.0$ is plotted, for better visibility

At first glance, the load step response is the same as in Chapter 3 for the PI-P cascade. But the setpoint step responses show an erratic behavior not experienced so far.

For the large initial setpoint step $r_1 = 0.98$ (see bottom left for details), and for the first sampling period, the control variable u moves to the upper saturation u_{hi} as one would expect. But for the second sampling period, surprisingly it moves to the lower saturation u_{lo}. From there on, it slowly increases up to u_{hi} again (here in about 12 T_s). This "lack of u" produces a transient decrease of y_2 instead of the monotonic increase one would expect as regular behavior. Then y_1 will also decrease transiently, although this is not visible here.

For the small setpoint step $r_2 = 0.02$ applied at $t = 15$ (bottom right), this effect is even more pronounced. Here, y_1 distinctly moves in the opposite direction first (in this case to approximately $-r_2$) before settling at the final setpoint. *Such responses are obviously not acceptable.*

Note that the PI-P structure does not show such effects; see Fig. 5.15

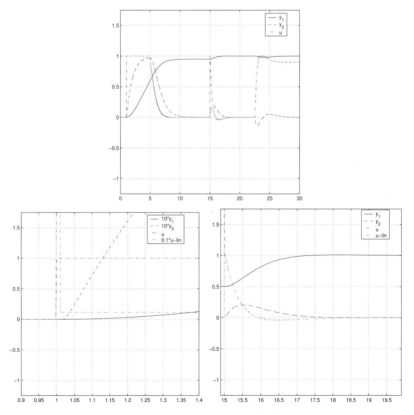

Fig. 5.15. Responses with linearly equivalent PI(aw)-P controller, and deadbeat awf gain

The Cause

Starting from the generic form of the input constraint controller of Chapter 2
with the linear parts R_1 and R_2

$$R_1 = k_p \left(1 + sT_i + s^2 T_i T_d \right)$$
$$R_2 = \frac{1}{sT_i} \qquad\qquad (5.24)$$

for the discrete-time version in difference equation form is:

$$\Delta u[k] = k_p \left(\frac{T_s}{T_i} e[k] + (e[k] - e[k-1]) + (e[k] - 2e[k-1] + e[k-2]) \frac{T_d}{T_s} \right)$$
$$u_{lin}[k] = u[k-1] + \Delta u[k];$$

where

$$u[k] = SAT(u_{lin}[k]); \qquad \text{with} \quad u_{lo}, u_{hi} \quad \text{for} \quad SAT \qquad (5.25)$$

For the reference step input at $k = 1$ of height r and considering both $y_1 = 0$
and its time derivative $dy_1/dt = 0$ to be still at their initial steady state values
for $k = 1, 2, 3, 4$, then

$$e[k] = 0 \quad \text{for} \quad k = -3, -2, -1, \ 0$$
$$e[k] = r \quad \text{for} \quad k = 1, \ 2, \ 3, \ 4 \qquad (5.26)$$

yields

$$\Delta u[1] = k_p \left[\frac{T_s}{T_i} r + (1 - 0)\, r + (1 - 2 \times 0 + 0)\, r \frac{T_d}{T_s} \right]$$
$$= k_p \left(\frac{T_s}{T_i} + 1 + \frac{T_d}{T_s} \right) r$$
$$\Delta u[2] = k_p \left[\frac{T_s}{T_i} r + (1 - 1)\, r + (1 - 2 \times 1 + 0)\, r \frac{T_d}{T_s} \right]$$
$$= k_p \left(\frac{T_s}{T_i} - \frac{T_d}{T_s} \right) r$$
$$\Delta u[3] = k_p \left[\frac{T_s}{T_i} r + (1 - 1)\, r + (1 - 2 \times 1 + 1)\, r \frac{T_d}{T_s} \right]$$
$$= k_p \frac{T_s}{T_i} r$$
$$\text{and} \quad \Delta u[4] = \Delta u[3] \qquad\qquad (5.27)$$

Note that

$$\frac{T_s}{T_i} \ll 1 \quad \text{and} \quad \frac{T_d}{T_s} \gg 1 \qquad (5.28)$$

In other words, for the large setpoint step $r_1 \approx 1$:
  - $\Delta u[1]$ will be $\gg 0$, whereas

- $\Delta u[2]$ will be $\ll 0$, due to $-r(T_d/T_s)$, and
- $\Delta u[3], \Delta u[4] = \epsilon$ will be > 0, but small.

Therefore

$$u[1] = u_{hi} \ ; \quad u[2] = u_{lo} \ ; \quad u[3] = u_{lo} + \epsilon \ ; \quad u[4] = u[3] + \epsilon \ ; \quad \cdots \quad (5.29)$$

which explains the transient of u in Fig. 5.14 (bottom left).

This "deficit on u"-effect will be visible for decreasing r values (all other parameters being constant), until

$$u_{hi} + \Delta u[2] = u_{lo} \quad i.e. \quad \Delta u[2] = -\,(u_{hi} - u_{lo}) = k_p \left(\frac{T_s}{T_i} - \frac{T_d}{T_s} \right) r \quad (5.30)$$

which determines the threshold value r_ℓ of r

$$r_\ell = \frac{u_{hi} - u_{lo}}{k_p} \frac{T_s}{T_d} \quad (5.31)$$

where for the current benchmark

$$k_p T_d = (3\Omega T_2 - a_2)\, T_1 \quad (5.32)$$

Inserting the current numerical values yields

$$r_\ell = \frac{2}{(4.5 - 1)5} 0.01 = 1.143 10^{-3} \quad (5.33)$$

which is far below the smaller setpoint step size $r_2 = 0.02$ from the benchmark, see Fig. 5.14 (bottom right). The slower recovery of u is due to the smaller increment ϵ from above. This explains why the deflection of y_1 in the opposite direction is so much stronger than with the large setpoint step, Fig. 5.14 (bottom left).

Consider now a reset of u to 0 instead of u_{lo}. This will suppress the "deficit on u" effect. In order to obtain this, the setpoint step size must be further reduced to r_{ℓ_0}

$$r_{\ell_0} = \frac{u_{hi} - 0}{k_p} \frac{T_s}{T_d} \ ; \quad i.e. \text{ for the benchmark:} \quad r_{\ell_0} = 0.5 r_\ell = 0.571 \cdot 10^{-3}$$
$$(5.34)$$

Consider finally reducing r_2 further to r_{lin} such that the initial increment $\Delta u[1]$ is below $u_{hi} - 0$, and therefore the subsequent response is linear:

$$r_{lin} = \frac{u_{hi}}{k_p \left(\frac{T_s}{T_i} + 1 + \frac{T_d}{T_s} \right)} \ ; \quad \text{and using Eq. 5.28:} \quad r_{lin} < \frac{u_{hi}}{k_p T_d} T_s = r_{\ell_0}$$
$$(5.35)$$

Exercise
- Check and visualize this by simulations.
- Investigate the effect of a first-order filtered derivative action
   by simulations first,
- and then more analytically.

5.2.4 Some Solutions

It is obvious that such erratic behavior must be eliminated. Again, many different approaches are possible. We shall focus on three such solutions here.

Avoiding the Derivative Action

This behavior can be eliminated by simply avoiding the D action.
And this can be achieved very efficiently by using the linearly equivalent[4] PI(aw)-P cascade instead; see Fig. 5.15. Such a cascade has further practical advantages. It is easy to tune and to handle during start-up of the control system. It is also less prone to negative effects from high-frequency measurement noise (such as actuator wear).
Note however that this requires an additional measured signal y_2 to be available, where y_2 contains at least a strong component proportional to dy_1/dt. It may also be useful to consider an observer.

A "two degrees of freedom" Structure

An inherent property of the standard "text book" PID controller of Eq. 5.18 is, that both the measured variable $y(t)$ and the setpoint signal $r(t)$ are differentiated, due to $e = r - y$. In other words, the transfer function of r is the same for both inputs $-y$ and $+r$, and therefore this structure is said to be of "one degree of freedom".

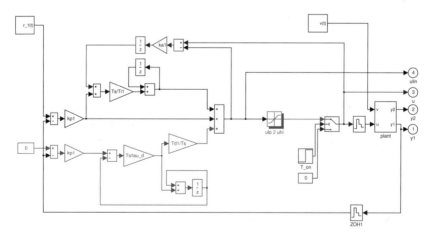

Fig. 5.16. "Two degrees of freedom" structure; with first-order filtered derivative action on the y-path only

---

[4] It has the same characteristic equation.

However, closed-loop stability, and performance of responses to initial conditions and plant disturbances, will need dy/dt in the feedback only. The part of dr/dt is a feedforward path to u, which only contributes to the performance of the reference response $r(t) \rightarrow y(t)$.

So, in order to eliminate the cause for the "deficit on u" effect, one would simply suppress this feedforward path of dr/dt; see Figs. 5.16 and 5.17.

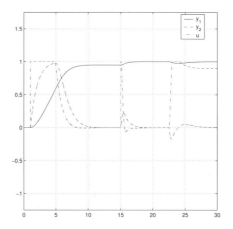

Fig. 5.17. Response of system in Fig. 5.16 to the benchmark, with deadbeat awf gain

In industrial process control systems, such a modified controller structure is not always standard. But it may be available as an option, called "derivative action on the measured variable only" or similarly. Note that this will slightly slow down the linear reference response $r(t) \rightarrow y(t)$. But also notice that for very small sizes of r, see Eq 5.35, the saturation on u will become active and diminish any such speedup effect from dr/dt.

Now the transfer function of the controller is different for both inputs, which contributes an additional degree of freedom for the design. This concept may be expanded to the proportional path of R as well. Then the reference step response will slow down further. However, this also suppresses the overshoot on the linear setpoint step response, which is so typical for PI control, and without any noticeable increase in loop settling time.

Lower awf gain

This alternative is based to the standard one degree of freedom form.

The basic idea is not to suppress the negative $\Delta u[2]$ from the differentiation of r, but to reduce its offset effect on $u[2]$, which is produced by the very strong awf action, as a consequence of the (high) deadbeat gain. In other words, one attempts to reduce k_a sufficiently.

But this will increase the transient windup, and thus tends to reduce performance and affect stability. But as has been shown above, there is no significant reduction as long as k_a stays above a certain threshold value k_a^*, the compensating one. Note that for the current benchmark: $k_a^* = 3$.

Consider u_{lin} while operating along the upper saturation u_{hi} (see Chapter 2):

$$u_{lin} = r\, k_p \left(1 + sT_i + s^2 T_i T_d\right) \frac{1}{sT_i + k_a} + \frac{sT_i k_a}{sT_i + k_a} u_{hi} \qquad (5.36)$$

Then let

1. $u_{hi} = const.$ $\rightarrow$ $\dfrac{sT_i k_a}{sT_i + k_a} u_{hi} = 0$

2. $T_d = T_i/4$;   as in the Ziegler–Nichols and Chien–Hrones–Reswick rules
   ($i.e.$ lower than $T_d \approx T_i/3$ from pole assignment)

3. $k_a = 2$ (also lower than $k_a^* = 3$ from the design above)   (5.37)

Inserting into Eq. 5.36 yields

$$u_{lin} = r\, k_p \left(1 + 2s\frac{T_i}{2} + s^2 \frac{T_i^2}{4}\right) \frac{1}{2} \frac{1}{s\frac{T_i}{2} + 1} = r\, k_p \frac{1}{2}\left(1 + s\frac{T_i}{2}\right) \qquad (5.38)$$

Applying an r-step to this will produce *no undershoot* on u_{lin} after the first overshoot, and therefore eliminates the "deficit on u" effect. Furthermore, this is valid for r of arbitrary size. So there is no such limit on r, as found in the deadbeat awf gain case.

In case of the first-order filtered derivative action with time constant $\tau_d \ll T_d$ (here $\tau_d = 0.1T_d$), note that

$$1 + sT_i + s^2 T_i T_d \frac{1}{1 + s\tau_d} \approx 1 + s\frac{T_i}{2}\left(1 + \frac{1}{1 + s\tau_d}\right) + \left(s\frac{T_i}{2}\right)^2 \frac{1}{1 + s\tau_d}$$

$$= \left(1 + s\frac{T_i}{2}\right)\left(1 + s\frac{T_i}{2}\frac{1}{1 + s\tau_d}\right)$$

$$i.e. \quad u_{lin} = r\, k_p \frac{1}{2}\left(1 + s\frac{T_i}{2}\frac{1}{1 + s\tau_d}\right) \qquad (5.39)$$

Again, the step response to r will produce no undershoot. This is confirmed by the simulations on the current benchmark in Fig. 5.18.

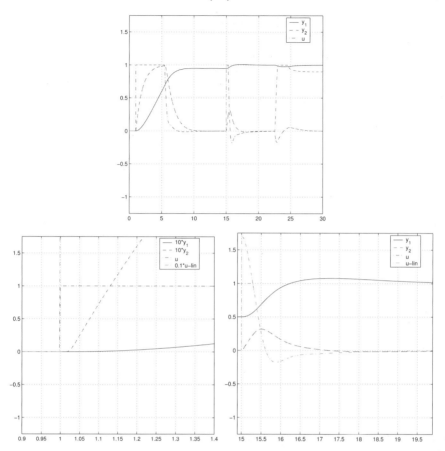

Fig. 5.18. Responses of the control loop structure in Fig. 5.14, but with first-order filtered derivative action, $\tau_d = 0.20T_d$, and with compensating awf gain, $k_a = 3.0$

Remarks

As stated before, there are many more than the three alternatives discussed so far.

One other possibility would be to use the low-pass filter G_a in the awf path together with the deadbeat gain k_a, in order to suppress the undershoot in the tracking loop.

Another strategy would be to clip the derivative action to the working range of u by a separate saturation block. But the stability analysis is more complicated, as there are now two nonlinear blocks.

A third approach would be to consider the filtered D action as to contribute a second state variable (besides the I action) to the controller, which may also "wind up". In other words the erratic behavior is attributed to wind up, and thus an awf to this second state variable is indicated. Note however, that the time constants in this awf loop are very short compared with all others, which would enforce a very short sampling time. Also, it seems that the case with the non-filtered D action is not covered directly.

5.2.5 Stability Analysis

The general approach of Chapter 2 directly applies here.

The Linear Subsystem

The Nyquist contour is

$$F + 1 = \frac{1 + k_a G_a R_2}{1 + R_1 R_2 G} = \frac{s^2 \left(s + \frac{a_2}{T_2}\right)}{(s + \Omega)^3} \frac{s + \frac{k_a}{T_i}}{s} \tag{5.40}$$

PI(aw) and D with low-gain awf

Using $k_a = k_a^* = 3$, *i.e.* $k_a / T_i = \Omega$

$$F + 1 = \frac{\frac{s}{\Omega} \left(\frac{s}{\Omega} + \frac{a_2}{\Omega T_2}\right)}{\left(\frac{s}{\Omega} + 1\right)^2} \tag{5.41}$$

- For the special case of $\frac{a_2}{\Omega T_2} = 1$

$$F + 1 = \frac{\frac{s}{\Omega}}{\frac{s}{\Omega} + 1} \tag{5.42}$$

which evolves in the right-hand half plane for $\omega > 0$, and thus indicates asymptotic stability in the large by the on-axis circle test.
- In the special case of $\frac{a_2}{\Omega T_2} = 1$, *i.e.* $a_2 = 3$,

$$F + 1 = \frac{\frac{s}{\Omega}}{\left(\frac{s}{\Omega} + 1\right)} \frac{\left(\frac{s}{\Omega} + \frac{a_2}{\Omega T_2}\right)}{\left(\frac{s}{\Omega} + 1\right)} \tag{5.43}$$

the Nyquist contour will start for $\omega = 0$ at the origin and evolve along the positive imaginary axis, *i.e.* with a vertical tangent there. This again indicates asymptotic stability in the large by the off-axis circle test or by the Popov test.
- For $a_2 = 0$ the Nyquist contour will start horizontally from the origin to the left. This leads to a finite radius of attraction; see Chapter 4.

Fig. 5.19 (left) illustrates this for the current benchmark system. The continuous equivalent is used.

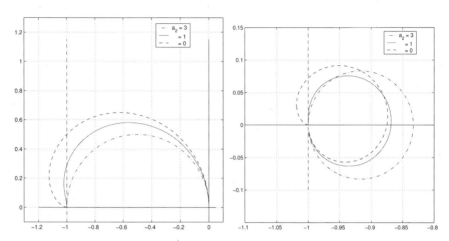

Fig. 5.19. Nyquist contours for the continuous equivalent to the current benchmark system, with non-filtered D action:
(left)   with low-gain awf (compensating, $k_a = 3.0$);
(right) with high-gain awf (deadbeat equivalent)

PI(aw) with D action on the measured variable and with deadbeat-gain awf

Here

$$F + 1 = \frac{\left(\frac{s}{\Omega}\right)^2 \left(\frac{s}{\Omega} + \frac{a_2}{\Omega T_2}\right)}{\left(\frac{s}{\Omega} + 1\right)^3} \frac{1}{\frac{s}{\Omega}} = \frac{\frac{s}{\Omega}}{\frac{s}{\Omega} + 1} \frac{\frac{s}{\Omega} + \frac{a_2}{\Omega T_2}}{\frac{s}{\Omega} + 1} \frac{1}{\frac{s}{\Omega} + 1}$$

$$= (F+1)_{k_a = k_a^*} \frac{1}{\frac{s}{\Omega} + 1} \qquad (5.44)$$

The shape from above is modified by a unity gain first-order low pass in series. So this will not alter the Nyquist contour for low frequencies $0 < \omega < \Omega$. Thus the stability properties are not altered significantly for $\omega \to 0$, and the deformation of the contour is towards the right half plane, *i.e.* in the favorable direction for stability.

Fig. 5.19 (right) illustrates this for the continuous equivalent of the current benchmark system.

PI(aw)-P Control

The Nyquist contours are the same for the same awf gains.
(Show this as a short **exercise**.)

The Nonlinear Subsystem

As shown in Chapter 2 for the graphic test, the nonlinear part determines the position of the straight line on the negative real axis:

$$-\frac{1}{b} = -1 - \frac{u_{hi}}{u_{lin_{max}} - u_{hi}} \qquad (5.45)$$

for $\bar{u} = 0$ and for the case $u_{lin_{max}} > u_{hi}$.

This needs a good estimate of $u_{lin_{max}}$. It can be determined from the setpoint step size as in Chapter 2, but only if there is no "lack of u" effect, and therefore no undershoot of $y_1(t)$, $y_2(t)$.

For the one degree of freedom PID structure with $T_d = 0.25T_i$ and a low-gain awf $k_a = 2 < k_a^*$

$$u_{lin} = r\frac{k_p}{k_a}\left(1 + s\frac{T_i}{k_a}\right) \quad \rightarrow \quad u_{lin_{max}} \approx rk_p\frac{1}{k_a} \qquad (5.46)$$

Then, taking into account that for the graphic test (see Chapter 2), k_a cancels on both the nonlinear subsystem side and on the Nyquist contour, and from Chapter 4 for the on-axis circle test, for the non-trivial case $a_2 = 0$

$$\frac{1}{b} > \frac{1}{8} \quad \rightarrow \quad \frac{u_{lin_{max}}}{u_{hi}} < 9 \quad i.e. \text{ finally } \quad r_{max} < 9\frac{u_{hi}}{k_p} \qquad (5.47)$$

And for the two degrees of freedom PID structure and deadbeat awf, the contribution of the D action to u is 0, as long as the outputs $y_2(t)$, $y_1(t)$ do not start to move. Then again for the graphic test, k_a cancels on both the nonlinear subsystem side and on the Nyquist contour, and again for $a_2 = 0$, the final result is the same: $r_{max} < 9\frac{u_{hi}}{k_p}$.

5.2.6 Summary

The main finding has been that the standard awf strategy (which has never failed so far) does not always work well with derivative action, if used in a standard (one degree of freedom) PID structure. It *fails* for high-gain awf, but *succeeds* for low-gain awf (with the compensating value). But in a two degrees of freedom PID structure, high-gain awf also performs well.

Similar effects will appear on a PI(aw) control loop, if r is not moved stepwise to a new steady state, but instead as a *pulse* of finite length, such that the awf loop has settled due to its short closed-loop time constant, but the controlled output y has barely started to move. Another such situation is measurement noise on y, to be investigated in the next section.

This indicates some "limits of applicability" (and a *"use best for..."* region) for those simple and effective awf design methods from Chapter 2.

5.3 PI(aw) Control with Measurement Noise

5.3.1 Introduction

Intuitively, the best design choice for the awf gain k_a would be to make it as high as possible, *i.e.* on its deadbeat value in time-discrete implementations or to near infinity in continuous forms. This will produce the best possible tracking of the controller output to the constraints, *i.e.* to minimum transient windup, and thus intuitively to minimal performance degradation.
However, such design rules must be used with caution, because there are always limits to their applicability one may not be aware of.

One such limit has been found for the "PI(aw) and D" case. On the standard one degree of freedom structure, the high-gain awf has produced a significant degradation in closed-loop response for small setpoint steps. But the two degrees of freedom structure provides a simple work-around, which allows to go on using the high-gain awf.

A more severe limitation will be addressed in this chapter. This limitation appears if the measured signal $y_1(t)$ contains some persistent (quasi-stationary) high-frequency disturbances. This may be measurement noise, *e.g.* due to turbulence, or residual interference from the power line frequency, *etc.* It is assumed in the following, that this cannot be filtered further without degrading the closed-loop performance in the linear working range.

Then, depending on the operating point, a control error $\bar{e}_1(t) \neq 0$ can appear, where $\bar{e}_1(t)$ shall denote the median of $e_1(t)$ taken over the linear closed-loop settling time. $\bar{e}_1$ may be slowly time varying or can even turn into a steady state offset. This shall be denoted as the *offset effect*.

In contrast, if no such persistent high-frequency disturbances are present, then ultimately $\bar{e}_1(t) \to 0$ due to the integral action, as shown in the previous chapters.

This offset effect was reported by Rundqwist in his PhD thesis [42]. As it is caused by the awf, it can be reduced by lowering the awf gain (among other alternatives). For this "low-gain awf" approach, Rundqwist [42] presents design rules such as

$$T_t = \min\left[T_i, \max\left(\sqrt{T_i T_d}, \frac{T_i}{2}\right)\right] \tag{5.48}$$

with T_t as the awf tracking time constant.
The design rule in Eq. 5.48 has been derived on a specific case, *i.e.* for a continuous implementation of a PI(aw)D controller on a second order process (positioning of a DC-servo), subjected to both pulse and high-frequency sinusoidal disturbances. Since then not much seems to have been published on the subject, *e.g.* [52].

Similar effects have been experienced by the authors while commissioning antiwindup loops in industrial plants, for instance with digital temperature control loops using PI(aw)D controllers. Here heat-up turned out to be significantly slower than predicted by simulations, which had been performed without such high-frequency disturbances.

Another such experience will be described in Section 5.3.2.

The *aim* is to explain the mechanisms with a simple model and to predict the offset effect quantitatively. This is done for three typical situations:
- PI(aw) control on a plant of dominant first order,
- PI(aw)D control on a plant of dominant second order, and
- PI(aw)-P cascade control on the same plant.

First, a motivating example is presented ("Louis' problem"). A suitable measurement noise model is introduced and discussed next. Then the three typical situations from above are investigated, using the same framework of
- specification by a benchmark,
- analysis of the steady state control error $\lim_{t \to \infty} \bar{e}_1(t)$, and
- simulations to check performance and the results.

Finally, some alternatives for reducing or eliminating the offset effect are discussed.

5.3.2 Louis' Problem

In the early 1980s electronic analog control, *e.g.* [24], had become the standard equipment in the hydropower industry (see also John's case, Sect. 2.1). At the same time, the first industrial microprocessor-based control systems with comparable prices were available, although only for comparatively slow thermal processes.

At that time Louis was a young control engineer with the R&D department of a major supplier of hydroturbines and associated control equipment. His project was to show the technical feasibility of microprocessor-based control for hydroturbines. He had to put together a prototype based on a suitable commercial microcomputer system, implement the main functionality of its analog counterpart, test it by simulations, install it (in parallel to an analog controller) at a power station in the Swiss Alps, commission it, and obtain some indicative experience from several months of continuous operation.

All went reasonably well. But one day, during the continuous operation period, the plant manager informed him that the synchronization took much longer than customary with the analog controller, and that this needed fixing.

Observations quickly revealed that the actual turbine speed was not equal to the speed reference (*i.e.* grid frequency) as one would expect with PI control. It was consistently higher and outside the control error gate of the synchronization logic for most of the time. Also, this offset was slowly fluctuating, and from time to time fell below this error gate. Only then was the unit switched into the grid, but with a noticeable bump.

Recordings then documented a typical narrow-band stochastic variation of the discrete time control error, which originated from the grid, and with peak to peak values of approximately ±0.1 Hz. A second element was that Pelton turbine units typically operate at the speed-no-load steady state quite close to the fully closed actuator, *i.e.* here at $\bar{u} - u_{lo} \approx 0.01$ with $u_{lo} = 0.0$.

This induced him in the design phase to reduce any windup to the minimum, in order to avoid any speed overshooting. Therefore, he set the awf gain to its deadbeat value and the lower opening limitation at the PI controller output directly to u_{lo}.

Trying to resolve the synchronization problem, Louis looked more closely at the analog counterpart implementation. He found a small difference: it had a supplementary tracking offset on the opening limitation of approximately ±0.02, *i.e.* the output of the controller was tracking to approximately 0.02 *outside* the actual actuator position. This had been introduced earlier on an intuitive basis to have the servomotor press the gate vanes with a nonzero force into its end position and thus provide better closure. Louis then demonstrated by simulations that this windup had a negligible effect on transients. So he also introduced the supplementary tracking offset on the digital version. And the synchronization problem disappeared! But it reappeared when he introduced some small derivative action on the speed controller to improve on initial overshoot robustness.

The reasons for all this were not fully understood at the time.

Exercise

This is a further Case Study:

After having studied the material of this chapter, investigate Louis' case further and try to reproduce his findings by simulations using:

- the model of Sect. 2.1.1,
- the control structure of Sect. 2.1.3, with a sampling time of 0.010 s,
- and with a (weak) derivative action on the speed controller,
- the awf gain at deadbeat and compensation values,
- first without and then with a tracking offset as described above,
- and the information about the high-frequency disturbance level
  from below.

5.3.3 The High-Frequency Disturbance Model

Two types of such high-frequency disturbance may be distinguished.

The first one stems from residual grid frequency interference. It will be periodic at the grid frequency (here at nominal 50 Hz) and possibly some multiples for harmonics, and will have approximately constant amplitude and zero mean. This is similar to what Rundqwist [42] assumed.

The second type will be some stochastic process, such as the grid frequency fluctuations in Louis' case. We may assume it to be stationary and of zero mean. Its power spectrum shall have its dominant part closely below the sampling frequency. This will produce the narrow-band noise characteristics observed by Louis. As the sampling frequency is much higher than both the open-loop bandwidth of the plant and of the closed loop, this also justifies the attribute "high frequency".

The disturbance signals are continuous and additive on y. They are then sampled and quantized in the AD-converter. To give some quantitative values in Louis' case: using a standard 10-bit converter, *i.e.* a 'Least Significant Bit' (LSB) of approximately 10^{-3}, and a typical sensor span of 45 to 55 Hz as 1.0, this would lead to 1 $LSB := 10$ mHz. And short-term grid frequency variations may easily attain ± 100 mHz.

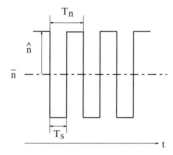

Fig. 5.20. The "rectangular wave" disturbance model and its parameters

Then set $T_s = 0.010$ s, *i.e.* $2T_s$ to one nominal grid period, and consider the first type of disturbance only. Then the output of the AD-converter plus zero-order-hold due to the disturbance signal will be a rectangular wave, with
- period $T_n := 2T_s$,
- zero mean,
- and peak-to-peak value $2\hat{n}$ (which defines $\hat{n}$).

Fig. 5.20 illustrates this model of $n(t)$.

If the second type of disturbance is considered, then the analysis gets much more involved, as the awf effect is highly nonlinear and depends on the *actual* time course $n(t)$. So a usual RMS-based modelling with Gaussian distribution is not applicable. A detailed analysis of the general case seems not to be available to date.

Therefore, a conservative approximation is used here, where the actual noise process is replaced by the "rectangular wave" from above, and $2\hat{n}$ is set to the maximum observed peak-to-peak value. If no such observations are available, then one has to resort to an educated guess first, and refine it later from experiments whenever possible.

We shall now consider the following three cases of
- PI(aw) control,
- PI(aw)D control, and
- PI(aw)-P cascade control.

5.3.4 PI(aw) Control

The Control Problem

The following set of specifications is used as a benchmark

(a) The *plant* is given by a first-order transfer function

$$y(s) = \frac{b}{sT + a} \left(u(s) - v(s) \right) \tag{5.49}$$

There are two inputs, the control variable u and a load disturbance variable v. For simplicity, the delay e^{-Ds} is omitted here.
Numerical values to be used are: $b := 1.0; \quad T := 1.0; \quad a := 0.0;$

(b) As *controllers*, both P and PI(aw) types shall be considered.

The *P controller* is

$$u(s) = k_p \left(r(s) - y(s) \right) + u_v \tag{5.50}$$

where the additional offset input u_v is used to compensate the steady state control errors due to persistent loads $\bar{v} \neq 0$.

And the *PI controller* is

$$R(s) = k_p \left(1 + \frac{1}{sT_i} \right) \quad \rightarrow \quad R(z) = k_p \left(1 + \frac{1}{T_i} \frac{T_s}{1 - z^{-1}} \right) \tag{5.51}$$

where $T_s := 0.010$ s.

(c) The *controller settings* k_p and T_i are obtained by pole assignment for the continuous-time case with the first-order plant dynamics $G_u(s)$ and $R(s)$. From the closed-loop characteristic equation

$$0 = 1 + R(s)G_u(s) = 1 + k_p \left(1 + \frac{1}{sT_i} \right) \frac{b}{sT + a} = s^2 + s\frac{a + bk_p}{T_i} + \frac{bk_p}{T_iT}$$

$$\stackrel{!}{=} (s + \Omega)^2$$

$$\rightarrow \quad bk_p = 2\Omega T - a; \quad \frac{1}{T_i} = \frac{\Omega^2 T}{bk_p} \tag{5.52}$$

and for *numerical values* $b = 1$ and $a = 0$: $\rightarrow \quad k_p = 2\Omega T; \quad \frac{1}{T_i} = \frac{\Omega}{2}.$

The closed-loop bandwidth is set to $\Omega = 5$ rad/s. This provides a sufficient margin with respect to the sampling frequency.

That is, finally $k_p = 10$; $T_i = 0.40$ s.

(d) As *awf structure* we shall consider only the standard control conditioning form **D** of Chapter 2, with the awf gain k_a as design parameter.

Note that in [42] a slightly different controller structure is used, with a tracking time constant T_t instead of the awf gain k_a. However, by comparing both structures:

$$T_t \overset{!}{=} T_i/k_a$$

(e) The following *test sequence* shall be applied to the closed loop, where r is the setpoint and v is the load as specified in item (a):

- Initially (index $_0$) the loop is to be at "standstill" conditions: $r_0 = 0$ and $v_0 = 0$;   *i.e.* $\bar{y}_0 = 0$;   and $\bar{u}_0 = 0$.

- At time T_1, a setpoint step to $r_1 = 1$ is applied while there is still no load, *i.e.* $v_1 = 0$. Then, at equilibrium, a plant input $\bar{u}_1 = ar_1 = 0$ results.

- Then, at time T_2, a load step v_2 is applied ($v_2 = +0.90$), while the setpoint is kept constant, $r_2 = r_1 = 1$.

(f) The *actuator saturation* is: $u_{low} \overset{.}{=} -1.0$ and $u_{high} = +1.0$

(g) For the *measurement disturbance* signal, the model of the previous section is used, with the numerical value $\hat{n} = 0.015 = 1.5$ % .

Note that both the Ziegler–Nichols and the Chien–Hrones–Reswick tuning rules would also allow a derivative action. But if high-frequency measurement disturbances are present, then the derivative action is usually set to zero at commissioning in order to avoid excessive high-frequency movement of the control signal.

In Chapter 2 this system was analyzed for its transient response (although without $n(t)$) and its stability properties. This has led to the concept of "compensating" awf gain $k_a^* = 2$.

Using the notation of [42]

$$T_t := T_i/k_a$$

then yields

$$T_t = T_i/2$$

which nicely confirms the result in Eq. 5.48 from [42] for PI(aw) control (*i.e.* $T_d = 0$).

Transient Response

The aim is to illustrate by simulations, how the response can be affected by such a high-frequency disturbance. Fig. 5.21 shows the transient responses to the benchmark test sequence with P control, (left) without and (right) with $n(t)$ applied, and Fig. 5.22 the same for PI(aw) control.

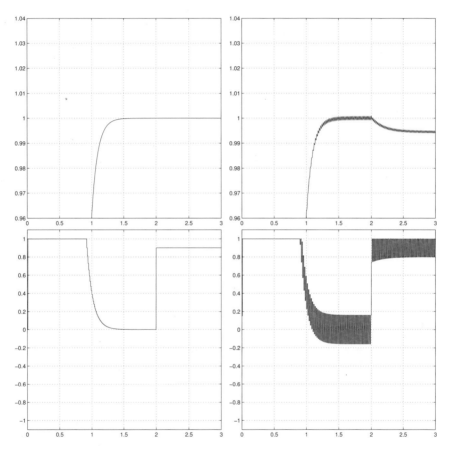

Fig. 5.21. Structure and responses with P control,
(left) without and (right) with $n(t)$;   top row: $y(t)$; bottom row: $u(t)$

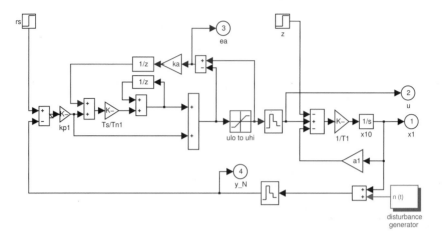

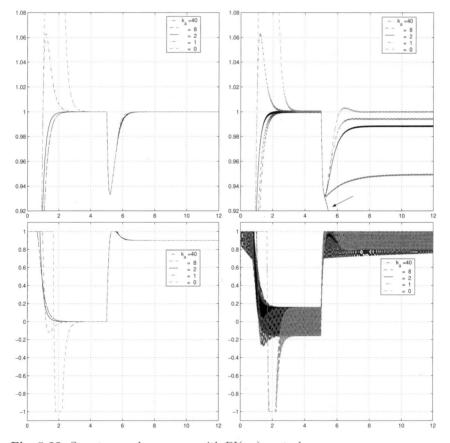

Fig. 5.22. Structure and responses with PI(aw) control,
(left) without and (right) with $n(t)$;   top row: $y(t)$; bottom row: $u(t)$

Remarks

For the *P control*, there is no significant effect of $n(t)$ in phase 1, *i.e.* on run-up overshoot and on the subsequent steady state. In phase 2 (the load swing), for $n = 0$ there is no steady state offset on $y(t)$ due to the fact that $u_v = v$ has been switched in at the same instant as the step on v. But for $n \neq 0$, there appears a steady state control error $\bar{e} := r_1 - \bar{y} \neq 0$. The numerical value read from the experiment is $\bar{e} \approx +0.0055$.

And for the *PI(aw) control*, $n(t) \neq 0$ has again no significant effect in phase 1. For instance, there is no perceptible overshoot of $y(t)$ for $k_a = k_a^* = 2$, and the overshoot for $k_a = 1$ is the same ($y_{max} \approx 1.062$).
In phase 2 there again appears a steady state control error $\bar{e} \neq 0$, which now increases with increasing awf gain k_a. The deadbeat value of $k_a = T_i/T_s = 40$ leads to $\bar{e} \approx 0.454$ after a settling time of about 50 s. This is not fully shown in Fig. 5.22 (top right), but the trace is indicated by the small arrow.

Offset Analysis

The aim now is to calculate this steady state control error as function of the main parameters for the experiment.

Linear Operating Conditions

Consider first the operating conditions at the end of run-up phase 1. Applying the square wave sequence $u(z)$ with $\hat{u}$ to the plant $G_u(z)$ produces an amplitude $\hat{y}$ on the output

$$\hat{y} = \frac{1}{2} \frac{T_s}{T} \, \hat{u} \qquad (5.53)$$

Then for the controller input

$$\hat{e} = \hat{y} + \hat{n} \qquad (5.54)$$

For the P- and the PI-controller respectively

$$\hat{u}_P = k_p \hat{e} \quad \text{and} \quad \hat{u}_{PI} = k_p \left(1 + \frac{T_s}{T_i}\right) \hat{e} \qquad (5.55)$$

yielding

$$\hat{u}_P = \frac{k_p}{1 - k_p \frac{1}{2}\frac{T_s}{T}} \, \hat{n} \quad \text{and} \quad \hat{u}_{PI} = \frac{k_p \left(1 + \frac{T_s}{T_i}\right)}{1 - k_p \left(1 + \frac{T_s}{T_i}\right) \frac{1}{2}\frac{T_s}{T}} \, \hat{n} \qquad (5.56)$$

and with the numerical values for the benchmark

$$\hat{u}_P \approx 0.158 \quad \text{and} \quad \hat{u}_{PI} \approx 0.162 \qquad (5.57)$$

Operating Conditions Close to the Actuator Saturation

Consider now the operating conditions in phase 2. For steady state operation, a mean control input value

$$\bar{u} = -v = 0.90 \tag{5.58}$$

is required, which is close to the upper saturation $u_{hi} = 1.0$. If now the disturbance signal $n(t)$ is applied, then the actuator input signal $u_{lin}(t)$ will periodically saturate; Fig. 5.23.

Then the steady state operation of the closed loop requires

$$\frac{\overline{d}}{dt}y \overset{!}{=} 0 \tag{5.59}$$

i.e. from Fig. 5.23

$$\hat{u} = u_{hi} - \bar{u} \tag{5.60}$$

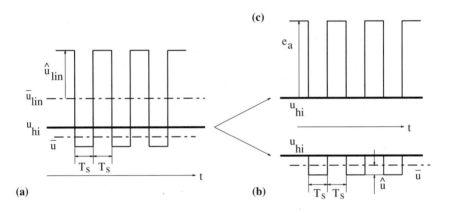

Fig. 5.23. The signals (a) at the input to the saturation, (b) at its output, *i.e.* the input to the plant, (c) the input to the awf, $e_a(t)$

For the *P controller*:

$$u_{lin}(t) = k_p e(t)$$
$$\bar{u}_{lin} = k_p \bar{e}$$
$$\hat{u}_{lin} = k_p \hat{e}$$
$$\text{where} \quad \hat{e} = \hat{n} + \hat{y}$$
$$\text{and} \quad \hat{y} = \frac{1}{2}\frac{T_s}{T}\hat{u} = \frac{1}{2}\frac{T_s}{T}(u_{hi} - \bar{u}) \tag{5.61}$$

By inspection of Fig. 5.23 for steady state conditions

$$\bar{u}_{lin} - \hat{u}_{lin} \overset{!}{=} \bar{u} - u_{hi} \tag{5.62}$$

i.e.

$$\bar{u}_{lin} = k_p \bar{e}_P = \hat{u}_{lin} + \bar{u} - u_{hi} = k_p \hat{e}_{lin} + \bar{u} - u_{hi} \tag{5.63}$$

yielding the main result for P control:

$$\bar{e}_P = \hat{e} - \frac{u_{hi} - \bar{u}}{k_p} \;=\; \hat{n} - \frac{u_{hi} - \bar{u}}{k_p}\left(1 - k_p \frac{1}{2}\frac{T_s}{T}\right) \tag{5.64}$$

i.e. a nonzero steady state control error $\bar{e}_P$ appears for $\hat{n}$ large enough such that u_{lin} periodically saturates.
And the offset $\bar{e}_P$ will be zero if

$$\hat{n} < \frac{u_{hi} - \bar{u}}{k_p}\left(1 - k_p \frac{1}{2}\frac{T_s}{T}\right) \tag{5.65}$$

i.e. the high-frequency disturbance level $\hat{u}$ must be low enough compared with the residual control working range $u_{hi} - \bar{u}$.

For the *PI(aw) controller*:
For steady state conditions the awf action has to reset the integrator output u_I in Fig. 5.23 (a) to the same value as one period of the square wave before. During this time interval u_I will move upward due to the non-zero error input $\bar{e}_{PI}$ in both half periods:

$$\Delta u_{I\ up} = 2k_p \frac{T_s}{T_i}\,\bar{e}_{PI} \tag{5.66}$$

and u_I is moved downward at the end of the second half period due to the awf error e_{aw} in the second half period by

$$\Delta u_{I\ dn} = k_a \frac{T_s}{T_i}\,e_{aw} \tag{5.67}$$

For steady state conditions

$$\Delta u_{I\ up} \overset{!}{=} \Delta u_{I\ dn} \quad \rightarrow \quad 2k_p \bar{e}_{PI} = k_a e_{aw} \tag{5.68}$$

Furthermore, from Fig. 5.23 (b) and (c),

$$e_{aw} = 2\hat{u}_{lin} - 2\left(u_{hi} - \bar{u}\right) \tag{5.69}$$

that is

$$k_p \bar{e}_{PI} = \frac{1}{2}k_a e_{aw} = k_a\left[\hat{u}_{lin} - \left(u_{hi} - \bar{u}\right)\right] \tag{5.70}$$

To determine $\hat{u}_{lin}$, consider the discrete-time awf loop in Fig. 5.22:

$$\hat{u}_{lin} = \frac{1}{1 - \frac{1}{2}\frac{T_s}{T_i}k_a} \, k_p \left(1 + \frac{1}{2}\frac{T_s}{T_i}\right) \hat{e} \tag{5.71}$$

Inserting this into Eq. 5.70

$$\bar{e}_{PI} = \frac{k_a}{1 - k_a\frac{1}{2}\frac{T_s}{T_i}} \left[\left(1 + \frac{1}{2}\frac{T_s}{T_i}\right) \hat{e} - \frac{u_{hi} - \bar{u}}{k_p}\right] \tag{5.72}$$

and using

$$\hat{e} = \hat{n} + \hat{y}$$

with $\hat{y}$ from Eq. 5.61 produces the main result for PI(aw) control:

$$\bar{e}_{PI} = \frac{k_a}{1 - k_a\frac{1}{2}\frac{T_s}{T_i}} \left[\left(1 + \frac{1}{2}\frac{T_s}{T_i}\right) \hat{n} - \frac{u_{hi} - \bar{u}}{k_p} \left(1 - k_p\frac{1}{2}\frac{T_s}{T}\right)\right] \tag{5.73}$$

Note that the offset $\bar{e}$ will be zero if:
- $k_a = 0$, i.e. for the no-awf case - or if the second term [...] is zero.

This means

$$\left(1 + \frac{1}{2}\frac{T_s}{T_i}\right) \hat{n} < \frac{u_{hi} - \bar{u}}{k_p} \left(1 - k_p\frac{1}{2}\frac{T_s}{T}\right) \tag{5.74}$$

That is, the high-frequency disturbance level $\hat{u}$ must be low enough compared with the residual control working range $u_{hi} - \bar{u}$:

$$\hat{n} < \frac{u_{hi} - \bar{u}}{k_p} \frac{1 - k_p\frac{1}{2}\frac{T_s}{T}}{1 + \frac{1}{2}\frac{T_s}{T_i}} \; ; \quad \text{or for short} \quad \hat{n} < \frac{u_{hi} - \bar{u}}{k_p} \tag{5.75}$$

Comparison

P control
The main result in Eq. 5.64 evaluates to

$$\bar{e}_P = 1.5\% - 1.0\% \, (1 - 0.05) \quad = \quad 0.55\% \tag{5.76}$$

which agrees well with the simulation result $\sim 0.55\%$ from Fig. 5.21.

PI(aw) control
Inserting the numerical values from $k_a = 0$ up to the deadbeat value $k_a = 40.0$:

	k_a	0	1	2	8	40
Evaluation:	$\bar{e}_{PI}$ [%]	0	0.576	1.167	5.06	45.5
Simulation:	$\bar{e}_{PI}$ [%]	0	0.55	1.18	5.10	45.4

Again, the results from the simulation in Fig. 5.22 agree well with the values predicted by the main result in Eq. 5.73.

5.3.5 PI(aw) Control with Derivative Action

The case discussed so far of PI(aw) control on a dominant first-order plant covers a large area of industrial applications. But if a PI(aw) controller is used on a plant of dominant second order, the phase margin will be low and linear range closed loop performance will be poor. Then it is very effective to add a derivative action. However the high frequency disturbance contribution from the derivative action to $\hat{u}$ will become large. This will produce an even larger control error offset $\bar{e}$. Low pass filtering of the derivative may reduce this, but at the cost of reducing the phase margin again.

The Control Problem

This shall be stated as in Sect. 5.3.4 by modifying the relevant items.

(a) The *plant* model is replaced by

$$G_u(s) = \frac{1}{(sT_1 + a_1)(sT_2 + a_2)} \tag{5.77}$$

again with $a_1 = 0$ and $a_2 = 0$, and also $T_1 := T_2 := T$, and with the load transfer function $G_v(s) = -G_u(s)$.

(b) As *controller*, a standard PID type shall be considered with the linear transfer function

$$R(s) = k_p \left(1 + \frac{1}{sT_i} + sT_d \frac{1}{s\tau_d + 1} \right) \quad \rightarrow$$

$$R(z) = k_p \left(1 + \frac{1}{T_i} \frac{1}{\frac{1-z^{-1}}{T_s}} + T_d \frac{1-z^{-1}}{T_s} \frac{z^{-1}}{\frac{1-z^{-1}}{T_s}\tau_d + z^{-1}} \right)$$

(c) The *controller settings* shall be calculated using pole assignment with $G_u(s)$ for $a_1, a_2 = 0$, and $R_d(s) = k_p \left(1 + \frac{1}{sT_i} + sT_d \right)$, where the filter with τ_d is omitted. The poles shall be placed at

$$(s + \Omega)^3 = 0 \tag{5.78}$$

leading to

$$k_p = 3\,(\Omega T)^2 \;\; ; \quad \frac{1}{T_i} = \frac{\Omega}{3} \; ; \quad T_d = \frac{1}{\Omega} \tag{5.79}$$

For numerical values, Ω shall be set such that the same value for k_p results as for the dominant first-order system, *i.e.*

$$\Omega = \sqrt{2 \cdot 5/3} \approx 1.826 \quad \rightarrow \quad k_p := 10 \; , \quad \frac{1}{T_i} := \frac{1}{1.643} \; , \quad T_d := 0.548 \tag{5.80}$$

Using the common design rule $\tau_d := 0.1T_d$ (e.g. see [2]) yields $\tau_d = 0.0548$; and as $T_s = 0.010$, therefore set

$$\tau_d := 5T_s. \tag{5.81}$$

and items (d), (e), (f), and (g) shall be the same as in Sect. 5.3.4

Stability analysis

Again, the awf tracking loop will attain its dead beat response at $k_{a_{max}} = \frac{T_i}{T_s}$; and with the numerical values for the benchmark $\quad k_{a_{max}} = \frac{3}{\Omega T_s} \approx 164.3$.

The nonlinear stability analysis for the main loop shall be conducted in its continuous equivalent form. With

$$G_d(s) = \frac{1}{(sT_1 + a_1)(sT_2 + a_2)} \quad \text{and} \quad R_d(s) = k_p \left(s^2 T_d T_i + sT_i + 1 \right) \frac{1}{sT_i}$$

$$
\begin{aligned}
F(s) + 1 &= \frac{(sT_1 + a_1)(sT_2 + a_2)\, sT_i}{\left[s^3 T_1 T_2 T_i + s^2 T_1 T_i(a_1 + a_2 + k_p T_d) + sT_i(a_1 a_2 + k_p) + k_p \right]} \\
&\quad \times \left(1 + k_a \frac{1}{sT_i} \right) \\
&= \frac{(s/\Omega)^2 + (s/\Omega)\left[(a_1/T_1\Omega) + (a_2/T_2\Omega) \right] + (a_1/T_1\Omega)(a_2/T_2\Omega)}{\left[(s/\Omega) + 1 \right]^2} \\
&\quad \times \frac{\left[(s/\Omega) + (k_a/T_i\Omega) \right]}{\left[(s/\Omega) + 1 \right]}
\end{aligned}
\tag{5.82}
$$

Global asymptotic stability can be shown, if $a_1 > 0$ and $a_2 > 0$ and $k_a > 0$, using the off-axis circle criterion.

Again, for the awf gain

$$k_a := k_a^* = \Omega T_i$$

one pole in $F + 1$ cancels with the zero from the awf loop. And inserting $a_1 = a_2 = 0$ in the controller design

$$\frac{1}{T_i} = \frac{\Omega}{3} \rightarrow k_a^* = 3 \tag{5.83}$$

$$\text{and} \quad (F + 1)^* = \frac{(s/\Omega)^2}{\left[(s/\Omega) + 1 \right]^2} \tag{5.84}$$

Note that $k_a^* = 2$ was found for the comparable PI(aw) case.

For $k_a^* < k_a \le k_{a_{max}}$ the second fraction in $F(s) + 1$ in Eq. 5.82 is a low-pass element. Therefore, it shifts the phase of the Nyquist contour in the direction, which is favorable for stability.

If, however, $0 < k_a \le k_a^*$ then the second fraction in $F(s) + 1$ is a high-pass element, which reshapes the contour unfavorably. In other words, if $k_a < k_a^*$ then windup effects will again become pronounced.

Finally, note that the contour for Eq. 5.83 is such (it starts from the origin along the negative real axis) that only a finite radius of attraction may be shown. So some overshoot must be expected for the run-up response. But this can be reduced by increasing $k_a > k_a^*$.

Transient Responses

Again, we shall compare the cases with $n(t)$ to those without. The amplitude parameter $\hat{n}$ has been reduced such that the amplitude $\hat{u}_{lin}$ on the control signal in the linear range has about the same size as with the PI(aw) case, i.e. $\sim 16.0\,\%$.

Fig. 5.24 documents the structure of the control loop used in Fig. 5.25.

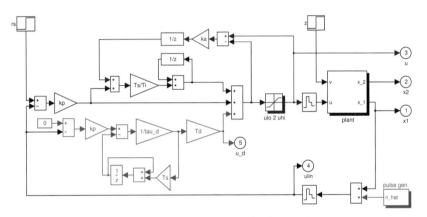

Fig. 5.24. Structure with PI(aw)D control

Consider Fig. 5.25 (top left). Using the compensating awf gain value $k_s = k_a^* = 3$ produces an overshoot of $\sim 2.75\%$. This is in contrast to the finding for the first-order plant. But it is to be expected from the stability analysis, which indicates that k_a should be increased. From the simulations $k_a := 3k_a^* = 9$ reduces the run-up overshoot to about 0.07%, and for the dead-beat gain $k_a = T_i/T_s \approx 164.3$ to about 0.035%. Note that this is still > 0.

Note that the level of disturbance $\hat{n}$ is so low that it is not visible in the traces of either $x_1(t)$ or $x_2(t)$, but only shows up on $u(t)$.

Comparing the left and right columns in Fig. 5.25 reveals that $n(t)$ has no significant effect on either run-up or stabilization in phase 1 of the test sequence.

However, in phase 2, there appears a non-negligible steady state offset on the control error.

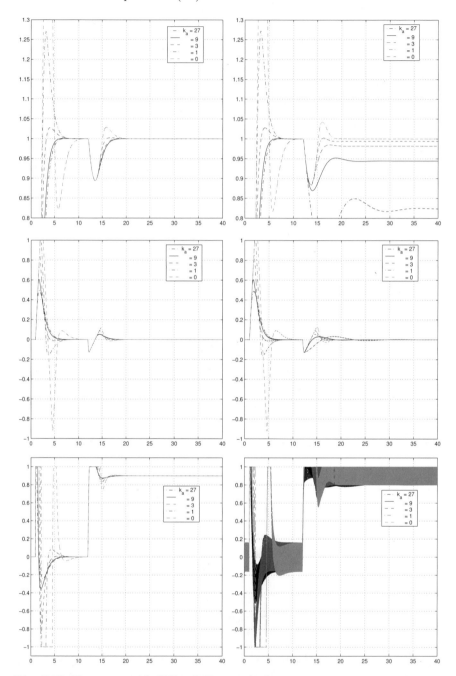

Fig. 5.25. Responses with **PI(aw)-D - control**,
top $x_1(t)$, center $x_2(t)$, bottom $u(t)$,   left: $n(t) = 0$, right: $n(t) = 1.5 \%/12.35$

Offset Analysis

The approach from the PI(aw) case is used.

The residual amplitude $\hat{x}_1$ from $\hat{u}$ is reduced by a factor of $(T_s/T)^2$ and may well be neglected, *i.e.*

$$\hat{e} = \hat{n} + \hat{x}_1 \approx \hat{n}$$

The input to the saturation element is (for $\tau_d := \kappa\, T_s$ with integer $\kappa \geq 1$ and for the $\hat{n}$-period $T_{p_n} = 2T_s$):

$$\hat{u}_{lin} \approx k_p \left(1 + \frac{1}{2}\frac{T_s}{T_i} + \frac{\frac{1}{\tau_d}}{1 - \frac{1}{2}\frac{1}{\tau_d}T_s}T_d \right) \hat{e} = k_p \left(1 + \frac{1}{2}\frac{T_s}{T_i} + \frac{1}{1 - \frac{1}{2}\frac{T_s}{\tau_d}}\frac{T_d}{\tau_d} \right) \hat{e}$$

Also

$$e_{aw} = \hat{u}_{lin} - (u_{hi} - \bar{u})$$

and for steady state at the I(aw) part of the controller

$$k_p \left(1 - k_a \frac{1}{2}\frac{T_s}{T_i} \right) \bar{e} = k_a e_{aw}$$

finally resulting in the main result for PI(aw)D control:

$$\bar{e}_{PID} \approx k_a \frac{1}{1 - k_a \frac{1}{2}\frac{T_s}{T_i}} \left[\left(1 + \frac{1}{2}\frac{T_s}{T_i} + \frac{1}{1 - \frac{1}{2}\frac{T_s}{\tau_d}}\frac{T_d}{\tau_d} \right) \hat{n} - \frac{u_{hi} - \bar{u}}{k_p} \right] \quad (5.85)$$

Inserting the numerical values of the benchmark

$$\hat{u}_{lin} = k_p \left(1 + \frac{1}{2}\frac{T_s}{T_i} + \frac{1}{1 - \frac{1}{2}\frac{T_s}{\tau_d}}\frac{T_d}{\tau_d} \right) \hat{n} = 10\,(1 + 0.00304 + 12.1955)\,1.5\%$$

$$= 198\%\ (!)$$

shows that the contribution to $\hat{u}_{lin}$ from the derivative action is now the dominating one. It is by a factor of ~ 12.2 larger than the contribution of the proportional action.

Such a large offset is not tolerable. It must be reduced by acting on the disturbance level $\hat{n}$. A reasonable assumption would be that $\hat{u}_{lin}$ attains the same level as for the PI(aw) case, *i.e.* 16.0%. This requires a reduction factor ρ

$$\rho = \frac{197.98\%}{16.0\%} = 12.35$$

as has been used in the simulations in Fig. 5.25.

Comparison

Evaluating $\bar{e}_{PID}$ by using Eq. 5.85 first and then taking it from the simulations in Fig. 5.25 (top right) yields

	k_a	0	1	3	9	27	164.3 (deadbeat)
Evaluated:	$\bar{e}_{PID}[\%]$	0	0.599	1.82	5.55	17.6	198
Simulated:	$\bar{e}_{PID}[\%]$	0	0.6	1.82	5.56	17.6	197.2

where the deadbeat case has been simulated, but is not shown in Fig. 5.25. So Eq. 5.85 seems to predict the experimental results well enough.

To summarize, it is *definitely not recommended* to use a PI(aw)-D controller in the presence of such high frequency disturbances. Other solutions must be sought.

5.3.6 PI(aw)-P cascade control

In industrial applications a PI-P cascade structure is often used instead of a single PID controller. It avoids calculating the sampled derivative of $y_1(t) + n_1(t)$ by using a second feedback variable $y_2(t)$. However, this needs an additional sensor and will introduce its own high-frequency disturbance component $n_2(t)$.

This alternative shall be investigated next, and compared with the PI(aw)D version.

Specifications

- In item (b) of the specifications in section 5.3.4, the controller is replaced by

$$\text{the ``slave'' controller} \quad R_2(z) = k_{p_2}$$

$$\text{and the ``master'' controller} \quad R_1(z) = k_{p_1}\left(1 + \frac{T_s}{T_{i_1}}\frac{1}{1 - z^{-1}}\right)$$

- In item (c) the controller settings are calculated as before using the continuous-time equivalent and $a_1 = a_2 = 0$. This results in

$$k_{p_2} = 3\Omega T_2; \quad k_{p_1} = \Omega T_1; \quad \frac{1}{T_{i_1}} = \frac{\Omega}{3}$$

- In item (d) the awf shall act on the integral action input as before. Its awf gain shall be denoted as k_a', as k_{p_2} is in the awf loop here.

- In item (g) the high-frequency disturbance signals $n_1(t)$ and $n_2(t)$ shall be of the same form and frequency as $n(t)$ from Sect. 7.3. Furthermore $n_1(t)$ and $n_2(t)$ shall be "in phase", which is the most pessimistic assumption.

For *numerical values*, again choose $\hat{n}$ such that $\hat{u}_{lin} :\approx 16\%$, which leads to

$$\hat{n} := 1.0\% \tag{5.86}$$

i.e. a reduction factor $\rho \approx 1.5$ with respect to the PI(aw) case. In other words, the high-frequency disturbance level can be higher by a factor of ~ 8 compared with the PI(aw)D case.

Fig. 5.26 shows the structure of the control loop.

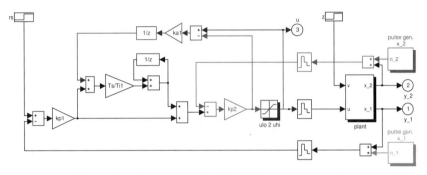

Fig. 5.26. Structure with PI(aw)-P cascade control

Stability Analysis

The awf tracking loop now contains the "slave" regulator gain k_{p_2} as well. To obtain the same properties of the awf loop as before, set:

$$k'_a := k_a/k_{p_2}$$

Then $F(s) + 1$ is the same as in Sect. 5.3.5, *i.e.* the PI(aw)-P cascade has the same stability properties as the PI(aw)D structure.

Offset Analysis

This is investigated using the same approach and previous results. The input amplitude $\hat{u}_{lin}$ to the saturation from Fig. 5.26 is generated by two components from $\hat{n}_1 + \hat{y}_1$ and $\hat{n}_2 + \hat{y}_2$.

As above $\qquad\qquad\qquad\qquad \hat{y}_1 := 0$

Then for the inner loop $\qquad \hat{u}_{lin} = \dfrac{1}{1 - \frac{1}{2}k_{p_2}\frac{T_s}{T_2}}(\hat{u}_2 + \hat{u}_1)$

where $\qquad\qquad\qquad \hat{u}_2 = k_{p_2}\hat{n}_2; \text{ and } \hat{u}_1 = k_{p_2}k_{p_1}\left(1 + \dfrac{1}{2}\dfrac{T_s}{T_{i_1}}\right)\hat{n}_1$

and also $\qquad k_{p_2}k_{p_1}\bar{e} = \dfrac{k_a}{1 - \frac{1}{2}k_a\frac{T_s}{T_{i_1}}}[\hat{u}_{lin} - (u_{hi} - \bar{u})]$

finally resulting in

$$\bar{e} = \frac{k_a}{1 - \frac{1}{2}k_a \frac{T_s}{T_{i_1}}} \left\{ \frac{1}{1 - \frac{1}{2}k_{p2}\frac{T_s}{T_2}} \left[\frac{1}{k_{p_1}}\hat{n}_2 + \left(1 + \frac{1}{2}\frac{T_s}{T_{i_1}}\right)\hat{n}_1 \right] - \frac{u_{hi} - \bar{u}}{k_{p2}k_{p_1}} \right\}$$

$$(5.87)$$

Transient Response

The simulation results in Fig. 5.27 compare the cases with $n(t)$ to those without, using $\hat{n} = 1.0\%$. Considering Fig. 5.27 (top left), the compensating awf gain value $k_s = k_a^* = 3$ produces an overshoot of $\approx 3.5\%$. And for $k_a := 3k_a^* = 9$ the run-up overshoot is suppressed.

Again, $n(t)$ has no significant effect on either run-up or stabilization in phase 1 of the test sequence. And in phase 2 there is a steady state offset on the control error as predicted.

Comparison

Evaluating Eq. 5.87 for $\hat{n}_1 = \hat{n}_2 = 1.0\%$ agrees well enough with the results from the simulations in Fig. 5.27:

	k_a	0	1	3	9	27	164.3 (deadbeat)
Evaluated:	$\bar{e}_{PI\text{-}P}$ [%]	0	0.59	1.79	5.48	17.4	195
Simulated:	$\bar{e}_{PI\text{-}P}$ [%]	0	0.58	1.75	5.35	17.1	191

5.3.7 Some Solutions

The basic move is to avoid derivative action strictly in the controllers. And if such additional feedback is needed for loop stabilization or performance, then the derivative action should be replaced by a cascade (or equivalent state feedback) structure. Then there are several possible solutions to this offset problem.

Low-Gain awf

From Eq. 5.73 (also 5.85) and 5.87, we find that for $k_a = 0 \rightarrow \bar{e} = 0$, but obviously this is not an acceptable solution, as the windup effect is then at its maximum. So the second move is to avoid high-gain or deadbeat awf, and choose k_a as low as possible. In other words, a design compromise has to be made between offset and overshoot. This seems to work rather well with the basic dominant first-order plant and PI(aw) control, but deteriorates quite markedly for the dominant second-order plant and PI(aw)-P control.

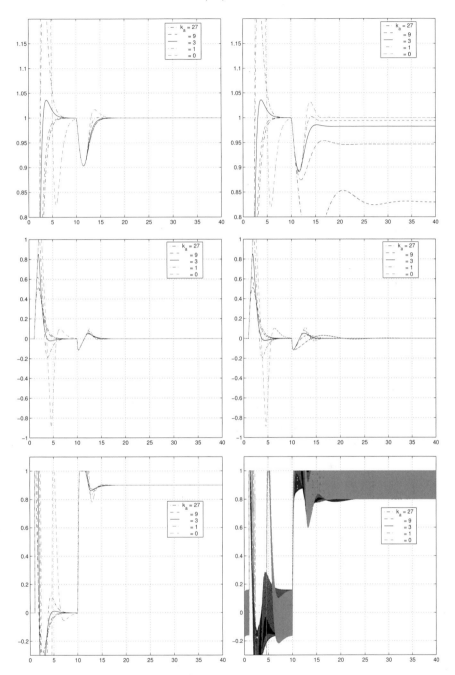

Fig. 5.27. Responses with PI(aw)-P cascade control.
Top: $x_1(t)$; center: $x_2(t)$; bottom: $u(t)$;   left: $n(t) = 0$; right: $n(t) = 1.0\%$

Actuator Re-sizing

Again, from Eq. 5.73 and 5.87 also for 5.85:

$$\bar{e} \;\to\; 0 \quad \text{if} \quad \hat{u}_{lin} \;\le\; (u_{hi} - \bar{u}) \tag{5.88}$$

independent of the k_a value and the regulator type. In other words, the residual actuator maneuvering range $(u_{hi} - \bar{u})$ must be re-sized so as to accommodate the actual $\hat{u}_{lin}$-level.

That is, the actuator limit u_{hi} should be moved outward accordingly. In numerical values for the benchmark, Sect. 5.3.4, this would move the original value $u_{hi} = 100\%$ to the re-sized value $u_{hi_r} = (90 + 16.2)\% = 106.2\%$

Inserting a Controller Output Saturation

But note that this will not work if the actuator operates near its fully closed position $u_{lo} = 0$ (as in Louis' case).

His solution was to insert an additional saturation on the controller output with limits u_{lo-lo}, u_{hi-hi} outside the actual ones u_{lo}, u_{hi} in the actuator with the margins:

$$\Delta u_{lo} = u_{lo-lo} - u_{lo}; \qquad \Delta u_{hi} = u_{hi-hi} - u_{hi} \tag{5.89}$$

But this will add to integrator windup, and thus increase the overshoot. So the margins must be kept to the minimum. This shall be investigated further using the benchmark case of Sect. 5.3.4, with the following modifications (see Fig. 5.29 and 5.30):

$$r_1 = 1.0; \; a = 0.010; \; v = 0.0; \;\to\; \bar{u} = 0.010; \text{ and } \; u_{lo} = 0.0; \; u_{hi} = +0.20 \tag{5.90}$$

Several entries of $\hat{n}$ shall be considered in order to get an overview of the overshoot as a function of the awf gain k_a.

Offset Analysis

The aim is to determine the margin Δu_{lo}, such that the steady state offset on $\bar{e}$ is suppressed. In Fig. 5.28 (a)

$$\hat{u}_{lin} = k_p \left(1 + \frac{1}{2} \frac{T_s}{T_i} \right) \hat{n} \tag{5.91}$$

This is separated in the nonlinear element into Fig. 5.28 (c) and (b). From the steady state condition at the plant input (see Fig. 5.28 (c))

$$\overline{\frac{d}{dt} y_1} = 0 \;\to\; \hat{u} = \bar{u} - u_{lo}; \; i.e. \text{ in Fig. 5.28 (b): } v_{lo} := 2\hat{u}_{lin} - 2\hat{u} \tag{5.92}$$

Now the awf must not be activated, *i.e.* $e_a = 0$, or

$$|\Delta u_{lo}| \geq |v_{lo}| \quad \rightarrow \quad |\Delta u_{lo}| \geq 2k_p \left[\left(1 + \frac{1}{2}\frac{T_s}{T_i}\right)\hat{n} - \frac{\bar{u} - u_{lo}}{k_p} \right] \tag{5.93}$$

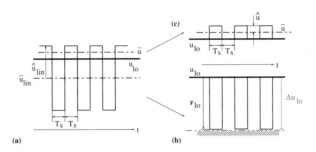

Fig. 5.28. Analysis for the controller output saturation larger than the actuator limits

Transient Response and Overshoot

Fig. 5.29 shows the PI(aw) control structure on the first-order plant, with an additional saturation u_{lo-lo} to u_{hi-hi} inserted at the controller output and then used for awf instead of u_{lo} to u_{hi}.

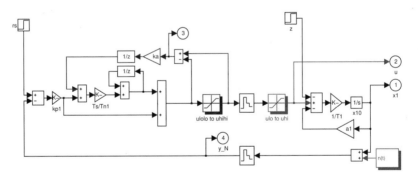

Fig. 5.29. Inserted controller output saturation

Fig. 5.30 documents the responses of $y(t)$, with e_{max} values for $k_a = 2$.

	Left		Right	
Top	$\hat{n} = 1.5\%$	$e_{max} \approx 1.8\%$	$\hat{n} = 1.0\%$	$e_{max} \approx 1.8\%$
Bottom	$\hat{n} = 0.5\%$	$e_{max} \approx 0.28\%$	$\hat{n} = 0.20\%$	$e_{max} \approx 0.04\%$

Note that the overshoot is not suppressed by increasing k_a beyond k_a^*. The only way out seems to be a small enough $\hat{n}$ (as in Sect. 5.3.2, where $\hat{n} \approx 0.1\%$).

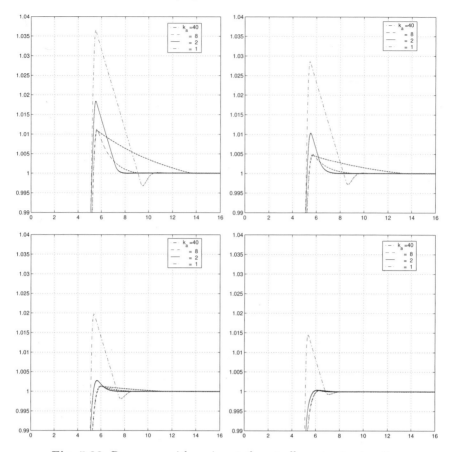

Fig. 5.30. Responses with an inserted controller output saturation

Comparison

For the numerical values of $\hat{n}$, Equation 5.93 evaluates to

	$\hat{n}$ [%]	1.5	1.0	0.5	0.20	0.10
From evaluation:	Δu_{lo} [%]	28.375	18.25	8.125	2.05	0.025
From simulation:	Δu_{lo} [%]	28.6	18.4	8.25	2.2	0.03

This is compared with values obtained from the simulation in Fig. 5.30. There, Δu_{lo} has been adjusted such that a reasonable speed of convergence on $y_1(t)$ is obtained for $k_a = 40$ (deadbeat).

Digital Filtering of Measured Signals

In the benchmark case of Sect. 5.3.5, the closed-loop bandwidth is set to $\Omega = 1.826$, and the frequency ω_n of $n(t)$ is 100π rad/s, *i.e.* $\omega_n/\Omega = 172$. In other words, there is a sufficient distance to insert digital filtering without disrupting closed-loop performance.[5] The structure is shown in Fig. 5.31.

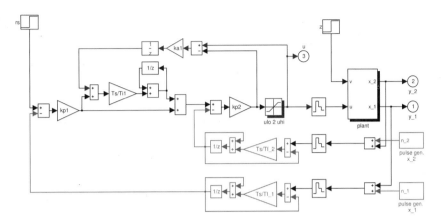

Fig. 5.31. Structure of Fig. 5.26 with digital first-order filtering of inputs

Simulation results are shown in Fig. 5.32.

For the left column, the filter bandwidth

$$\Omega_f := 43.5\Omega_1 \quad (\text{or} \quad \Omega_f := 0.253\omega_n)$$

has been chosen such that

$$\hat{u}_{lin} := u_{hi} - \bar{u} := 10\%$$

In other words $\qquad \bar{e}_{PI\text{-}P} = 0$ for all k_a

And for the right hand column,

$$\Omega_f := 12.6\Omega_1 (\quad \text{or} \quad \Omega_f := 0.040\omega_n)$$

such that

$$\hat{u}_{lin} := 2\%$$

Note that such filtering is only feasible if the gap ω_n to Ω_1 is large enough to allow inserting a filter at Ω_f with sufficient rejection at ω_n, as was the case here.

If not, then an observer may be used, *e.g.* see [52]. But this requires modules, which are usually non-standard within current industrial process control systems, and therefore lie outside the "Standard Techniques" environment.

---

[5] This also holds for the case of Sect.5.3.4, where $\omega_n/\Omega = 314.2/5 = 62.8$.

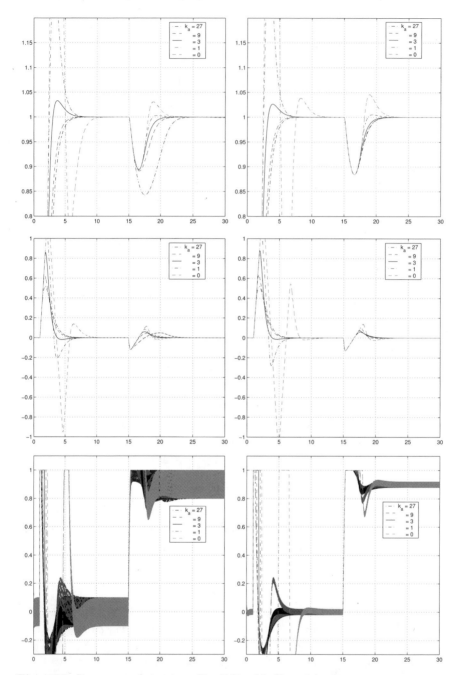

Fig. 5.32. Responses of structure Fig. 5.31 with filtered inputs.
Left: $\Omega_f = 43.5\Omega_1$; right: $\Omega_f = 12.6\Omega_1$;   top: x_1; center: x_2; bottom: u

Exercise
- These are experimental results. Confirm them by calculation.
- Consider the case of PI(aw) control, Sect. 5.3.4, along the same lines.
- Idem for PI(aw)D control, Sect. 5.3.5.

5.3.8 Conclusion

The most common structures for discrete-time controllers with integral action, control signal saturation, and static awf (PI(aw), PI(aw)D, PI(aw)-P cascade) have been investigated, while operating the control loop close to the saturation, and with stationary high-frequency measurement disturbances $\hat{n}$ at the sampling frequency.

Results are presented for the steady state control error offset $\bar{e}$ induced by such $\hat{n}$. They demonstrate the effect of design parameter choices. They are also compared with simulation results, with good agreement.

It is often postulated, at least implicitly, that awf must be "high-gain". But it has been shown in previous chapters that this is not necessarily so, and that "low-gain" awf does not significantly affect the stability properties and the run-up overshoot, if it is higher than a threshold, which depends on the plant transfer function type. This property is useful to reduce the offset, but cannot eliminate it.

It is also demonstrated that any derivative action should be avoided in this context. But if this were necessary for loop performance, then an equivalent PI(aw)-P cascade produces much less offset.

Several proposals for suppressing this offset have been investigated.

One is to redesign the actuator size such that $\hat{u}_{lin}$ stays in the linear actuator range for all steady state operating conditions. If this is not feasible due to physical restrictions, then inserting an additional saturation on the controller output is helpful, but only as long as $\hat{n}$ is small. Otherwise, this will lead to an increased overshoot.

Also, the effect of the high-frequency disturbance may be reduced by digital filtering of measured signals. But this requires a sufficiently large gap between closed-loop bandwidth and the sampling frequency. If this is not available, then an observer may be used instead. But this also has its limits regarding rejection of large load swings.

Finally, the controller structures with a low-pass filter $G_a(s)$ in the awf path from Chapter 2 have not been investigated here. They might perform better. But this remains to be investigated.

To *summarize:* Controllers with static awf are sensitive to such high-frequency measurement disturbances. So, from an overall design point of view, it is strongly recommended to make every effort to reduce as much as possible the level $\hat{n}$ out of the sensor subsystem first, and then use one or a combination of the measures proposed above to handle the residual $\hat{n}$.

Part II

Advanced Techniques

6

Generalized Antiwindup

6.1 Introduction

In the first part of this book, the focus has been on plants of dominant first order, where standard PI(aw) controllers are best suited. The integral action has been considered as a must, and not as an option. Such control systems can be directly implemented, as PI(aw)-controllers are available as standard modules in industrial process control systems.

The concepts have been quite successful so far. They are now extended as follows.

- Plants of dominant higher order n $(n \geq 2)$ are considered, where simple PI(aw) control would perform poorly in the linear range, if stabilizable at all.

- The second characteristic element is that we shall explicitly use state feedback controllers, and consider the integral action not as a must, but as an option. To simplify matters we shall consider all state variables to be measured, *i.e.* no observer is required. Note that such controllers may still be implemented on industrial process control systems by re configuring them into cascaded P controls.

- The third characteristic element is "plant windup": Internal state variables linked to the dominant part of the dynamics run so far off their equilibrium values that they cannot be brought back in the final approach phase in time due to the control saturation. So far, "windup" has been used exclusively for the integral action, or more generally to the state variable within the controller. Now "windup" becomes applicable to all state variables of the loop: it is *generalized*.

- And finally, nonlinear stability now becomes a major issue for practical design. It is more than just a nice property of academic interest.

We shall start with a motivating example (Peter's case), then state the control problem and design the linear range controller. Nonlinear stability analysis is performed next in a systematic way, using both the circle and Popov criteria and the describing function technique. Then a transient analysis is performed for two main cases, also to obtain some insight into the conservativeness of the stability analysis and corresponding hints for design.

The focus then moves to design. As a measure of best performance, the minimum-time (optimal) trajectories are given for the two main cases. Then a selection of typical design techniques is presented and considered in detail.

6.2 A Motivating Example: Peter's Case

In the 1980s, Peter and his colleagues were active in university research aimed at bringing state feedback design techniques closer to application. An important aspect therein was control-variable saturation. They found that the response of linearly designed control loops could be very sensitive to such input constraints. Their findings are given in the following. The presentation is slightly adapted to the framework and notation used here.

6.2.1 The Control Loop

As to the *plant*, they did not consider a specific application (as we did above in the various cases), but rather a more general form

$$G_u(s) = \frac{y}{u} = \frac{1}{(sT+1)^3} \tag{6.1}$$

They also did not consider any non-modeled fast dynamics, captured above by a short delay e^{-sD} in series. So there was no closed-loop bandwidth limitation for their feedback design.

As *controller* they considered an observer-based state feedback with constant coefficients. Here (as mentioned above) the state variables shall be considered as directly measurable without any lag. And G_u shall be inserted in its control canonical form; see Fig. 6.1. The gain in the setpoint path is such that the steady state control error tends to zero for setpoint steps. Also, there is no disturbance: $v(t) = 0 \quad \forall t \geq 0$.

The *controller parameters* are determined using pole assignment to $0 = (s + \Omega_1)^3$ yielding

$$b_u k_1 = (\Omega_1 T)^3 - a_1 ; \quad b_u k_2 = 3(\Omega_1 T)^2 - a_2 ; \quad b_u k_3 = 3(\Omega_1 T) - a_3 \tag{6.2}$$

with the design parameter Ω_1 (or with the 'relative pole shift' factor $\Omega_1 T$ instead).

The controller is implemented in its discrete- time form with sampling time $T_s = 0.01$, and one delay T_s to model the computational delay. And $b_u := 1$.

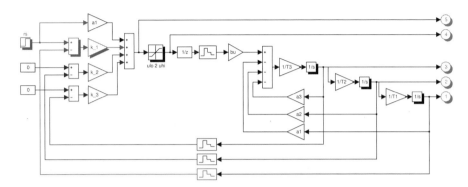

Fig. 6.1. The control system with the third-order plant

The *test sequence* is reduced to starting from equilibrium at $r_0 = 0$ and stepping the setpoint up to $r_1 = 1.0$ at $T_{r_1} = 1.0$. No further load swings v are applied. This produces a steady state control value

$$\bar{u}_1 = a_1 r_1; \quad i.e. \text{ with} \quad a_1 = 1.0 \quad \bar{u}_1 = 1.0 \tag{6.3}$$

The *saturation values* are set to

$$u_{lo} = -\Delta u + \bar{u}_1; \quad u_{hi} = +\Delta u + \bar{u}_1; \quad \text{with} \quad \Delta u = 1.0 \tag{6.4}$$

6.2.2 Transient Responses

Fig. 6.2 shows the simulation results. For $\Omega = 4$ there is no perceptible overshoot on the response of x_1. The actuator working range is well used by $u(t)$. So this would be a near-optimal response from an application point of view. For $\Omega = 8$ there is a considerable overshoot and the settling time increases from ~ 4 s to ~ 7 s. And for $\Omega = 16$ there is a limit cycle with a period of ~ 3 s and an amplitude $\hat{x}_1 \approx 0.08$.

6.2.3 Peter's Analysis

The limit cycle lead Peter to select as his analysis tool the "Describing Function Method" (DFM for short), see e.g. [12], and Appendix A.1.

As the limit cycle period is much larger than the sampling time T_s, we shall use the continuous approximation.

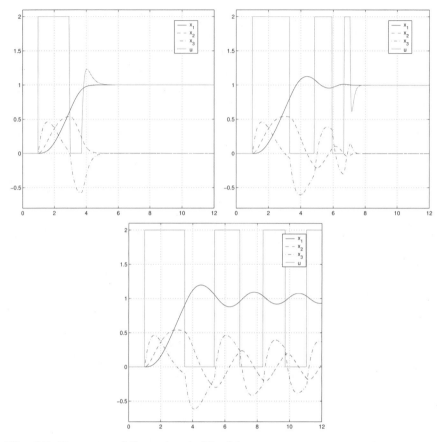

Fig. 6.2. Responses of the system in Fig. 6.1:
(top left) $\Omega = 4$;   (top right) $\Omega = 8$;   (bottom) $\Omega = 16$

Then the *linear subsystem* has the transfer function L

$$L(s) = R(s)G_u(s) \quad \text{with} \quad G_u(s) = \frac{1}{(sT)^3 + a_3(sT)^2 + a_2(sT) + a_1}$$

$$\text{and} \quad R(s) = k_3(sT)^2 + k_2(sT) + k_1 \tag{6.5}$$

The *describing function* N for the saturation element with input amplitude A, unity gain in the linear range and saturation values being symmetrical at $[-a, +a]$ to the final equilibrium $\bar{u}_1$ in the linear range, and where $\bar{u}_1$ is also shifted to zero, is:

$$0 \le A < a \quad N = 1.0$$

$$A \ge a \quad N = \frac{2}{\pi} \left[\arcsin\left(\frac{a}{A}\right) + \frac{a}{A} \sqrt{1 - \left(\frac{a}{A}\right)^2} \right] \tag{6.6}$$

The negative inverse of N is at $(-1, \ j0)$ for $0 \leq A \leq a$, and then for increasing $A > a$, proceeds along the negative real axis to $-\infty$.

From the graphic test performed in Fig. 6.3, limit cycles may appear, if the Nyquist contour $L(j\omega)$ for the linear subsystem crosses over the negative real axis between -1 and $-\infty$.

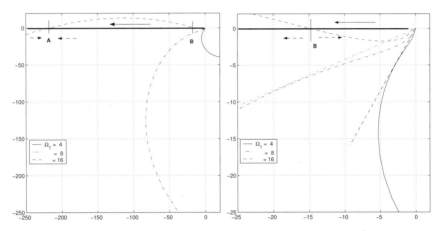

Fig. 6.3. Applying the DFM test to the system in Fig. 6.1 (in the discrete-time version) and using the 'dlinmod' and 'dnyquist' functions of Simulink. The right-hand plot is a zoom-in of the left-hand one

This is only the case with the contour for $\Omega = 16$. There are two intersection points A and B, where A evaluates to a stable limit cycle, and B to an unstable one. Small initial amplitudes (to the right of B) will decay to the linear range and thus to zero. Larger initial amplitudes (to the left of B) will increase and end up in A. And very large amplitudes (to the left of A) will decay to A.

Exercise
- How well does the DFM applied at the intersection point A predict quantitatively the limit cycle obtained from the simulation?
- Check the response around the unstable intersection point B (simulation).

There are no intersections for the cases of $\Omega = 8$ and $\Omega = 4$. But the DFM implicitly assumes that such an intersection exists. Only this allows the harmonic linearisation to be applied, as well as the Nyquist stability test for the linearized system. In other words, the DFM is not able to supply an answer here.

However,

- by inspection of those two Nyquist contours in Fig. 6.3 and the associated transients from Fig. 6.2 (top row), and also
- by carrying over the phase margin concept from linear design by defining a "phase margin sector" from the negative real axis to the dashed tangents on the Nyquist contours out of the origin, see Fig. 6.3 (right side),

Peter and his colleagues *conjectured*

- that a "phase margin sector" of $\sim 45^o$ is associated with an acceptably well-damped closed loop response (case of $\Omega = 8$), and

- that a "phase margin sector" of $\sim 60^o$ is associated with no perceptible overshoot and near minimum settling time (case of $\Omega = 4$).

This led them further to develop a "systematic design procedure" [41]:

> choose the closed-loop bandwidth Ω such that a "phase margin sector" of at least 45° results.

They reported testing it successfully on several other plants of the Hurwitz class as well.

The results of this conjecture seem to be good, but regrettably its theoretical basis is not very solid. A better founded approach (see also [40]) will be shown below. It will also include explicitly both plants and controllers being non-Hurwitz, and show in more detail the effect of the setpoint step size and of the saturation width (*i.e.* the actuator working range).

6.3 Problem Statement

This is to specify the control loop and its subsystems, as a basis for the subsequent stability analysis. Numerical values will be introduced later for the transient responses.

6.3.1 The Plant

The model of *dominant first order* has been discussed in Chapter 2. It will be listed here for completeness.

$$G_1(s) = e^{-sD}\frac{b}{sT + a_1} \qquad \text{with} \qquad -1.0 \le a_1 \le +1.0 \qquad (6.7)$$

The focus will be on the model of *dominant second order*

$$G_2(s) = e^{-sD}\frac{b}{(sT)^2 + a_2(sT) + a_1}$$
$$\text{with} \qquad -1.0 \le a_1 \le +1.0$$
$$\text{and} \qquad -1.0 \le a_2 \le +1.0 \qquad (6.8)$$

and on the model of *dominant third order*.

$$G_3(s) = e^{-sD} \frac{b}{(sT)^3 + a_3 (sT)^2 + a_2 (sT) + a_1}$$

$$\text{with} \quad -1.0 \leq a_1 \leq +1.0$$

$$-1.0 \leq a_2 \leq +1.0$$

$$\text{and} \quad -1.0 \leq a_3 \leq +1.0 \tag{6.9}$$

The state space representation of the "dominant dynamics" shall be in control canonical form.

Remarks

- Let $b := 1$. This implies an appropriate scaling.

- There is a common time scale T for the "dominant dynamics" part, a small delay D to cover the non-modelled fast dynamics, and the bounds on the a_i are to indicate, that the a_i may be zero or negative as well, but this within reasonable bounds.[1]

- Note that there are no zeros in the transfer functions $G_i(s)$. The focus is on the basic state feedback, without feed-through, and without a more general output feedback (Such cases have been investigated in e.g.[56]).

- A persistent load v shall be applied as above, but only at the plant control input, *i.e.* parallel to u, similar to Chapter 2.

- To provide some applications background, the model in Eq. 6.7 may be seen as speed control of a rigid body, and where the control input (the driving force) can be applied without significant lag. The second-order model in Eq. 6.8 can be associated to position control of a rigid body with "normal spring", "zero spring", or "negative spring" force feedback, plus a feedback force proportional to speed, which provides either "positive (dissipative)" damping, or "zero damping", or "negative damping". And in the third-order model in Eq. 6.9, the force or torque buildup is no longer instantaneous. A typical example would be an elevator position control with limits on the acceleration buildup ("jerk limits").

- Plants with dominant order larger than three shall not be considered here. This is mainly for practical reasons, as such cases tend to be difficult to control in the linear range, and a work-around (meaning a redesign of the plant) is strongly indicated anyway.

---

[1] Otherwise adapt T and b suitably.

6.3.2 Controller structures

Two versions shall be considered, the P$^+$ and the I(aw)-P$^+$ types. Both versions are implemented in their discrete time forms, with the sampling time T_s set short enough to allow using their continuous forms for analysis (as above).

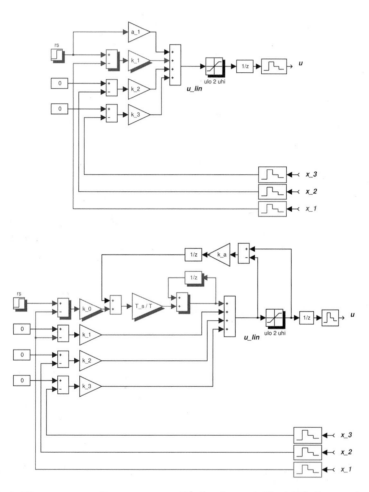

Fig. 6.4. The two controller structures, P$^+$ (top) and I(aw)-P$^+$ (bottom), for the model of dominant third order

The *P$^+$ controller* in Fig. 6.4 (top) consists essentially of a P controller for $y = x_1$ with gain k_1. In addition, there is a feedforward path from r to u_{lin} with weight a_1, in order to drive the steady state control error for setpoint steps to zero. Then there are state feedbacks $-k_2 x_2$ (and $-k_3 x_3$ for the case in Eq. 6.9). These motivate changing the notation from P- to P$^+$-control.

And the *I(aw)-P$^+$ controller* Fig. 6.4 (bottom) is basically an I(aw)-controller producing u_0 with additional state feedback u_{123}:

$$u_{123} = -(k_1 x_1 + k_2 x_2 + k_3 x_3) \quad \text{and} \quad u_{lin} = u_0 + u_{123} \tag{6.10}$$

This may also be seen as a special "two degrees of freedom" implementation of a PI(aw) controller, where the setpoint r is applied only to the integral action, and with additional state feedbacks $u_{23} = -(k_2 x_2 + k_3 x_3)$. The motivation for this is that, with this particular structure for R in Eq. 6.11, the overshoot on setpoint step responses (G_r) in the linear range produced by "one degree of freedom" PI controllers can be suppressed:

Let $\quad R_r = \dfrac{k_0}{sT} \quad$ and $\quad R_y = \dfrac{k_0}{sT} + k_1 + k_2(sT) + k_3(sT)^2$

Then

$$G_r(s) = \frac{R_r G_u}{1 + R_y G_u}$$

$$= \frac{k_0}{(sT)^4 + (a_3 + k_3)(sT)^3 + (a_2 + k_2)(sT)^2 + (a_1 + k_1)(sT) + k_0}$$

$$\rightarrow \frac{1}{(\frac{s}{\Omega_1} + 1)^4} \tag{6.11}$$

Thus G_r has unity gain and has no zero(s), as would be the case if $R_r := R_y$.

The second motivation is that $u(t)$ is not subjected to a large step by the proportional path (from $+k_1 r(t)$), and thus tends to saturate less.

Similarly, for the P$^+$ version

let $\quad R_r = k_1 + a_1 \quad$ and $\quad R_y = +k_1 + k_2(sT) + k_3(sT)^2$

then

$$G_r(s) = \frac{R_r G_u}{1 + R_y G_u}$$

$$= \frac{k_1 + a_1}{(sT)^3 + (a_3 + k_3)(sT)^2 + (a_2 + k_2)sT + (a_1 + k_1)}$$

$$\rightarrow \frac{1}{(\frac{s}{\Omega_1} + 1)^3} \tag{6.12}$$

Again, G_r has no zero(s), and it has unity gain.

6.3.3 Controller Parameters

These are determined as above by pole assignment for the dominant dynamics:

$$0 = (s + \Omega_1)^{n_p + n_c} \tag{6.13}$$

where n_p is the order of the dominant plant dynamics and n_c the order of the controller ($n_c = 0$ for P$^+$, and $n_c = 1$ for I(aw)-P$^+$). The damping ratio 2ζ in the linear range is set to $2\zeta = 2$ to avoid overshoot also in the nonlinear range (from previous experience; see Sect. 2.1).

The equations for the controller parameters are then particularly simple:

n_p	n_c	k_0	k_1	k_2	k_3
1	0	0	$(\Omega_1 T) - a_1$	0	0
	1	$(\Omega_1 T)^2$	$2(\Omega_1 T) - a_1$	0	0
2	0	0	$(\Omega_1 T)^2 - a_1$	$2(\Omega_1 T) - a_2$	0
	1	$(\Omega_1 T)^3$	$3(\Omega_1 T)^2 - a_1$	$3(\Omega_1 T) - a_2$	0
3	0	0	$(\Omega_1 T)^3 - a_1$	$3(\Omega_1 T)^2 - a_2$	$3(\Omega_1 T) - a_3$
	1	$(\Omega_1 T)^4$	$4(\Omega_1 T)^3 - a_1$	$6(\Omega_1 T)^2 - a_2$	$4(\Omega_1 T) - a_3$

where the closed-loop relative bandwidth $\Omega_1 T$ must be selected low enough for the delay D and the sampling time T_s not to interfere.

6.3.4 Test Sequence

The test sequence of Chapter 2 is slightly extended, by inserting the second item:

- standstill with $r = 0$ and $v = 0$,
- step-up of r to 0.99, while u is forced to 0, such that the awf loop stabilizes to $\Delta u_{lin_{max}}$,
- release of u and run-up to equilibration,
- a small setpoint step 0.01 up to $r = 1.0$ for near-linear behavior,
- a full load step v up to +0.90,
- and a full load reversal to $v = -0.90$.

6.3.5 Saturation Limits

In order to provide sufficient maneuverability around $r = 1$ (and centered at $v = 0$), they are set to

$$u_{lo} = -\Delta u + \bar{u}_1; \quad u_{hi} = +\Delta u + \bar{u}_1; \quad \text{with} \quad \Delta u = 1.0 \tag{6.14}$$

where $\bar{u}_1 = a_1 r_1$

6.3.6 Antiwindup Feedback

For the I(aw)-P$^+$ version, the standard static awf structure of Chapter 2 shall be used. Then

$$\text{deadbeat response will be produced by} \quad k_a := \frac{T}{T_s},$$
$$\text{and the compensating value is at} \quad k_a := \Omega_1 T.$$

6.4 Nonlinear Stability Analysis

The general result in Chapter 2 has been derived in the context of a dominant first order system with PI(aw) control. It has already been mentioned there that it is not restricted to this class, but may be applied to systems of arbitrary order.

So, the following is essentially a straightforward application of this general result. However, some refinements will be made here, which are not strictly necessary, but provide more insight.

Based on the experience from Chapter 2, the focus will be on the run-up phase, following the large setpoint step r_1 from 0 to 0.99): the largest "saturation overrun" $\Delta u_{lin_{max}} = u_{lin_{max}} - u_{hi}$ is to be expected there.

6.4.1 The P$^+$ Case

The control loop is visualized in Fig. 6.5

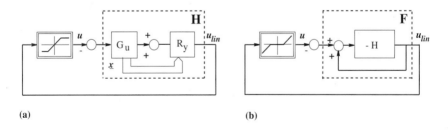

(a)   (b)

Fig. 6.5. The control loop with P$^+$-control and no awf:
(a) with the saturation nonlinearity;   (b) redrawn with the deadspan element

The Linear Subsystem

The general result was

$$F + 1 = \frac{1}{1 + R_y G_u} \tag{6.15}$$

that is

$$F + 1 = \frac{(sT)^{n_p} + a_{n_p}(sT)^{n_p-1} + \cdots + a_2(sT) + a_1}{(sT)^{n_p} + (a_{n_p} + k_{n_p})(sT)^{n_p-1} + \cdots + (a_2 + k_2)(sT) + (a_1 + k_1)}$$

where

$$a_1 + k_1 := (\Omega_1 T)^{n_p}$$

finally

$$F + 1 = \frac{(sT)^{n_p} + a_{n_p}(sT)^{n_p-1} + \cdots + a_2(sT) + a_1}{(s + \Omega_1)^{n_p}}; \quad \text{or} \tag{6.16}$$

$$= \frac{1}{a_1 + k_1} \frac{(sT)^{n_p} + a_{n_p}(sT)^{n_p-1} + \cdots + a_2(sT) + a_1}{\left(\frac{s}{\Omega_1} + 1\right)^{n_p}} \tag{6.17}$$

The numerator polynomial of $F+1$ is the denominator polynomial of the plant (without feedback control), whereas the denominator polynomial of $F + 1$ is the closed-loop polynomial (with feedback control). This is easy to memorize.

The Nonlinear Subsystem

The general result was

$$-\frac{1}{b_{max}} = -1 - \frac{\Delta u}{\Delta u_{lin_{max}} - \Delta u} = -1 - \Delta_{max} \tag{6.18}$$

which defines Δ_{max}. A conservative estimate is derived next.

Before the setpoint step is applied (*i.e.* for $t < 0$) is $r_0 = 0$. And all state variables $\underline{x} = [x_1, \ x_2, \cdots x_{n_p}]$ are zero from the equilibrium condition there.

Consider now the first time increment (at $t = +0$) after the setpoint step r_1 has been applied. The contribution to the control signal $u_{lin}(+0)$ from the plant feedback $-\underline{k}^T \underline{x}$ through R_y is zero, as all state variables $\underline{x}$ have not moved from zero.
And the contribution from the setpoint path R_r to the control variable is; see Fig. 6.4 (top):

$$\Delta u_{lin}(+0) = r_1 (a_1 + k_1) \tag{6.19}$$

The next step is to set as an estimate for the run-up transient

$$\Delta u^e_{lin_{max}} := \Delta u_{lin}(+0) = r_1 (a_1 + k_1) \tag{6.20}$$

This seems reasonable from past experience with the dominant first-order system, but should be confirmed by recording $\Delta u_{lin}(t)$ in simulations. Then

$$\Delta_{max} = \frac{\Delta u}{\Delta u_{lin_{max}} - \Delta u}$$

$$= \frac{1}{\Delta u_{lin_{max}}} \frac{\Delta u}{1 - \frac{\Delta u}{\Delta u_{lin_{max}}}} = \frac{1}{a_1 + k_1} \frac{\Delta u}{r_1} \frac{1}{1 - \frac{\Delta u}{r_1(a_1 + k_1)}} \quad (6.21)$$

where the last of the three factors is > 1. It tends to unity for $r_1(k_1 + a_1) \gg \Delta u$, which is the case for high-performance control ($\Omega_1 T \gg 1$).
And for the first factor:

$$a_1 + k_1 = (\Omega_1 T)^{n_p} \quad (6.22)$$

This leads to the conservative approximation, which is also easy to memorize

$$\Delta_{max}^a = \frac{1}{a_1 + k_1} \frac{\Delta u}{r_1} = \frac{1}{(\Omega_1 T)^{n_p}} \frac{\Delta u}{r_1} < \Delta_{max} \quad (6.23)$$

Stability Test

By inspection, Eq. 6.16 and 6.23 have the same factor $(k_1 + a_1) = (\Omega_1 T)^{n_p}$. If now k_1 increases, because $\Omega_1 T$ is increased for better linear control performance, then the plot of $F + 1$ and the position of the straight line at $-1 - \Delta_{max}$ will shrink equally.

To allow a more direct comparison of variations of design parameter $\Omega_1 T_1$, this shrinking can be *compensated* by

multiplying both $\quad F + 1 \quad$ and $\quad \Delta_{max}^a \quad$ by $\quad k_1 + a_1 := (\Omega_1 T)^{n_p} \quad (6.24)$

This may also be seen as *scaling* the complex plane accordingly, hence the index *sc*.

Then the graphic test is to be performed with

$$(F + 1)_{sc} = \frac{(sT)^{n_p} + a_{n_p}(sT)^{n_p - 1} + \cdots + a_2(sT) + a_1}{\left(\frac{s}{\Omega_1} + 1\right)^{n_p}}$$

and $\quad (\Delta_{max} + 1)_{sc}^a = \frac{\Delta u}{r_1} \quad (6.25)$

That is, the position of the straight line in the stability test is now invariant of the design parameter $\Omega_1 T$, and given by a very simple expression. $F(j\omega) + 1$ is discussed next.

The Shape of the Nyquist Contour

The shape of $F + 1$ is seen as being generated by

- a high-pass element with the transfer function $P_H(s)$ consisting of the numerator of Eq. 6.25

$$P_H(s) = (sT)^{n_p} + a_{n_p}(sT)^{n_p - 1} + \cdots + a_2(sT) + a_1 = p(s) \quad (6.26)$$

- connected in series to a low-pass filter $P_L(s)$ with unity gain and relative bandwidth $\Omega_1 T$.

$$P_L(s) = \frac{1}{\left(\frac{s}{\Omega_1} + 1\right)^{n_p}} = \frac{(\Omega_1)^{n_p}}{(s + \Omega_1)^{n_p}} = \frac{(\Omega_1)^{n_p}}{P(s)} \qquad (6.27)$$

- where $p(s)$ is the characteristic polynomial of the plant
- and $P(s)$ the closed-loop characteristic polynomial.

In other words, the shape of $P_H(s)$ in the complex plane is determined by the plant $(p(s))$ only and cannot be modified by changing the control feedback $P(s)$.

It can be plotted easily in the complex plane from

$$P_H(j\omega) = a_1 + ja_2(\omega T) - a_3(\omega T)^2 - j(\omega T)^3 \qquad (6.28)$$

as shown in Fig. 6.6 (left) for a third-order system with all $a_i > 0$.

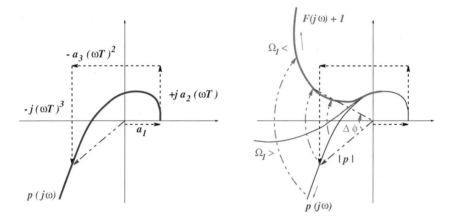

Fig. 6.6. Construction of the Nyquist contour for the high-pass element $P_H(s)$ (left); and of the Nyquist contour $F(j\omega) + 1$ (right) at the low-frequency end from $p(j\omega)$

Note that this is also the Nyquist contour of $F + 1$ for $\Omega_1 \to \infty$, *i.e.* the asymptote of the shape for a loop with infinitely high performance.

Now the shape of the low-pass element $P_L(s)$ is discussed. It is determined by the design of the closed loop only, *i.e.* by the choice of $P(s)$. From Chapter 2, the nonlinear stability properties are determined by the shape of the Nyquist contour at the low frequency end $0 \leq \omega \ll \Omega_1$.

In this *low-frequency* region $0 \leq \omega \ll \Omega_1$
and using the polar form for $P_L(j\omega)$

$$|P_L| \approx 1.0 \quad \text{and} \quad \Delta\varphi = -n_p \arctan\left(\frac{\omega}{\Omega_1}\right) \approx -n_p \frac{\omega}{\Omega_1} \tag{6.29}$$

i.e.
  - P_L will not change the length of $|P_H(j\omega)|$ significantly,
  - but will generate a progressive phase shift $\Delta\varphi$.
This will reshape the contour of P_H from Fig. 6.6 (left) as illustrated in Fig. 6.6 (right), drawn for a high closed-loop bandwidth (marked $\Omega_1 >$), and for a low bandwidth ($\Omega_1 <$).
In other words, decreasing the closed-loop bandwidth will improve the non-linear stability properties, as the Nyquist contour is pushed to the right. And thus it makes room for moving the straight line further to the right, *i.e.* for smaller Δ^a_{max}. This leads to
  - either a larger allowable size of r_1,
  - or smaller allowable size of Δu, *i.e.* lower u_{hi} and thus smaller actuators.
It also nicely explains the experimental findings in Fig. 6.2 for low Ω_1's, where the DFM is not applicable.

For the *high-frequency* region $\omega \gg \Omega_1$
the Nyquist contour converges to

$$\lim_{\omega \to \infty} (F(j\omega) + 1)_{sc} \rightarrow \lim_{\omega \to \infty} \left[\frac{(sT)^{n_p} + \dots}{\left(\frac{s}{\Omega_1}\right)^{n_p} + \dots}\right] = (\Omega_1 T)^{n_p} \tag{6.30}$$

i.e. the contour ends on the positive real axis at $(\Omega_1 T)^{n_p}$. So the end point moves out to the right for increasing relative bandwidth, and the size of the contour increases accordingly. But note that the result in Eq. 6.30 neglects any small delay D in the plant and finite sampling time T_s from the controller. The actual shape there can best be determined by using the functions 'dlinmod' and 'dnyquist' on the full Simulink model.

Finally, at the *bandwidth frequency* $\omega = \Omega_1$
the low-pass element $P_L(s)$ furnishes from the end point at $\omega \to \infty$ backwards to $\omega = \Omega_1$ a positive phase shift of

$$\Delta\varphi := +n_p 45^\circ \tag{6.31}$$

and a gain of

$$|L_P(j\omega)|_{\omega=\Omega_1} = \left(\sqrt{2}/2\right)^{n_p} \tag{6.32}$$

that is

$$|F(j\omega) + 1|_{\omega=\Omega_1} = |p(j\omega)|_{\omega=\Omega_1} \left(\sqrt{2}/2\right)^{n_p} \tag{6.33}$$

where $|p|$ is the length of the vector due to the high pass $P_H(s)$ as constructed in Fig. 6.6 (left).

The three elements

- Equations 6.26 or 6.28 for the low-frequency range,
- Eq. 6.30 for the high-frequency end, and
- Eq. 6.31, and 6.33 for the point at $\omega = \Omega_1$ in the intermediate frequency range

allow one to sketch the approximate shape of $F + 1$, and get a first idea on how critical a specific case is concerning nonlinear stability.

The Describing Function Test

On the other hand, the circle or sector test cannot show asymptotic stability if there is an intersection of the straight line and the Nyquist contour. This will appear for large Ω values and small Δ^a_{max}; but then limit cycles may appear, as the experiments in Sect. 6.2 have shown.

This suggests using the DFM as a supplementary tool for nonlinear stability analysis as well.

Consider the control loop in its redrawn form, Fig. 6.5 (b). For small enough initial conditions its response is linear and decays exponentially to the steady state. All variables shall be zero there, by adding offsets of appropriate size.

Then, for the *linear part*:

$$F = -\frac{H}{1+H} \tag{6.34}$$

where $H = G_u R_y$ is a low pass, and therefore F is a low pass as well. Furthermore, F is Hurwitz by design of the linear loop, even for unstable plants.

It is interesting to note the relation between F used here and the transfer function E used by Peter and his colleagues in the "phase margin sector" criterion, (Sect. 6.2). From there for $n_p = 3$:

$$E := R(s)G_u(s) \quad = \frac{k_3(sT)^2 + k_2(sT) + k_1}{(sT)^3 + a_3(sT)^2 + a_2(sT) + a_1} \tag{6.35}$$

Then introduce

$$E := -1 + E'$$

i.e.

$$E' = E + 1$$

and

$$E' = \frac{\left[(sT)^3 + a_3(sT)^2 + a_2(sT) + a_1\right] + \left[k_3(sT)^2 + k_2(sT) + k_1\right]}{(sT)^3 + a_3(sT)^2 + a_2(sT) + a_1}$$

$$= \frac{(sT)^3 + (a_3 + k_3)(sT)^2 + (a_2 + k_2)(sT) + (a_1 + k_1)}{(sT)^3 + a_3(sT)^2 + a_2(sT) + a_1}$$

$$= \frac{(s + \Omega_1)^3}{s^3 + \frac{a_3}{T}s^2 + \frac{a_2}{T^2}s + \frac{a_1}{T^3}} = \frac{(s + \Omega_1)^3}{s^3 + a_3's^2 + a_2's + a_1'} = \frac{P(s)}{p(s)}$$

finally

$$\frac{1}{E'} = \frac{1}{1 + E} \overset{!}{=} F + 1 \tag{6.36}$$

i.e.

$$\arg\{F(j\omega) + 1\} = -\arg\{E(j\omega) + 1\} \text{ and } |F(j\omega) + 1| = \left|\frac{1}{E(j\omega) + 1}\right| \tag{6.37}$$

The *nonlinear part* is the unity gain deadspan element of width $\pm a$, symmetrical to zero ($\bar{u} := 0$), and input amplitude $A := \widehat{\Delta u}_{lin}$. Then, the describing function is

$$0 \le A < a \quad N = 0$$

$$A \ge a \quad N = \left[1 - \frac{2}{\pi}\arcsin\left(\frac{a}{A}\right) - \frac{2}{\pi}\frac{a}{A}\sqrt{1 - \left(\frac{a}{A}\right)^2}\right] \tag{6.38}$$

and the plot of the negative inverse of N
- starts at $(-\infty, +j0)$ for $0 \le A \le a$
- runs to the right along the negative real axis
- and ends at $(-1.0, +j0)$ for $A/a \to \infty$.

In the *stability test*, the contour of $F + 1$ has been used for the sector test. To go on using this contour here, $-[(1/N) + 1]$ must be plotted. It will start at $-\infty$ as well, run along the negative real axis and end up at the origin for $A/a \to \infty$.

So, if $E(j\omega) + 1$ crosses over the negative real axis, which indicates limit cycles, then $F(j\omega) + 1$ does as well, and at the same ω-values.
Or also, if the Nyquist contour of $F + 1$ crosses over the negative real axis, then the DFM conditions are met. Note that the scaling on the negative real axis is different for both tests.
Fig. 6.7 shows two such cases. Note that intersection point **B** is also indicating the Δ_{max} for the off-axis circle criterion.

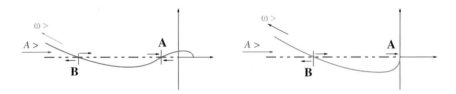

Fig. 6.7. Two cases from the describing function test:
(left) with two intersections, **B** indicating an unstable limit cycle, and **A** indicating a stable limit cycle with large bounded amplitudes A/a;
(right) with the stable limit cycle moving to infinite A/a

To *summarize*: the Nyquist contour plot can be used for both tests at the same time, and thus provide a sound answer for *both* situations, the "fast response without overshoot" one, which is usually specified and is the relevant design target, and the "limit cycle" one, which is more of theoretical interest, because it must be avoided in applications. But clearly, if $F(j\omega) + 1$ evolves across or even near the negative real axis, then this indicates a "demanding" case for practical design.

6.4.2 Stability Charts

The idea of drawing such stability charts is to visualize the effect of different dominant plant dynamics ($p(s)$) on the Nyquist contour of $F + 1$ in a systematic way, and thus on the nonlinear stability properties to be expected.

p(s) of first order

Variation of the single parameter a_1 will produce a one-dimensional sequence of corresponding plots (along the horizontal). Fig. 6.8 shows the charts for $F(j\omega + 1)_{sc}$ for three values of a_1.

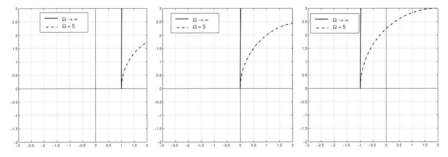

Fig. 6.8. Nyquist contours for the dominant first-order plant:
(left) $a_1 = +1$; (center) $a_1 = 0$; (right) $a_1 = -1$

Therefore, the stability properties are good for $a_1 \geq 0$, because $\Delta_{max_{sc}} \to 0$, *i.e.* the radius of attraction tends to infinity.
But the radius of attraction is bounded to

for $a_1 < 0$: $\Delta_{max_{sc}} = |a_1|$ independent of the choice of Ω ;

$$i.e. \text{ with }\quad \Delta_{max_{sc}} = \frac{\Delta \bar{u}}{r_{1_{max}}} ; \quad \to \quad r_{1_{max}} = \frac{\Delta \bar{u}}{|a_1|} \qquad (6.39)$$

which is the same as what is obtained by considering the equilibrium condition of the loop.

Exercise

Show this in more detail.

p(s) of second order

Variation of the two parameters a_1, a_2 will produce a two-dimensional chart, as shown in Fig. 6.9.

Discussion

- Only for the case depicted in the upper left corner (with $a_1 > 0, a_2 > 0$) can global stability be shown with the on-axis circle criterion, if Ω is selected small, such that $\Re(F+1) > 0$ $\forall$ ω.
  In other words, it has the best stability properties. Or, it is the least sensitive to saturations.

- For the case $a_1 > 0, a_2 = 0$, *i.e.* an oscillator with zero damping, the condition $\Re(F+1) > 0$ $\forall$ ω may also be attained, but then Ω must be selected even smaller.
  For the higher Ω value shown in Fig. 6.9 (left column, center row), the off-axis circle criterion is needed to show global stability.

- For the case $a_2 > 0, a_1 = 0$, Fig. 6.9 (top row, center column), *i.e.* an open integrator in series to a first-order lag, the off-axis circle criterion will show stability in the very large only.

- And for the case $a_2 = 0, a_1 = 0$, Fig. 6.9 (center row, center column), *i.e.* the two-open-integrator chain, the radius of attraction is restricted even more: the off-axis circle criterion will show stability in the large, but not in the very large.

- For the cases with $a_1 < 0$, $a_2 \geq 0$ (top and center row, right column), the radius of attraction is restricted to $|a_1|$, as with the case of first order p.

- Finally, the case with negative damping $a_2 < 0$ (bottom row, right column) has an even smaller radius of attraction.
  Or, it is the most sensitive to saturations.

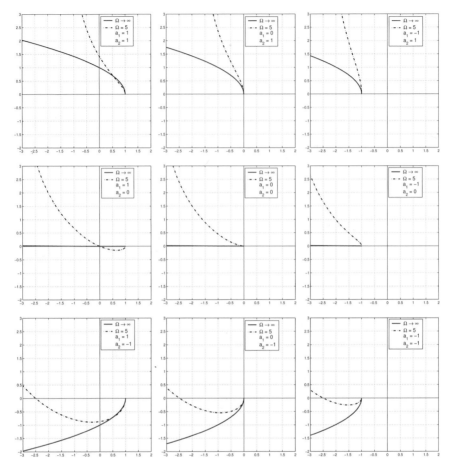

Fig. 6.9. Nyquist contours for the dominant second-order plant.
Rows: (top) $a_2 = +1$,   (center) $a_2 = 0$,   (bottom) $a_2 = -1$;
columns: (left) $a_1 = +1$,   (center) $a_1 = 0$,   (right) $a_1 = -1$

p(s) of third order

Variation of the three parameters a_1, a_2, a_3 will produce a 3-D chart.
A selection of $p(s)$ instantiations with $a_i \geq 0 \; \forall \, i$ is shown in Fig. 6.10, arranged
as follows:

	$p(s)$	
top left	$(s+1)^3$	1
top center	$(s+1)^2$	s
top right	$(s+1)$	s^2
bottom center	1	s^3

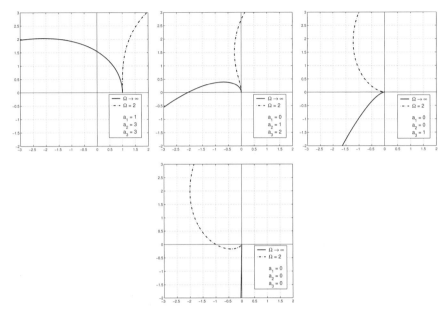

Fig. 6.10. A selection of Nyquist contours for the dominant third order plant; see legends for coefficients of $p(s)$ and Ω

Remarks

- Again for the case $a_1 = 1, a_2 = 3, a_3 = 3$ (top, left), *i.e.* three first-order lags in series (this is Peter's case), global stability can be shown with the on-axis circle criterion, if Ω is selected sufficiently low.
- If $a_1 = 0, a_2 = 1, a_3 = 2$ (top, center), *i.e.* one open integrator in $p(s)$, the situation is similar to that in Fig. 6.9 (top, center), and one can show stability in the very large using the off-axis circle criterion.
- If $a_1 = 0, a_2 = 0, a_3 = 1$ (top, right), *i.e.* a double integrator chain and one first-order lag in series, then the Nyquist contour ejects tangentially along the negative real axis from the origin. Thus, stability properties are similar to those of the double integrator chain in Fig. 6.9 (center, center).
- And for the triple integrator chain, the Nyquist contour ejects along the negative imaginary axis. Thus, there will always be an intersection of the Nyquist contour with the negative real axis at a finite distance from the origin, yielding a finite $\Delta_{max_{sc}}$, and thus a finite radius of attraction. By choosing Ω smaller, this distance $\Delta_{max_{sc}}$ can be reduced. But it will be zero only for $\Omega \to 0$, which is not a useful solution for design purposes.

To *summarize*: the stability properties of the control loop with input constraints (saturation) are dominated by the number N of open integrators in the plant. If this number is zero, then stability properties are very good. They are still good for $N = 1$. But they are significantly reduced for $N = 2$, and even more so for $N > 2$.

In other words, it is easy to obtain good performance for plants of $N = 1$ with input saturation, but it gets increasingly difficult for $N \geq 2$. And the situation quickly deteriorates with one (and then multiple) unstable poles.

Finally, what Peter and his colleagues investigated in their case (Sect. 6.2) turns out to be the least difficult situation. This also holds for the benchmark plant introduced by Rundqwist [42], which is of second order with $a_1 > 0$ and $a_2 > 0$, and which has been used by many other authors to illustrate their proposals for awf.

6.4.3 The I(aw)-P$^+$ Case

The Nonlinear Subsystem

The awf of Fig. 6.4 shall be used. Then, along the upper constraint

$$\Delta u_{lin_{max}} - \Delta u_{hi} = e_{a_{max}} \tag{6.40}$$

and for steady state in the awf loop while r_1 has been stepped up, but u is still blocked at zero for standstill:

$$r_1 k_0 := \bar{e}_a k_a := e_{a_{max}} k_a \tag{6.41}$$

That is, for the position Δ_{max} of the straight line on the negative real axis:

$$\Delta_{max} = \frac{\Delta u_{hi}}{\Delta u_{lin_{max}} - \Delta u_{hi}} = \frac{\Delta u_{hi}}{e_{a_{max}}} = \frac{\Delta u_{hi}}{r_1} \frac{k_a}{k_0} \tag{6.42}$$

The Linear Subsystem

The general result of Chapter 2 yields, with the closed-loop bandwidth Ω_I, and where $k_0 := \Omega_I^{n_p+1}$:

$$F + 1 = \frac{p(s)}{(s + \Omega_I)^{n_p}} \frac{s + k_a}{s + \Omega_I} = \frac{k_a}{k_0} \frac{p(s)}{(\frac{s}{\Omega_I} + 1)^{n_p}} \frac{\frac{s}{k_a} + 1}{\frac{s}{\Omega_I} + 1} \tag{6.43}$$

The Graphic Stability Test

Again Δ_{max} and $F + 1$ contain the common factor (k_a/k_0), which may be eliminated in both, and produces the "scaled" plot:

$$\Delta_{max_{sc}} = \frac{\Delta u_{hi}}{r_1} \quad \text{and} \quad (F + 1)_{sc} = \frac{p(s)}{(\frac{s}{\Omega_I} + 1)^{n_p}} \frac{\frac{s}{k_a} + 1}{\frac{s}{\Omega_I} + 1} \tag{6.44}$$

Thus, the position of the straight line
   - again does not depend on the closed-loop bandwidth design Ω_1,
   - and also is the same as for the P$^+$ cases.

Compensating-gain awf: $k_a = \Omega_1 T$

For this special case, the numerator and denominator of the second factor in Eq. 6.44 cancel, and

$$(F+1)_{sc} = \frac{p(s)}{(\frac{s}{\Omega_I}+1)^{n_p}} \tag{6.45}$$

which is the same result as for the P$^+$-controller. So one expects the same stability properties and the same transient response (see below).

High-gain awf: $\quad k_a = \frac{T_i}{T_s} \gg \Omega_I T$

From Eq. 6.44, with

$$k_a = \gamma \Omega_I \quad \text{with } 0 \le \gamma \le \frac{1}{T_s \Omega_I}, \quad \text{and } T_s \Omega_I \ll 1$$

$$\lim_{\gamma \to \infty} (F+1)_{sc} \;=\; \frac{p(s)}{(\frac{s}{\Omega_I}+1)^{n_p}} \cdot \frac{1}{(\frac{s}{\Omega_I}+1)} \;=\; \frac{p(s)}{(\frac{s}{\Omega_I}+1)^{n_p+1}} \tag{6.46}$$

For the *low-frequency* region $\quad 0 \le \omega \ll \Omega_I \quad$ again

$$|F+1| = |p(j\omega)| \quad \text{and} \quad \arg(F+1) = -(n_p+1)\arctan\left(\frac{\omega}{\Omega_I}\right) \tag{6.47}$$

that is, the phase shift into the favorable direction for the stability test is larger than for the P$^+$ controller. In other words, the bandwidth Ω_I with the I(aw)-P$^+$-controller can be increased (see Fig. 6.11 (right)).

For the *high-frequency* end $\quad \omega \gg \Omega_I$

$$|F+1| \to 0 \quad \text{and} \quad \arg(F+1) = -90^o \tag{6.48}$$

Keep in mind that this neglects the effect of the finite (small) T_s and any small delay D, and extrapolates the dominant dynamics to $\omega \to \infty$.

And at the *bandwidth frequency* $\quad \omega = \Omega_I$

$$|F+1| = |p(j\omega)|\left(\sqrt{2}/2\right)^{n_p+1} \quad \text{and} \quad \arg(F+1) = -(n_p+1)\,45^o \tag{6.49}$$

This again enables you to sketch the shape of the Nyquist contour by deformation of the shape for the P$^+$ controller case, and thereby see the consequences on the stability properties.

Fig. 6.11 (left) illustrates this deformation for Peter's case, Sect. 6.2
- $p(s) = (s+1)^3$,
- $\Omega_P = \Omega_I = 4.0$,
- discrete-time controllers from Fig. 6.4 with $T_s = 0.01$,
- and deadbeat awf gain $k_a = \gamma \Omega_I$ with $\gamma := 1/(\Omega_I T_s)$.

And in Fig. 6.11 (right), $\Omega_P = 4.0$ for the P^+-controller, whereas Ω_I has been manually adjusted ($\to 5.5$), such that the Nyquist contours for both controller types are nearly the same in the relevant low-frequency range.

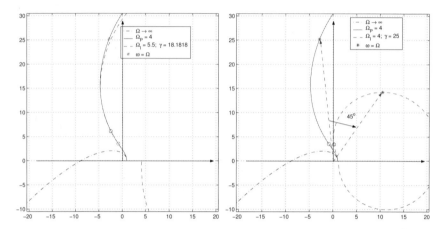

Fig. 6.11. Deformation of the Nyquist contour for the plant $p(s) = (s+1)^3$: (left) $\Omega_P = 4.0$, $\Omega_I = 4.0$, and deadbeat awf gain;   (right) $\Omega_P = 4.0$, $\Omega_I = 5.5$, such that the Nyquist contours are approximately the same

6.4.4 Another Approach to the Graphic Stability Test

Consider the graphic stability test in its *scaled* version in Eq. 6.25. There, variations of the closed-loop bandwidth Ω affect the shape of the Nyquist contour $(F+1)_{sc}$ quite strongly, whereas the position of the straight line $(\Delta_{max})^a_{sc}$ is not affected.

So if variations of Δu or r_1 are considered, and if stability is to be conserved, then Ω has to be adjusted by trial and error. This type of design problem may be solved more conveniently by choosing a different approach to scaling.

The **P^+ case** is considered first.

For the *linear subsystem* in Eq. 6.16, divide the numerator of $F+1$ by $a_1 + k_1 = (\Omega_1 T)^{n_p}$:

$$F+1 = \frac{\left(\frac{s}{\Omega}\right)^{n_p} + \frac{a_{n_p}}{\Omega_1 T}\left(\frac{s}{\Omega}\right)^{n_p-1} + \cdots + \frac{a_2}{(\Omega_1 T)^{n_p-1}}\left(\frac{s}{\Omega}\right) + \frac{a_1}{(\Omega_1 T)^{n_p}}}{\left(\frac{s}{\Omega_1}+1\right)^{n_p}} \qquad (6.50)$$

As $\Omega T \gg 1$, the influence of the coefficients a_k, $k = 1 \cdots n_p$, decreases with increasing k. In other words, the shape of

$$F+1 = \frac{\left(\frac{s}{\Omega}\right)^{n_p}}{\left(\frac{s}{\Omega_1}+1\right)^{n_p}} \qquad (6.51)$$

may be considered as the asymptotic one for $\Omega T \to \infty$. Note that for the chain with n_p open integrators, where $a_k = 0$, $k = 1 \cdots n_p$, the shape is equal to the asymptotic one. Also note that this shape now is *independent* of the choice of Ω_1.

For the *nonlinear subsystem* from Eq. 6.23 then

$$\Delta^a_{max} = \frac{\Delta u}{r_1 (\Omega_1 T)^{n_p}} \tag{6.52}$$

And for the *stability test*, the graphic configuration remains the same if

$$\frac{r_1 (\Omega_1 T)^{n_p}}{\Delta u} = const \tag{6.53}$$

or using a reference case denoted by index n:

$$\left(\frac{(\Omega_1)}{(\Omega_1)_r} \right)^{n_p} := \frac{\left(\frac{r_1}{\Delta u} \right)_r}{\left(\frac{r_1}{\Delta u} \right)} \tag{6.54}$$

and thus the stability properties are the same. Then one would expect the performance properties to be quite similar as well.

If $(\Omega_1 T)_r$ has been determined for the reference case r, either by stability analysis or by simulations, then Eq. 6.54 allows one to generate the relative bandwidth $\Omega_1 T$ as function of r_1 and Δu in a very transparent way.
Note that this will produce only a first approximation for plants with some or all $a_k \neq 0$. However the approximation gets better for decreasing $|a_k|$ and increasing relative bandwidth $\Omega_1 T$, and *vice versa*.

The **I(aw)-P$^+$ case** is considered next.

For the *linear subsystem* in Eq. 6.16, divide $F + 1$ by $k_0 = (\Omega_I T)^{n_p + 1}$

$$F + 1 = \frac{\left(\frac{s}{\Omega} \right)^{n_p} + \frac{a_{n_p}}{\Omega_I T} \left(\frac{s}{\Omega_I} \right)^{n_p - 1} + \cdots + \frac{a_2}{(\Omega_I T)^{n_p - 1}} \left(\frac{s}{\Omega_I} \right) + \frac{a_1}{(\Omega_I T)^{n_p}}}{\left(\frac{s}{\Omega_I} + 1 \right)^{n_p}} \quad \frac{\frac{s}{\Omega_I} + \frac{k_a}{\Omega_I T}}{\frac{s}{\Omega_I} + 1}$$

and for the *nonlinear subsystem* in Eq. 6.23

$$\Delta^a_{max} = \frac{\Delta u}{r_1} \frac{k_a}{k_0} = \frac{\Delta u}{r_1} \frac{1}{(\Omega_I T)^{n_p}} \frac{k_a}{\Omega_I T}$$

For the *graphic test*, it is useful to apply further scaling: both equations are divided by the common factor $k_a / \Omega_I T := \gamma$.

$(F+1)_{sc} =$

$$\frac{(\frac{s}{\Omega_I})^{n_p} + \frac{a_{n_p}}{\Omega_I T}(\frac{s}{\Omega_I})^{n_p-1} + \cdots + \frac{a_2}{(\Omega_I T)^{n_p-1}}(\frac{s}{\Omega_I}) + \frac{a_1}{(\Omega_I T)^{n_p}}}{(\frac{s}{\Omega_I}+1)^{n_p}} \quad \frac{(\frac{1}{\gamma})(\frac{s}{\Omega_I})+1}{\frac{s}{\Omega_I}+1}$$

and

$$(\Delta_{max})_{sc}^a = \frac{\Delta u}{r_1 \cdot (\Omega_I T)^{n_p}} \tag{6.55}$$

Again the position of the straight line is not affected by the choice of k_a, i.e. γ, and it is the same as for the P$^+$ case.

For the compensating awf gain, $\gamma := 1$, the graphic configuration will be the same as for the P$^+$ case, if $\Omega_I := \Omega_1$.
And for deadbeat awf, $\gamma \gg 1$, the scaled contour $(F+1)_{sc}$ converges to

$$(F+1)_{sc}\Big|_{\gamma \gg 1} \rightarrow$$

$$\frac{(\frac{s}{\Omega_I})^{n_p} + \frac{a_{n_p}}{\Omega_I T}(\frac{s}{\Omega_I})^{n_p-1} + \cdots + \frac{a_2}{(\Omega_I T)^{n_p-1}}(\frac{s}{\Omega_I}) + \frac{a_1}{(\Omega_I T)^{n_p}}}{(\frac{s}{\Omega_I}+1)^{n_p}} \quad \frac{1}{\frac{s}{\Omega_I}+1}$$

$$\tag{6.56}$$

which is again not affected by the choice of Ω_I, if the chain of n_p open integrators is considered.

To *summarize*: the same relation as Eq. 6.54 holds for variations of Δu, r_1 and ΩT.

6.5 Transient Analysis

So far, the elements for a given control loop with state feedback and input saturation have been assembled to analyze, whether the initial condition response is stable. Any one of the sector criteria can be used.
However, these tests are known to be conservative, *i.e.* the design to nonlinear stability will, in fact, produce a response with a significant "rate of convergence". Note that this is in contrast to linear systems, where meeting the Nyquist stability condition will produce a stationary oscillation, where the rate of convergence is zero. Note also that the "damping ratio" is defined for linear systems, and should be used exclusively there. Therefore, the term "rate of convergence" is considered more appropriate for this nonlinear context.
The analysis and design can be done computationally, using concepts of Robust Control theory and Linear Matrix Inequalities (LMIs), *e.g.* see [55, 53] and references therein.

Here, we shall resort to simulations. The aim is to generate some insight as to how the geometric situation in the graphic stability test relates to the properties of the transient response. This may be useful in a first short design phase, and it may help to prepare a second design phase, when such a computational approach to performance optimization is to be used.

However, the problem with simulations is the increasing number of parameters. In general, a very large number of experiments must be performed to obtain a reasonable overview. A careful selection has to be made to keep this number down to a feasible size. Any such selection is subjective, and may thus be viewed very critically. Also, the results must be restricted to the selection, and may not be generalized unduly.

The following assumptions are made here:

-   The *plant* shall be of order 2 and 3, and with the delay set to $D := 0$. This shall be further narrowed down to the double-open-integrator chain, and the triple one, with equal time parameters $T_k = 1.0, k = 1, 2, 3$, *i.e.* $a_1 = a_2 = a_3 := 0$. This selects the cases for $p(s)$ in the center of the stability charts, and also considers the cases which are being borderline manageable.

-   The *controller structures* are P^+ and $I(aw)$-P^+, and in their discrete-time form, with a short sampling time $T_s = 0.01$.

-   The *controller coefficients* shall be determined from the linear closed-loop bandwidth, *i.e.* from pole assignment to $(s+\Omega)^m = 0$, where $m = n_p + n_c$. Then Ω is the main control design parameter.

-   The test sequence shall focus on the run-up response from forced standstill $y = 0$ with u blocked at zero, while $r = 1.0$, up to nominal operation at $\bar{x}_1 = 1.0$. From previous experience, this is the most critical transient regarding stability (where $|\Delta u_{lin_{max}}|$ is largest).

-   The saturation on u shall be symmetrical and allow a dynamic working range of u up to ± 1.0.

-   The compensating awf gain case is trivial from the stability analysis, once the P^+-case has been analyzed, and will therefore not investigated separately. More interesting is the deadbeat awf gain. Also, some reduced value shall be inserted. This is inspired by the noise sensitivity , see Chapter 5.

An Overview of the Simulations

In the following figures, the graphic stability test for one particular experiment is always shown in the left-hand plot, while the associated run-up time response is given in the right-hand plot. One figure typically covers three values of the single parameter to be varied, with the value for the best stability properties always shown in the top row.

The experiments are listed in the following table:

Plant order N	$p(s)$	Controller type	Variations on	Fig.
2	$(sT)^2$	P$^+$	Ω_P	6.12
		I(aw)-P$^+$, deadbeat k_a	Ω_I	6.13
3	$(sT)^3$	P$^+$	Ω_P	6.14
		I(aw)-P$^+$, deadbeat k_a	Ω_I	6.15
2	$(sT)^2$	I(aw)-P$^+$, Ω_I fixed	k_a	6.16
3	$(sT)^3$	I(aw)-P$^+$, Ω_I fixed	k_a	6.17

For the group of Figures 6.12 to 6.15, the main design parameter Ω has been adjusted such that the minimum value of the real part of $F(j\omega) + 1$ is approximately

-1.0 for the top row, -2.0 for the center row, -4.0 for the bottom row.

Considering *performance*, the results from the simulations for the run-up settling time measured up to

$$|u(t)| < 0.05, \quad \forall \quad t > t_E$$

are as follows.

For the second-order case $p(s) = (sT)^2$ with P$^+$ control; see Fig. 6.12:

Ω_P	2.75	3.8	5.35	Min. time
$t_E - t_0$ [s]	2.7	2.15	3.3	2.0

and for the third-order case $p(s) = (sT)^3$ with P$^+$ control; consult Fig. 6.14:

Ω_P	1.575	1.975	2.47	Min. time
$t_E - t_0$ [s]	4.3	4.1	> 7	3.175

The second-order case performs well, and comparatively better than the third-order case, which still performs acceptably.

These performance results will serve as a base line for the design methods.

Discussion

For the group of Figs. 6.12 to 6.15, the main parameter is the linear closed-loop bandwidth (ΩT).

The top row in all four figures corresponds to the situation, where the on-axis circle test indicates stability for the specific transient (the "run-up"). The transients show consistently that the response is not at the stability limit in the linear systems sense, but that there is still a considerable "margin". This illustrates that the circle test is quite conservative.

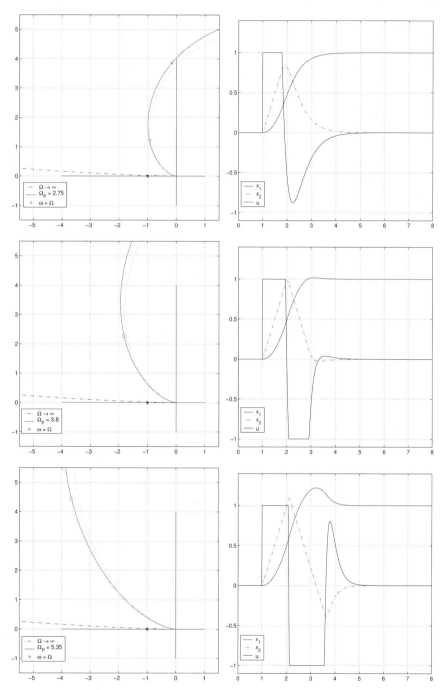

Fig. 6.12. Plant: $p(s) = (sT)^2$, P^+ control, effect of closed-loop bandwidth Ω_P

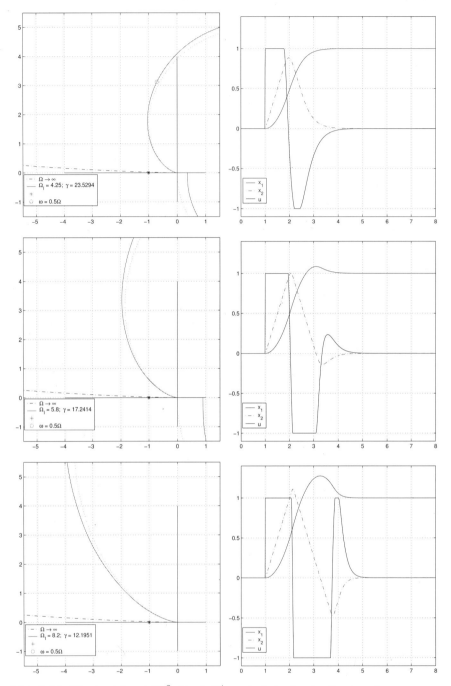

Fig. 6.13. Plant: $p(s) = (sT)^2$, I(aw)-P$^+$ control, k_a for deadbeat awf, effect of closed-loop bandwidth Ω_I

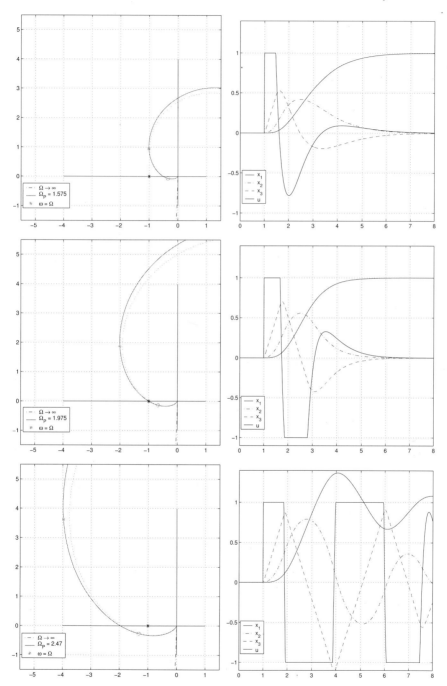

Fig. 6.14. Plant: $p(s) = (s\mathbf{T})^3$, P^+ control, effect of closed-loop bandwidth Ω_P

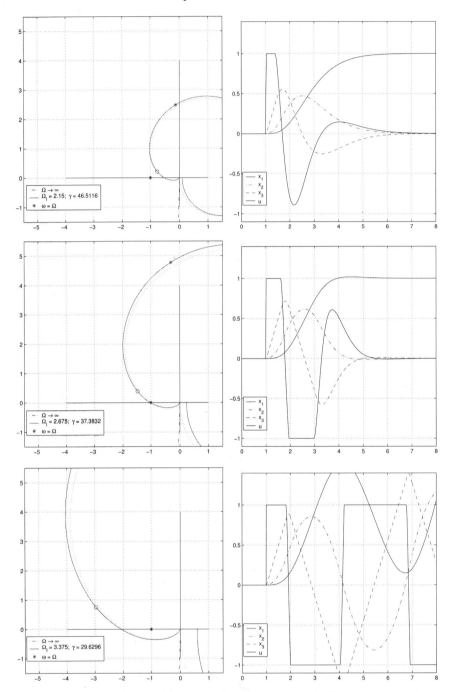

Fig. 6.15. Plant: $p(s) = (\mathbf{s}\mathbf{T})^3$, I(aw)-P$^+$ control, k_a for deadbeat awf, effect of closed-loop bandwidth Ω_P

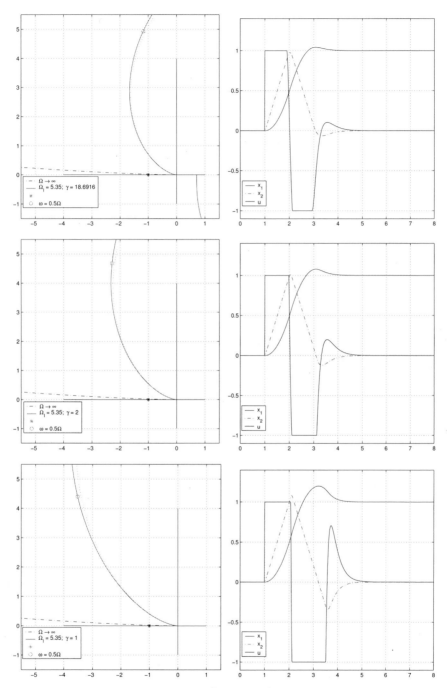

Fig. 6.16. Plant: $p(s) = (\mathbf{sT})^2$, I(aw)-P$^+$ control, variation of k_a

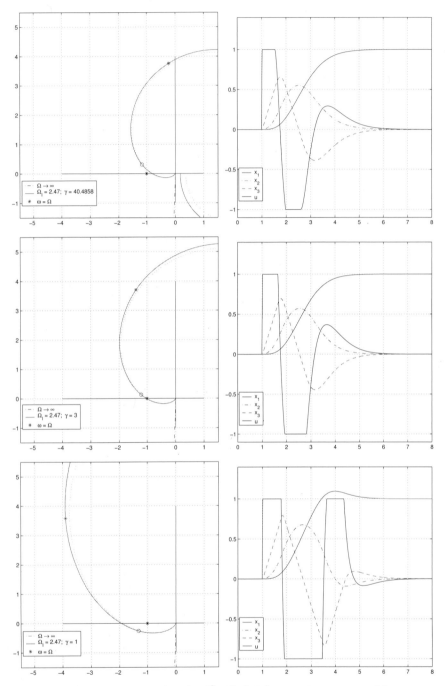

Fig. 6.17. Plant: $p(s) = (\mathbf{sT})^{\mathbf{3}}$, I(aw)-P$^+$ control, variation of k_a

But it also shows that the on-axis circle test produces a value of ΩT and an associated run-up response that is at least 'acceptable' from a practical point of view. It may be considered rather slow, but it also provides some margin for uncertain parameters. And it produces at least a reasonable starting point for subsequent tuning.

For the center row (and the bottom one as well) the situation is such that stability of the run-up transient can no longer be shown by the on-axis circle criterion. But, at least for the center row, it may still be shown with the off-axis circle or the Popov tests. The transients are such that there is a small overshoot on the main control variable. The settling time is close to the minimum-time response one. From a practical point of view, this is near to the optimal response. But it is only to be used if robustness is not an issue.

For the bottom row, where ΩT is adjusted to $\Re(F(j\omega)+1)\big|_{min} \approx -4$, the run-up response is not acceptable any more. There is a considerable overshoot for the $N = 2$ case, and a weakly damped oscillation for the $N = 3$ case.

The run-up responses, in particular $x_2(t)$, also illustrate the "plant windup" effect.

The second group of Figures 6.16 and 6.17, illustrates the effect of reducing the awf gain from the deadbeat value (top row) to the compensating one (bottom row). The bandwidth parameter Ω has been set to the value where $\Re(F(j\omega) + 1)\big|_{min} \approx -4$ for the P$^+$ controller. Note that

$$\gamma := \frac{k_a}{k_a^*} = \frac{k_a}{\Omega T} \quad \text{and} \quad \gamma_{db} = \frac{T}{T_s}\frac{1}{\Omega T} = \frac{1}{\Omega T_s} \; ; \quad \text{with} \quad 1 \le \gamma \le \gamma_{db}$$

From the stability test situation for $N = 2$, the effect of k_a is strong from $\gamma = 1$ up to $\gamma = 2$, and much less from $\gamma = 2$ to γ_{db}. And for $N = 3$, the effect is strong up to $\gamma = 3$.

6.6 Design Methods: An Overview

So far the **analysis** has been discussed. Now the synthesis or **design** shall be addressed. Eight design approaches or methods are selected from a still growing pool of proposals and ideas, and are investigated in more detail. They are bundled into three groups:

- de-tuning the linear loop (Sect. 6.7), and "nested loops" (Sect. 6.8),
- first-order dynamic awf (Sect. 6.9), trajectory generators with awf (Sect. 6.10), and the "continued states" concept (Sect. 6.11),
- and add-on output constraints implemented by cascade-limiters (Sect. 6.13), selection for approach speed (Sect. 6.14), and selection for lower-bandwidth control (Sect. 6.15).

The selection for lower-bandwidth control method is closely related to recent results in MPC. These links are discussed in Sect. 6.16.

The common aim is to produce design solutions for which nonlinear stability may be shown (at least by plausibility) within the specified range of input sizes, or equivalent initial conditions. Performance shall be judged by comparison of the simulation to the minimum-time run-up.
Further aims, such as formalized optimality, robustness, *etc.*, are not postulated here. This is an area, where research is currently very active.

6.7 Design by De-tuning Ω

The linear closed-loop bandwidth in absolute (Ω) or relative (ΩT) form is considered as the primary design parameter. It is very often determined by the requirements of the actual application, such as the disturbance rejection rate, *etc.*, and may not be altered at will.

If this is the case, then the shape of the Nyquist contour is given. Also the maximum size of the setpoint step is considered as given. Then the only remaining design parameter to comply with the stability test is Δu, *i.e.* the actuator working range. It has to be sized accordingly, which may lead to a very costly installation.

Then, as an engineering compromise, one may think of reducing the bandwidth nevertheless, such that stability may be shown with the existing actuator (and the setpoint step size as specified). And then one is interested in lowering Ω not more than necessary to comply with the stability test. Therefore, this is called the "de-tuning" design approach.

Then the *design rules* follow directly from the findings in the stability and transient analysis of the previous sections:

1. Use the on-axis circle test with $\Re(F(j\omega) + 1)|_{min} \approx -(\Delta u/r_1)$ to obtain a first value of Ω. This is conservative, but not excessively so.
2. Use the on-axis test with $\Re(F(j\omega) + 1)|_{min} \approx -2\,(\Delta u/r_1)$ to obtain a value of Ω, which produces a near minimum-time response. Some small additional tuning will be required.
3. For I(aw)-P$^+$ controllers set the awf gain to its compensating value: $\gamma := N$, for $N = 2, 3$. This reduces the admissible Ω value somewhat, but also lessens sensitivity to any high-frequency measurement disturbances indexSensitivity to!measurement noise.
4. If deadbeat awf gain γ_{db} can be used, then Ω_I can be substantially increased from the value Ω_P for the P$^+$ controller.

Note that these rules have been obtained on (and thus are valid only for) the double- and triple-open-integrator plants.

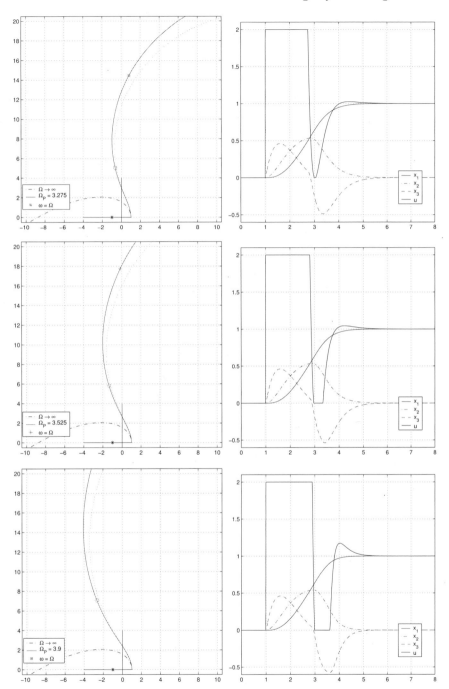

Fig. 6.18. Plant: $p(s) = (\mathbf{s} + \mathbf{1})^{\mathbf{3}}$, P$^+$ control: checking the design rules

Therefore, they are checked next on Peter's case, with $p(s) = (s+1)^3$ and for the P^+ controller only; see Fig. 6.18. This shows

- that adjusting Ω for $\Re(F(j\omega) + 1)|_{min} \approx -(\Delta u/r_1)$ produces again a "good" initial value.
- And nearly minimum-time run-up is produced by adjusting Ω for $\Re(F(j\omega) + 1)|_{min} \approx -4\,(\Delta u/r_1)$,
- instead of previously $-2\,(\Delta u/r_1)$ for the triple-open-integrator chain. This is to be expected from the stability analysis.

Exercises

Check the *de-tuning design rules* in adjacent areas
- for other $N = 3$ plants, $p(s) = s\,(s+1)^2$ and $p(s) = s^2\,(s+1)$,
- and for $N = 2$ plants, $p(s) = (s+1)^2$, and $p(s) = s\,(s+1)$,
- and for I(aw)-P^+ controllers, with high- and low-gain awf,
  as in Figs. 6.16 and 6.17.

- Investigate the effect of different setpoint step sizes,
  and then use Eq. 6.54 to adjust Ω.

- Apply the full test sequence of Sect. 6.3, and
- check also the assumption about $u_{lin_{max}}$ by simulations.

6.8 The "Nested Loops" Method

Looking again at the stability charts in Section 6.4.2 reveals that plants having their own internal negative feedback will yield much better closed-loop stability properties than those without. This is due to the influence of the feedback coefficients a_j, $j = 1 \cdots n_p$ being > 0 on the shape of $F(j\omega) + 1$. The basic idea of the nested loops method is to provide the plant with some weak negative feedback first, and then to implement the full design feedback in a second step; see Fig. 6.19.

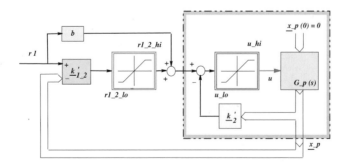

Fig. 6.19. The basic idea of the "nested loops" method

For the first step, the feedbacks k_{j_2}, $j = 1 \cdots n_p$, are determined by pole assignment

$$(s + \Omega_2)^{n_p} = 0 \; ; \quad \text{where} \quad 0 < \Omega_2 \ll \Omega_1 \qquad (6.57)$$

with Ω_2 as a first design parameter.

The next design step is to introduce the saturation element with its values marked in Fig.6.19 by $[\, r1\_2\_hi, \; r1\_2\_lo \,]$, as the second set of design parameters.

Finally the outer feedback is added, with the gains

$$k_{j_{1\_2}} := k_{j_1} - k_{j_2}, \; j = 1 \cdots n_p \qquad (6.58)$$

where the total gains k_{j_1} are given by the specified closed-loop bandwidth Ω_1 and by the pole assignment $(s + \Omega_1)^{n_p} = 0$.
Note that the outer feedback may again be split up into two cascaded feedbacks with an additional saturation in the outer loop. This method has been proposed by Bühler [51].

Now the design parameters must be entered.

-   The feedforward weight b is chosen such that the steady state effect of the inner feedback is canceled, *i.e.*

$$\text{from} \quad br_1 - (a_1 + k_{j_1})r_1 := 0 \quad \text{and with} \quad a_1 = 0 \quad \rightarrow \quad b := k_{j_1} \quad (6.59)$$

This lets the additional saturation element operate around its center in steady state.
-   The parameters for the additional saturation are intuitively set to the same values as with the control saturation

$$r1\_2\_hi := u_{hi} \; ; \quad r1\_2\_lo := u_{lo} \qquad (6.60)$$

Exercise
Clarify this.
Develop and check other ideas for entering the feedforward from r_1.

-   And Ω_2 is determined by tuning first. For this, the following trend is useful: increasing Ω_2 also increases the positive effect on the numerator shape of $F(j\omega) + 1$, and therefore will allow to increase the value of Ω_1, for approximately the same closed-loop stability properties. We shall show later how the value for Ω_2 can be derived more concisely by stability considerations.

Note that the basic idea and the layout in Fig. 6.19 may be seen differently: the input $r1\_2$ to the inner control loop is clipped to a size, which results in a stable trajectory for the inner loop. Then this method belongs also to the group of input size modifiers in Sect. 6.8 to 6.10.

6.8.1 Transient Response

The case $G(s) = 1/(sT)^3$ is considered. The control structure is shown in Fig. 6.20

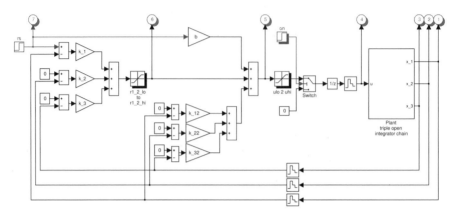

Fig. 6.20. The "nested loops" method applied to the plant $G(s) = 1/(sT)^3$

For the tuning of Ω_2 an intuitive first value would be $\Omega_2 := 1.0$. And from the specifications $\Omega_1 := 5.0$. This pair produces the response in Fig. 6.21 (top row). The response is stable (in contrast to the original one without a nested loop), but the oscillatory component is not acceptable.

The trend mentioned above indicates that this can be improved either by reducing Ω_1, or by increasing Ω_2.

In Fig. 6.21 (center) Ω_1 is reduced by tuning to 2.76 such as to suppress the oscillatory component, while the inner loop is kept at $\Omega_2 := 1.0$.

There are two distinct phases in the transient response. In the first phase, the outer loop controller output saturates at $r1\_2\_hi$, i.e. the inner loop "sees" a constant setpoint at $+1.0$. Thus it will generate a run-up transient, determined by the gains k_{j_2} due to Ω_2 and the control saturation $[u_{hi}, u_{lo}]$.

In the second phase, the inner loop controller output $u(t)$ does not saturate any more, i.e. the constraints $[u_{hi}, u_{lo}]$ can be omitted; that is, the plant with the weak feedback is now linear, and constitutes the modified plant (with all $a_j > 0$, $j = 1 \cdots n_p$), and only the outer loop constraints $[r1\_2\_hi, r1\_2\_lo]$ are active.

In Fig. 6.21 (bottom) both Ω values are increased by manual tuning: Ω_2 is raised to 1.575, while $\Omega_1 := 5.0$ has achieved its initial specification.

Again the two phases described above can be clearly distinguished. Note that $u(t)$ does transiently saturate for a short time interval (~ 0.2 s) in the second phase.

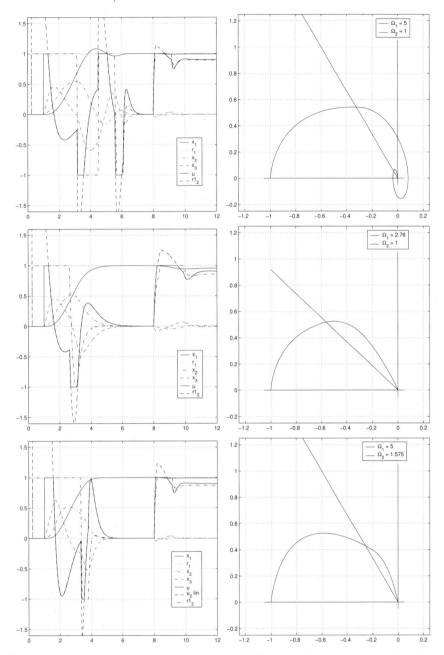

Fig. 6.21. Transient responses versus time (left) and in the phase plane (right):
(top) $\Omega_2 := 1.0$, $\Omega_1 := 5.0$;
(center) $\Omega_2 := 1.0$, $\Omega_1 := 2.76$;
(bottom) $\Omega_2 := 1.575$, $\Omega_1 := 5.0$

Performance is checked by reading t_E from the simulations for the three cases of Fig. 6.21, such that

$$|u(t)| < 0.05 \quad \forall \quad t > t_E$$

and comparing to the minimum-time run-up:

Ω_2	1.0	1.0	1.575	Min. time
Ω_1	5.0	2.76	5.0	
$t_E - t_0$ [s]	5.7	4.4	3.8	3.175

Exercise

Derive $t_E - t_0\big|_{min}$ for this benchmark case.

6.8.2 Stability Properties

From the time and the phase plane plots, the run-up trajectory may be seen as two consecutive parts, with a first part

> starting at $\quad x_1(0) = -r_1,$ $\quad$ (virtually) ending at the origin
> with one saturation $\quad [u_{hi}, \ u_{lo}]$;
> the plant being $\quad G(s) = 1/(sT)^3$;
> and the bandwidth $\quad \Omega_2 \ (\ll \Omega_1)$

leading up to a second trajectory part

> with input size $\quad r_1;$ $\quad$ ending at the origin
> with one saturation $\quad [r1\_2\_hi, \ r1\_2\_lo] := [u_{hi}, \ u_{lo}]$;
> the plant being $\quad G^+(s) = \dfrac{1}{[(s/\Omega_2) + 1]^3};$
> and the bandwidth $\quad \Omega_1$

The graphic stability test for both subsystems in Fig. 6.22 shows the Nyquist contours for $F + 1|_{G; \ \Omega_2}$ and $F + 1|_{G^+; \ \Omega_1}$. It also indicates the positions of the straight line for the circle test at $u_{hi}/[r_1(\Omega_2 T)^3]$ and $u_{hi}/[r_1(\Omega_1 T)^3]$. Therefore stability is indicated by the circle test for both subsystem trajectories. This now suggests a design procedure for Ω_2:

> select the inner loop bandwidth Ω_2 such that the circle test indicates stability for the subsystem with $F + 1|_{G; \ \Omega_2}$, and $u_{hi}/[r_1(\Omega_1 T)^3]$.

And this is just what the "de-tuning" method does. Therefore, the "nested loops" method has been bundled with the "de-tuning" method.

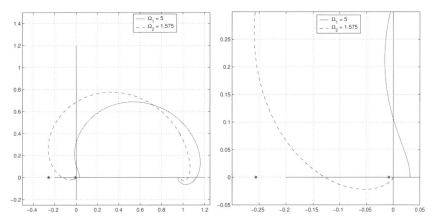

Fig. 6.22. Stability test for the two subsystems $F+1|_{G;\ \Omega_2}$ and $F+1|_{G+;\ \Omega_1}$; dots indicating the positions $u_{hi}/(r_1[\Omega_2 T]^3)$ and $u_{hi}/[r_1(\Omega_1 T)^3]$ of the straight lines, right: zoom-in to the relevant area

6.8.3 Checking the Design Procedure

The design procedure from above shall now be checked for the other benchmark case, *i.e.* $G(s) = 1/(sT)^2$. Let
- $\Omega_1 := 10.0$, as per specifications,
- $b := k_{12}$,
- $[r1\_2\_hi,\ r1\_2\_lo] := [u_{hi},\ u_{lo}]$, and
- $\Omega_2 := 2.75$ from the "de-tuning method" for this case (see Fig. 6.12).

Fig. 6.23 shows the stability properties for the two subsystems. The run-up response obtained for Ω_2 from the stability test is acceptable. Note, however, that for $\Omega_1 = 10$ the saturation $[u_{hi},\ u_{lo}]$ in the inner loop is also encountered in the second trajectory part, *i.e.* both saturations are active, and the separation into two subsystems with one saturation each (as has been assumed for the stability test) is no longer valid. But this can be attained by reducing Ω_1 to ~ 6.

Performance is again measured by the settling time to $|u| < 0.05$

Ω_2	2.75	2.75	Min. time
Ω_1	6.0	10.0	
$t_E - t_0$ [s]	2.25	2.50	2.0

To *summarize*: The "nested loops" method produces a simple structure, which is not difficult to implement. However, both the transient response and the stability test are not very transparent in the general case, but are better accessible in the special case of sufficiently low bandwidth Ω_1. Performance seems to be good. Note that this is also a useful concept for pre-stabilizing plants with unstable poles.

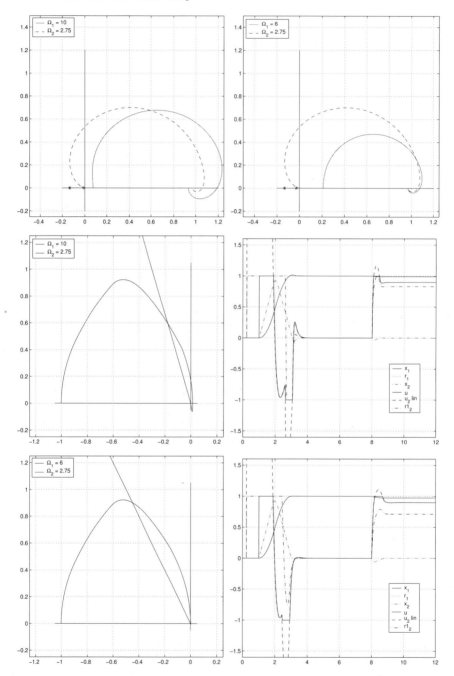

Fig. 6.23. Stability analysis for the two subsystems in $G(s) = 1/(sT)^2$:
(top left) $\Omega_2 := 2.75$ and $\Omega_1 = 10.0$;   (top right) $\Omega_2 := 2.75$ and $\Omega_1 = 6.0$;
and responses (center row) $\Omega_1 = 10.0$;   (bottom row) $\Omega_1 = 6.0$

6.9 First-order Dynamic Antiwindup Feedback

6.9.1 The Design Procedure

This method has been proposed by Peter and his colleagues as a solution to the stability problems encountered in his case in section 6.2. Their basic idea is

- to transplant the idea of control error conditioning awf to reference conditioning awf and from the case with integral action to the case without integral action,
- to stabilize the control system by inserting a dynamic block $G_a(s)$ in the awf loop,
- where $G_a(s)$ is always a unity gain first-order filter with time constant T_a,
- and to use both k_a and T_a as additional design parameters,
- thereby deforming the Nyquist contour of the linear subsystem, such that it stays outside a cone of say 60° from the negative real axis,
- where appropriate values for k_a and T_a have to be determined iteratively (by systematic variation and inspection of the resulting shape),
- but no indications are furnished on how to do this search systematically.

Fig. 6.24 shows the structure of the control loop.

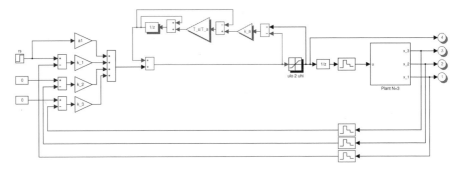

Fig. 6.24. The control system with dynamic awf, parameters k_a, T_a

This design procedure is now applied to Peter's case, Sect. 6.2, where $p(s) = (sT + 1)^3$, $T = 1$, and with $\Omega_P = 16$; see Fig. 6.2 (bottom).

Fig. 6.25 shows the result, where k_a has already been searched and found (at $k_a := 30$), and where T_a is still to be selected (three values are entered). This seems to suggest that $T_a := T$ may be a good first choice.

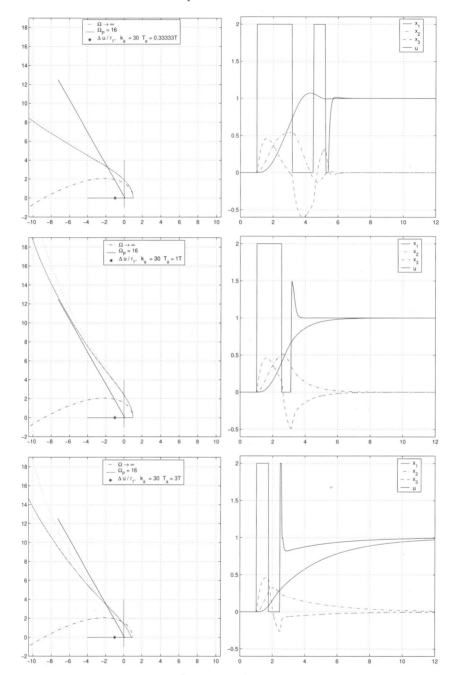

Fig. 6.25. Plant: $p(s) = (sT+1)^3$, $T = 1$, P$^+$-control, $\Omega = 16.0$: Peter's method, k_a constant, T_a variable. The "phase margin sector" of 60° is indicated

Looking at the bottom row plot reveals that the run-up response consists of two distinct phases. The second one is an approach to the final steady state, which seems to be formed by the response of $G_a(s)$ and its time constant T_a. And the first phase seems to be a nonlinear approach to the second phase trajectory. By further fine tuning of k_a, T_a from the values used in the center row plot, the second phase may be reduced to near zero; in other words, a near-minimum time response can be obtained.

In summary, this procedure is not very transparent, it is strongly tied to the specific plant, and the results are not that convincing. We shall try to improve on this now.

The first step will be discussion of stability properties, which has been very helpful so far to channel the parameter variations. Then the benchmark cases of the double- and triple-open-integrator chain are investigated in the graphic stability test and corresponding transient response. This will finally yield a more general design procedure for such a first-order dynamic awf, and show its limitations more clearly. And this will motivate the next following design method; Sect. 6.10.

6.9.2 Stability Analysis

Again, the general result of Chapter 2 applies.

For the *linear subsystem* recall:

$$F + 1 = \frac{1 + k_a G_a R_2}{1 + R_1 R_2 G} \tag{6.61}$$

where R_2 is the part of the controller transfer function contained within the awf loop. By inspection of Fig. 6.24

$$R_2 := 1 \tag{6.62}$$

Then, using the continuous approximation, *i.e.* considering the dominant dynamics of the plant and a very short sampling time T_s:

$$
\begin{aligned}
F + 1 &= (1 + k_a G_a) \frac{1}{1 + RG} \\
&= \frac{sT_a + 1 + k_a}{sT_a + 1} \frac{(sT)^{n_p} + \cdots + a_2(sT) + a_1}{(sT)^{n_p} + \cdots + (a_2 + k_2)(sT) + (a_1 + k_1)} \\
&= \frac{k_a + 1}{a_1 + k_1} \frac{s\frac{T_a}{1+k_a} + 1}{sT_a + 1} \frac{(sT)^{n_p} + \cdots + a_2(sT) + a_1}{\left(\frac{s}{\Omega} + 1\right)^{n_p}}
\end{aligned} \tag{6.63}
$$

For the *nonlinear subsystem* consider the awf loop transfer function from Fig. 6.24 in continuous form

$$\frac{e_a(s)}{r_1(s)} = (a_1 + k_1)\frac{1}{1 + k_a\frac{1}{sT_a+1}} = (a_1 + k_1)\frac{sT_a + 1}{sT_a + (1 + k_a)}$$

$$\rightarrow \quad \bar{e}_a = \frac{a_1 + k_1}{1 + k_a}r_1 := e_{a_{max}} \tag{6.64}$$

that is

$$\Delta_{max} + 1 = \frac{\Delta u}{\Delta u_{lin_{max}} - \Delta u} = \frac{\Delta u}{e_{a_{max}}} = \frac{\Delta u}{r_1}\frac{1 + k_a}{a_1 + k_1} \tag{6.65}$$

For the *graphic test*, again scaling is applied to both Eq. 6.63 and 6.65:

$$(F + 1)_{sc} = \frac{s\frac{T_a}{1+k_a} + 1}{sT_a + 1}\frac{(sT)^{n_p} + \cdots + a_2(sT) + a_1}{\left(\frac{s}{\Omega} + 1\right)^{n_p}}$$

$$= \frac{s\frac{T_a}{1+k_a} + 1}{sT_a + 1}\frac{p(s)}{\left(\frac{s}{\Omega} + 1\right)^{n_p}}$$

$$\text{and} \quad (\Delta_{max} + 1)_{sc} = \frac{\Delta u}{r_1} \tag{6.66}$$

Observe that the second term in $(F + 1)_{sc}$ is the same as for the system without dynamic awf.

And the first term from the dynamic awf loop is a low-pass filter, as $k_a > 0$. It therefore re-shapes the original Nyquist contour in the favorable direction for stability. But obviously the range of deformation is restricted, due to admitting only first-order blocks G_a.

And the position of the straight line is the same as without dynamic awf. So the plots can be drawn in the same window for easy comparison.

Compensating awf Gain

Now consider the special case for k_a' as follows:

$$\frac{T_a}{k_a' + 1} := \frac{1}{\Omega} \quad \text{yielding} \quad k_a' := \Omega T_a - 1 \tag{6.67}$$

Then the numerator contribution of the awf loop cancels with one of the closed-loop pole factors, and

$$(F + 1)_{sc} = \frac{1}{sT_a + 1}\frac{p(s)}{\left(\frac{s}{\Omega} + 1\right)^{n_p-1}} \quad \text{and} \quad (\Delta_{max} + 1)_{sc} = \frac{\Delta u}{r_1} \tag{6.68}$$

Consider further Peter' case, where $p(s) = (sT + 1)^3$.

If then $T_a := T$ is selected, the denominator contribution of the awf loop $(sT_a + 1)$ cancels with one factor of $p(s)$, and

$$(F + 1)_{sc} = \frac{(sT + 1)^{n_p-1}}{\left(\frac{s}{\Omega} + 1\right)^{n_p-1}} = \frac{(sT + 1)^2}{\left(\frac{s}{\Omega} + 1\right)^2}; \quad \text{while} \quad (\Delta_{max} + 1)_{sc} = \frac{\Delta u}{r_1} \tag{6.69}$$

which is equivalent to the case $N = 2$ with a P$^+$ controller (and no awf).
So this amounts to a reduction of equivalent system order by one, for instance
from $N = 3$ of the original system to $N = 2$ for the system with dynamic
awf, if k_a and T_a have been selected as just prescribed. This reduction of N
markedly improves the stability properties; see the stability charts in Figures
6.8 to 6.10.

Note that such cancelation requires of $p(s)$ to contain at least one such
term $(sT + 1)$; the rest of $p(s)$ may be arbitrary.

Note, further, that the integrators associated to the canceling zero and
pole are not physically the same. So there may be unwanted transient effects
due to non equal initial conditions, such as shown in [47].

Deadbeat awf Gain

In the continuous approximation this would lead to

$$k'_a \to \infty \quad i.e. \quad \frac{T_a}{k'_a + 1} \to 0 \tag{6.70}$$

Then there is no such cancelation as above, and

$$(F + 1)_{sc} = \frac{1}{sT_a + 1} \frac{p(s)}{\left(\frac{s}{\Omega} + 1\right)^{n_p}} ; \quad \text{and} \quad (\Delta_{max} + 1)_{sc} = \frac{\Delta u}{r_1} \tag{6.71}$$

If, again, Peter's case $(sT + 1)^3$ is considered and if also $T_a := T$ is selected,
then

$$(F + 1)_{sc} = \frac{(sT + 1)^{n_p - 1}}{\left(\frac{s}{\Omega} + 1\right)^{n_p}} = \frac{(sT + 1)^2}{\left(\frac{s}{\Omega} + 1\right)^2} \frac{1}{\left(\frac{s}{\Omega} + 1\right)}$$

$$\text{while} \quad (\Delta_{max} + 1)_{sc} = \frac{\Delta u}{r_1} \tag{6.72}$$

which is equivalent to the case of $N = 2$ with an I(aw)-P$^+$ controller and
high-gain awf.
In other words, the favorable phase shift on the Nyquist contour is even
stronger, and the stability properties will improve further.

For the discrete-time implementation, k'_a has an upper limit at the deadbeat response of the awf loop. This requires a unity open awf loop gain in Fig.
6.24:

$$1 := (k'_{a_{db}} + 1)\frac{T_s}{T_a} \quad \to \quad k'_{a_{db}} = \frac{T_a}{T_s} - 1 \tag{6.73}$$

However, using such a high gain will have an adverse effect if high-frequency
measurement disturbances are present (they always are in real plants). So it is
recommended to use the lower compensating value, and accept the somewhat
reduced stability properties.

6.9.3 Checking Peter's Method

The design procedure shall now be checked by simulations and the corresponding stability test plots for the double- and triple-open-integrator chain plants.

The Case $p(s) = (sT)^2$

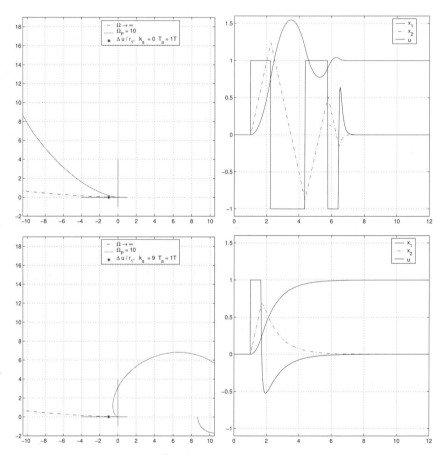

Fig. 6.26. Plant: $p(s) = (\mathbf{sT})^2$, P^+-control, $T_a = T_1$:
(top) $k_a = 0$;   (bottom) $k_a = k_a^*$

The stability analysis for the case without dynamic awf yields:

$$(F+1)_{sc} = \frac{(sT)^2}{\left(\frac{s}{\Omega}+1\right)^2} = (\Omega T)^2 \left[\frac{\frac{s}{\Omega}}{\frac{s}{\Omega}+1}\right]^2 \; ; \text{while} \; (\Delta_{max}+1)_{sc} = \frac{\Delta u}{r_1}$$

(6.74)

and with dynamic awf, letting $T_a := T$, and with compensating awf gain k_a:

$$(F+1)_{sc} = \frac{(sT)^2}{\left(\frac{s}{\Omega}+1\right)^2} \frac{\frac{s}{\Omega}+1}{sT+1} = (\Omega T)^2 \left[\frac{\frac{s}{\Omega}}{\frac{s}{\Omega}+1}\right]^2 \frac{\frac{s}{\Omega}+1}{sT+1}$$

(6.75)

where the value for $(\Delta_{max}+1)_{sc}$ is the same.

As $\Omega T \gg 1$, the last multiplicative factor in Eq. 6.75 is a low pass, which reshapes the Nyquist contour in the favorable direction for stability. From Fig. 6.26 this effect is considerable, and sufficient for practical purposes. There is no need to revert to deadbeat awf gain.

Exercise
Investigate the effect of varying T_a.

The case $p(s) = (sT)^3$

The stability analysis for the case without dynamic awf yields:

$$(F+1)_{sc} = (\Omega T)^3 \left[\frac{\frac{s}{\Omega}}{\frac{s}{\Omega}+1}\right]^3 \; ; \quad \text{while} \quad (\Delta_{max}+1)_{sc} = \frac{\Delta u}{r_1}$$

(6.76)

and with dynamic awf, setting $T_a := T$, and with compensating awf gain $k_a = k_a^*$:

$$(F+1)_{sc} = (\Omega T)^3 \left[\frac{\frac{s}{\Omega}}{\frac{s}{\Omega}+1}\right]^3 \frac{\frac{s}{\Omega}+1}{sT+1}$$

(6.77)

or with deadbeat awf gain

$$(F+1)_{sc} \rightarrow (\Omega T)^3 \left[\frac{\frac{s}{\Omega}}{\frac{s}{\Omega}+1}\right]^3 \frac{1}{sT+1}$$

(6.78)

with the same value of $(\Delta_{max}+1)_{sc}$ for both cases of k_a.
Again, the last multiplicative factor in Eq. 6.77 is a low pass, which reshapes the Nyquist contour in the favorable direction for stability. Now, two values of closed-loop bandwidth Ω are considered.
In Fig. 6.27 $\Omega := 2.5$ is chosen such that the transient response without dynamic awf is oscillatory, but still converging. Then, the reshaping effect is sufficient to yield a response within specifications.

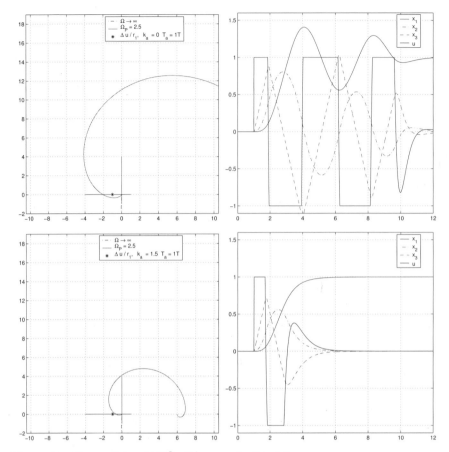

Fig. 6.27. Plant: $p(s) = (\mathbf{sT})^3$, P^+-control, $\Omega = 2.5$,
Peter's Method, $T_a = T_1$: (top) $k_a = 0$; (bottom) $k_a = k_a^*$

For Fig. 6.28 ($\Omega = 5.0$) this effect is not sufficient any more; k_a has to be increased to the deadbeat value (Eq. 6.78).

If Ω needs to be even higher from linear design specifications, then the procedure fails to achieve a stable response. This is because the reshaping capability of the first-order filter G_a is not sufficient for the shape of $(F + 1)$ produced by such high ΩT values.

This weakness will be addressed with the following design method.

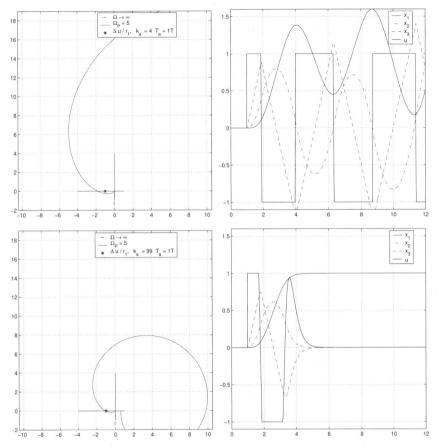

Fig. 6.28. Plant: $p(s) = (s\mathbf{T})^3$, P^+-control, $\Omega = 5.0$,
Peter's method, $T_a = T_1$: (top) k_a compensating; (bottom) k_a deadbeat

6.10 Trajectory Generators with Antiwindup Feedback

Observe that the main cause for the insufficient stability properties (and rate of convergence) is the large initial "overload" of $\Delta u_{lin_{max}}$ versus Δu_{hi}, where $\Delta u_{lin_{max}} \approx r_1(\Omega T)^{n_p}$ is generated by the large setpoint step r_1 with its near-infinite slope.

One work-around to reduce $\Delta u_{lin_{max}}$ by reducing ΩT has already been investigated above.

Here, the idea is to generate a reference trajectory $r_{d_1}(t)$ from the reference step r_1 by an appropriate filter, which produces less "overload" on $\Delta u_{lin_{max}}$, but does not lengthen the settling time unduly, and also leads to $\bar{e}_1 = 0$.

The filter shall be denoted as a "trajectory generator", or TG for short.

The design procedure is given first. It leads to a suitable controller structure. Then the stability properties are investigated, confirming the designed structure and yielding the parameter values. And the performance is checked by simulations on the triple-open-integrator chain plant.

6.10.1 The Design Procedure

- The plant shall be given in a state space representation of control canonical form, such as derived directly from the transfer function representation of (dominant) degree n_p. All state variables $(x_j)_1$, $j = 1 \ldots n_p$, are directly accessible for feedback.

- The controller shall be of P^+-form as above, and its gains $(k_j)_1$, $j = 1 \ldots n_p$, determined by pole assignment: $(s + \Omega_1)^{n_p} = 0$.

- The order of the TG shall be equal to n_p. This will produce for each state variable $(x_j)_1$, $j = 1 \ldots n_p$, of the plant one state variable $(x_j)_2$, $j = 1 \ldots n_p$, in the TG, which then serves as its reference time function $(r_d)_j(t)$, $j = 1 \ldots n_p$.

- The reference trajectory $\mathbf{r}_d(t)$ is the vector of these individual time functions.

- It is best to duplicate the plant model in the TG. Then the reference variables from the filter correspond one-to-one to the plant variables, and no additional scaling factors, *etc.* need to be introduced, *i.e.*
$\mathbf{r}_d(t) := \mathbf{x}_2(t)$.

- Then this plant model is augmented with its own P^+-state feedback with gains $k_{j_2}, j = 1 \ldots n_p$, such that its poles are assigned to $(s + \Omega_2)^{n_p} = 0$, where $\Omega_2 T$ is the TG design parameter. See to it that the steady state gain of the TG is unity.

- Note that $\Omega_2 T$ may be selected very low, such that the plant with its state feedback control can follow the generated trajectory without its control variable $u(t)$ ever saturating. This design strategy may be in order for some practical cases, but it is trivial, and shall not be investigated further.

- Here, $\Omega_2 T$ shall be chosen such that $u(t)$ will transiently saturate. This will reduce the settling time. A typical target response may be near minimum-time, such as found above in the de-tuning approach.
  Then the windup error $e_a(t)$ shall be injected by an awf into the TG, as shown in Fig. 6.29. This introduces one additional design parameter k_a, which is to be determined.

- An alternative is a structure with separate awf paths to each integrator in the TG. But it is more complex and will not be considered further here.

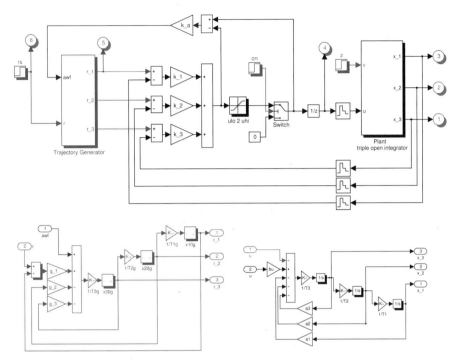

Fig. 6.29. Control Structure:
(lower left)   TG with awf k_a, P$^+$-feedback, Ω_2;
(lower right) plant: $p(s) = (sT)^3$, $(a_1 = a_2 = a_3 = 0)$, P$^+$-control, Ω_1

Fig. 6.28 also illustrates the main feature of the TG design method. It allows one to separate the reference response ("servo" response) with dynamics according to Ω_2 from the disturbance response ("regulator" response) with dynamics determined by Ω_1, and where $\Omega_1 \gg \Omega_2$ is feasible without deteriorating stability properties. This shall be shown next.

6.10.2 Stability Properties

Linear Subsystem

Owing to the assumptions on the awf structure in Fig. 6.29, the main result of Chapter 2 applies here as well.

$$F(s) + 1 = \frac{1 + k_a G_a(s) R_2(s)}{1 + R_1(s) R_2(s) G(s)} \quad \text{with} \quad G(s) = \frac{1}{p(s)}$$

where according to the design procedure, item 5, the TG incorporates the plant $G(s)$, *i.e.* contains $p(s)$ as the denominator of the "plant" $G_g(s)$ with

$$p_g(s) := p(s)$$

to which is added the TG state feedback $g(s)$. Then with Fig. 6.29

$$G(s) := \frac{1}{(sT)^{n_p} + a_{n_p}(sT)^{n_p-1} + \cdots a_2(sT) + a_1}$$

$$G_a(s) = \frac{1}{p_g(s) + g(s)} := \frac{1}{p(s) + g(s)}$$

$$= \frac{1}{(sT)^{n_p} + (a_{n_p} + g_{n_p})(sT)^{n_p-1} + \cdots + (a_2 + g_2)(sT) + (a_1 + g_1)}$$

$$R_2(s) = k_{n_p}(sT)^{n_p-1} + \cdots + k_2(sT) + k_1$$

$$R_1(s) = 1$$

and with the awf design parameter k_a to be determined.
Then

$$F + 1 = \frac{1}{1 + R_1(s)R_2(s)G(s)} \; (1 + k_a G_a(s)R_2(s))$$

$$= \frac{p(s)}{(sT)^{n_p} + \cdots + (k_1 + a_1)}$$

$$\times \frac{(sT)^{n_p} + (sT)^{n_p-1}(k_a k_{n_p-1} + a_{n_p-1} + g_{n_p-1}) + \cdots + (k_a k_1 + a_1 + g_1)}{(sT)^{n_p} + \cdots + (a_1 + g_1)}$$

$$= \frac{(sT)^{n_p} + \cdots + a_1}{(s + \Omega_1)^3}$$

$$\times \frac{(sT)^{n_p} + (sT)^{n_p-1}(k_a k_{n_p-1} + a_{n_p-1} + g_{n_p-1}) + \cdots + (k_a k_1 + a_1 + g_1)}{(s + \Omega_2)^{n_p}}$$

and then with d_{n_p-j} being the binomial coefficients for degree n_p

$$F + 1 = \frac{(sT)^{n_p} + \cdots + a_1}{[s + \Omega_1]^{n_p}}$$

$$\times \frac{s^{n_p} + (d_{n_p-1})s^{n_p-1}\Omega_1\left[k_a + \left(\frac{\Omega_2}{\Omega_1}\right)\right] + \cdots + \Omega_1^{n_p}\left[k_a + \left(\frac{\Omega_2}{\Omega_1}\right)^{n_p}\right]}{s^{n_p} + (d_{n_p-1})s^{n_p-1}\Omega_1 + \cdots + \Omega_1^{n_p}}$$

$$\tag{6.79}$$

This strongly suggests setting $\mathbf{k_a := 1}$ as a design rule

If, furthermore, $\Omega_2/\Omega_1 \to 0$,
then the coefficients of the numerator and denominator polynomial of the
second term in Eq. 6.79 tend to the same values, and

$$\lim_{\Omega_2/\Omega_1 \to 0} [F + 1] = F_2 + 1 = \frac{p(s)}{(s + \Omega_2)^{n_p}} = \frac{(sT)^{n_p} + \cdots + a_1}{(s + \Omega_2)^{n_p}} \tag{6.80}$$

which defines the asymptotic shape $F_2 + 1$.

- In other words, the Nyquist contour now is dominated by the choice of $\Omega_2 T$ for the TG, and the influence of the linear range closed-loop bandwidth Ω_1 tends to zero. This documents the separation effect of "servo" and "regulator" responses mentioned above.
- However, in usual design situations $0 < \Omega_2/\Omega_1 < 1$. Then the second term in Eq. 6.79 will be a lead-lag element, as the numerator coefficients are larger than the corresponding denominator coefficients. This will produce in the relevant ω-region a deformation of the contour of $F_2(j\omega) + 1$ to the left, and thus adverse to the stability properties.
- Equations 6.79 and 6.80 show the strength of this method. By designing the TG as prescribed in the procedure, one is able to *replace* the Nyquist contour for Ω_1 by one with a reduced Ω_2, and thus much better nonlinear stability properties. This is clearly more effective than *reshaping* the Nyquist contour by one first-order filter only.
- And Ω_2 can be directly taken over from the "de-tuning" method.
- Note that by admitting individual awf gains $(k_a)_j$, $j = 1, \ldots n_p$, instead of a common one as in Fig. 6.29, an exact compensation could also be attained for the usual design situation $0 < \Omega_2/\Omega_1$. However, the examples in Section 6.10.3 indicate that the improvement in stability properties is too small to merit the additional complication of the structure.

The Nonlinear Subsystem

Consider the system Fig. 6.29 at steady state with u forced to zero by the run-up enabling switch, but with the setpoint already set to its operating value $r_1(:= 1.0)$. Then for the TG to have unity gain, the reference input r must be weighted by $(a_1 + g_1)$. The awf loop with the TG will reach its equilibrium $\overline{e_a}$ at

$$r_1 \, (a_1 + g_1) = k_a \overline{e_a} \quad \text{where} \quad e_a = \Delta u_{lin} - \Delta u_{hi} \qquad (6.81)$$

Then inserting $k_a := 1$ yields for the position of the straight line (while assuming that this steady state also produces $e_{a_{max}}$)

$$\overline{e_a} := e_{a_{max}} = \Delta u_{lin_{max}} - \Delta u_{hi}$$

$$\Delta_{max} + 1 = \frac{\Delta u_{hi}}{\overline{e_a}} = \frac{\Delta u_{hi}}{r_1 \, (a_1 + g_1)} = \frac{\Delta u_{hi}}{r_1} \frac{1}{(\Omega_2 T)^{n_p}} \qquad (6.82)$$

The Stability Test

With

$$(F_2 + 1)_{sc} = \frac{(sT)^{n_p} + \cdots + a_1}{\left(\frac{s}{\Omega_2} + 1\right)^{n_p}}$$

then $\quad (F + 1)_{sc} = (F_2 + 1)_{sc} \dfrac{\left(\frac{s}{\Omega_1}\right)^{n_p} + \cdots + \left[1 + \left(\frac{\Omega_2}{\Omega_1}\right)^{n_p}\right]}{\left(\frac{s}{\Omega_1}\right)^{n_p} + \cdots + 1} \qquad (6.83)$

Note the similarities to the previous results for the design by "de-tuning" and Peter's method.

6.10.3 Checking the Transient Response

The cases of open integrator chains from above shall be used again, see Fig. 6.30 for $N = 2$ and Fig. 6.31 for $N = 3$.

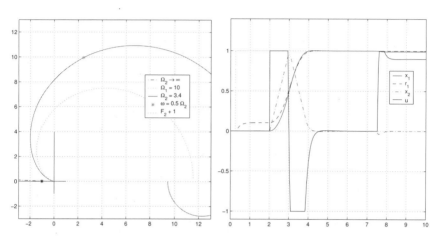

Fig. 6.30. Response of the system in Fig. 6.29 with $p(s) = (sT)^2$, $T = 1$, and with $\Omega_1 = 10.0$ and $\Omega_2 = 3.4$

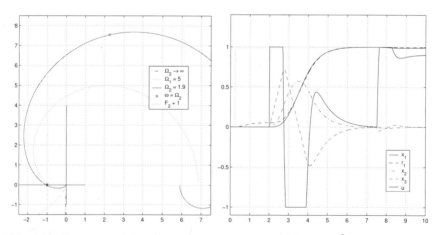

Fig. 6.31. Response of the system in Fig. 6.29 with $p(s) = (sT)^3$, $T = 1$, and with $\Omega_1 = 5.0$ and $\Omega_2 = 1.9$

Considering *performance*, the results for the run-up settling time
to $|u(t)| < 0.05$, $\forall$ $t > t_E$ are as follows:

Plant $p(s)$	Ω_1	Ω_2	$t_E - t_0$ [s]	Min. time
$(sT)^2$	10	3.4	2.3	2.0
$(sT)^3$	5	1.9	5.3	3.175

The second-order case performs well.
But there is a considerable loss of performance for the third-order case. This
is due to the slow trailer generated by the TG with its low Ω_2. Here, some
re-tuning (increasing Ω_2 further from its stability-based design value) may
be tried, but will not have a significant impact.

Discussion

- The results from the simulations nicely confirm the results from the "design
  for stability" process.
- Applying the on-axis criterion leads to a conservative solution, but not
  very much so. A near-minimum transient is obtained by using the off-axis
  criterion and by inserting Ω_2 such that $\Re F(j\omega) + 1)|_{min} \approx 2\,\Delta u/r_1$.
- If r_1 were reduced, but with Δu_{lin} still evolving into the awf region, then
  the TG with awf will produce in its linear final phase a slow trajectory,
  which follows from the low value of Ω_2 (see Fig. 6.31 for $t \geq 4.0$).
- But the stability analysis indicates that $(r_1(\Omega_2 T)^{n_p}/\Delta u)$ should be
  constant as a first approximation. This strongly suggests a simple "gain
  scheduling" type of adaptation of Ω_2 as function of the sizes of r_1 and
  Δu, such as proposed in [46].

Exercise
- Develop the TG design method further to incorporate an I(aw) action.
- Check the design procedure for other plants, for instance
  in Peter's case with $a_3 = 3$, $a_2 = 3$, $a_1 = 1$.
- How do the design methods "dynamic awf" and "TG with awf" relate
  to the standard "setpoint rampers" from process control?

To *summarize*: The "Trajectory Generator with awf"-design has a transparent
structure, which makes it easy to implement. Both transient response and
stability analysis are straightforward, as there is strictly one nonlinearity, the
u saturation. This also holds for the design of the control parameters.
Its main weakness seems to be the comparatively slow "finals" trajectory. This
is due to the low Ω_2 value needed for stability, and which is not suppressed or
switched out as in some other related methods; see Sect. 6.8, 6.14, and 6.15.

6.11 The "Continued States" Concept

This method was developed on a quite different track, [49]. We shall show that it is closely related to the TG method from the previous section.

The starting point was that the state variables in the plant $G_p(s)$ behave differently in the control saturated regime than what the linear state feedback controller would expect. To alleviate this, the method installs a parallel model $G_m(s)$ driven by the awf error $-e_a$ as its input; see Fig. 6.32.

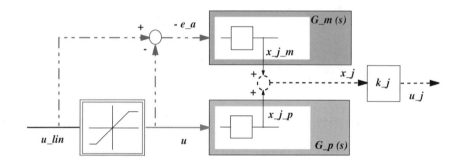

Fig. 6.32. The "continued states" idea

Assume $G_m = G_p$. Then for each state variable x_{j_p} in G_p, a corresponding one x_{j_m} is added by the model to the system.

Assume further adding corresponding state variables

$$x_j = x_{j_p} + x_{j_m} \qquad \forall \ j$$

Finally, assume both systems at rest at the origin, *i.e.* all initial conditions in G_p and G_m at zero.

that is, $u = u_{lin} \ \forall \ t > 0$, and $-e_a = 0$, *i.e.* the additional states x_{j_m} stay at zero, and

$$x_j = x_{j_p} + 0 \ \forall \ j \quad \text{and} \quad u(t) = -(k_1 x_{1_p} \cdots + k_j x_{j_p} \cdots) = u_{lin}(t)$$

In the *saturating regime* of u, the corresponding state variables $x_{j_m}(t)$ run off zero, and supplement the part of $x_{j_p}(t)$ which is intuitively "missing", as perceived from the linear control law.

But note that along the upper control constraint, $-e_a(t)$ is always positive, *i.e.* it drives the states x_{j_m} outward. Any inward movement, *i.e.* any decay of the states x_{j_m}, must therefore be caused by

- either u_{lin} running into the opposite saturation, *i.e.* $-e_a(t)$ being negative,
- or by $-e_a(t)$ still being positive and decaying to zero, in combination with negative internal feedback in $G_m(s)$.

The first situation is associated with the "bang-bang" type of response for second or higher order plants with very large inputs. And the second situation is what will happen with $-e_a(t)$ for any system transient as it returns to the linear control regime around the final steady state.

So G_m *must* be Hurwitz; and, from $G_m := G_p$, the method is strictly applicable only to plants which are Hurwitz. This is a severe restriction for practical applications. Fig. 6.33 visualizes the structure.

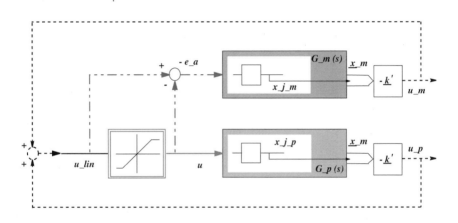

Fig. 6.33. The "continued states" method: the system structure

The work-around for *non-Hurwitz* plants is now to resort to an approximation, where G_m consists of an exact replica of G_p, but with an additional state feedback with gains g_j, $\forall\ j$, such as to obtain a Hurwitz G_m^{mod}. Thus the states x_{j_m} will not be a strict continuation in the above sense, but an approximative one.

This may be carried one step further, by not only using the g_j to ensure that G_m is Hurwitz, but to actively shape the dynamics of G_m, such that a desired overall system response results. Note that, in the initial "continuation of states" design method, the question of desired system response or performance has not been addressed. It only comes in now.

Fig. 6.34 shows the modified structure.

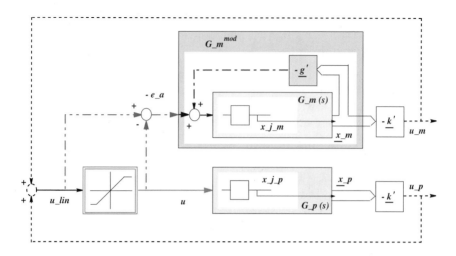

Fig. 6.34. The "continued states" method: the system structure with active feedback in G_m, *i.e.* $G_m \neq G_p$

As a next step, the system Fig. 6.34 is re-drawn into Fig. 6.35. Note the change of signs.

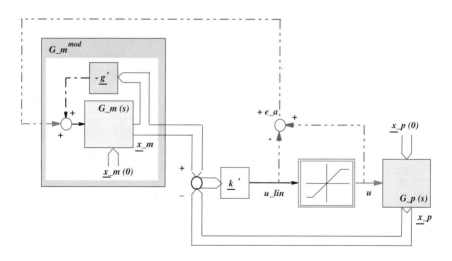

Fig. 6.35. The "continued states" method: the system structure with active feedback in G_m, redrawn

But this is now the structure from the TG method, with one difference: initial conditions appear instead of the reference input r_1 to the TG. But this is very minor for two reasons.

First, that setpoint step response from a given initial steady state up to the final steady state may always be generated as an initial conditions response with appropriately shifted initial conditions. And secondly, the stability-based design of the TG parameters g_j is in fact for the initial conditions response.

Also note the assumption that $G_m := G_p$, which has been used there as well, and the assumption that the corresponding initial conditions in G_m must be the same as in G_p. Otherwise the "continuation of states" concept will not even hold at the first instant. And if the corresponding initial conditions are not the same, then a fast equilibration transient will be generated, *e.g.* see [45].

On the other hand, the g_j of the "continuation of states" can be designed in a transparent way using the stability-based TG design.

To *summarize*: Starting from a very different point, the "continued states" concept leads to a control structure that is very nearly the same as obtained by the TG and awf method, and therefore does not bring new elements to the design. But it clarifies assumptions.

By comparative characteristics, such as complexity of structure, tuning, stability properties and performance, both methods score equally well.

6.12 The Group of "Add-on Output Constraints"

Generally speaking, the dynamic problems of large overshoot or stability problems are caused by state variables within the plant running off their equilibrium values, and not being brought back in time due to the saturation on u ("plant windup").

This suggests the next design approach. It considers such excessive buildup of state variables as a violation of operational constraints on these state variables, *i.e.* as additional output constraints. They have no direct physical meaning, such as was the case in Chapters 3 and 4, but they only help to improve the dynamic properties of the main control loop. Their contribution is to constrain the initial conditions for the finals approach phase of the R_1 loop with its u saturations, as it has been discussed at length in Chapter 4.

More precisely, the constraining action is only needed in some finite time interval before the control is transferred to R_1 for the "finals" phase. The length of this time interval is what is needed by the constraint feedback loop (R_2 loop) to attain those transfer conditions from its own earlier initial conditions, where it has started from. And this is determined by the R_2-loop bandwidth, and also its own control saturations effects.

Further backwards in time the constraining feedback should not restrict u further, as this increases the overall settling time.

Clearly this leads to constraint setpoints $r_{2_{hi}}$, $r_{2_{lo}}$ that are no longer constant, but mutate into functions of the actual state $\mathbf{x}(t)$, in the sense that they increase with the distance from the origin, but still end up in the region of appropriate transfer conditions to R_1-SAT-control around the origin. However, such state-dependent constraint setpoints add to the complexity of the control structure and of the design process. This suggests proceeding by successive approximations: start with the version having constant setpoints, and then extend it to non-constant setpoints wherever the better performance merits the additional complexity.

Considering *implementation* of such add-on output constraints, two structures were discussed in Chapter 3, the cascade-limiter and the selector methods. The cascade-limiter method is related to the "nested loop" method, and shall be investigated first. Then two structures with Max-Min-selectors are looked into. Note that the Max-Min-selectors may be replaced directly by the equivalent deadspan nonlinear-adding element. As will be seen, this may lead to systems with multiple nonlinearities. Thus, the structure gets more complex than in Chapter 4 for the general case of dominant plant order n_p, and the stability test tends to be more complex as well. However, a conservative approximation shall be presented, which is still manageable. As plant models, the open integrator chains of order $n_p = 2$ and $n_p = 3$ shall be used. This makes the basic idea, the design procedure and the control structure more transparent than if the most general case of $G(s)$ were considered.

Exercise
Consider the extension to the general case.

6.13 The Cascade Limiter Method

6.13.1 Problem Specification

As stated above, the *plant model* is assumed to be

$$G(s) = \frac{1}{(sT)^{n_p}} \; ; \quad n_p = 3 \qquad (6.84)$$

with T as a suitably selected time scaling factor. The corresponding state space representation is then in control canonical form. Note that for the more general case it is recommended to use this assumption as well.

The *linear controller* is designed by state feedback as in the preceding sections, yielding the feedback gains k_j. The closed-loop bandwidth is set to $\Omega_1 = 5$ as above, which produces an unstable response for $r_1 = 1.0$.

But then the structure is redrawn in a cascaded form; see Fig. 6.36. This is a linear transformation, yielding as gains h_j in the nested loops (by inspection of Fig. 6.36)

$$h_{n_p} = k_{n_p}$$

$$h_{n_p-1} \cdot h_{n_p} = k_{n_p-1} \quad \rightarrow \quad h_{n_p-1} = \frac{k_{n_p}}{k_{n_p-1}}$$

$$h_1 \cdot h_2 \cdots h_{n_p-1} \cdot h_{n_p} = k_1 \quad \rightarrow \quad h_1 = \frac{k_1}{k_2} \qquad (6.85)$$

Note that this uses the assumption on the special form of $G(s)$.

And let the *control saturation* be symmetrical to the final equilibrium value $\overline{u}$

$$\Delta u_{hi} = u_{hi} - \overline{u} = -\Delta u_{lo} = u_{lo} - \overline{u} \qquad (6.86)$$

6.13.2 The Design Procedure

This is developed by induction.

The case $N = 1$ has been discussed in Chapter 2, with the controller being of P type (no I(aw) action). There is no such plant internal state which can wind up, and no action is required.

The case $N = 2$ has been discussed in Chapter 4 in the phase plane, with x_1 proportional to the controlled variable y_1. Now there is one such plant internal state variable x_2, which can wind up. As

$$x_2(t) = T\frac{d}{dt}x_1(t)$$

the "plant windup" is an excessive buildup of approach speed. It must therefore be constrained such that it may be reduced to zero in time by the saturated effect of braking.

This has been discussed in detail in Chapter 4, see Eq. 4.9, yielding speed constraint setpoints $r_{2_{hi}}^*$, $r_{2_{lo}}^*$ as upper limits, such that the subsequent final approach does not overshoot unduly. These values may be adopted directly as saturation settings on r_2.

Note that due to the assumption in Eq. 6.86 and on $G(s)$ the constraint setpoints are symmetrical as well and the pair shall be denoted as $r_{2_{SAT}}$.

In practical applications, a safety factor β should be inserted for robustness

$$r_{2_{SAT}} := \beta \, r_{2_{SAT}}^* \quad \text{with} \quad 0 \ll \beta < 1$$

The case $N = 3$ is depicted in Fig. 6.36.

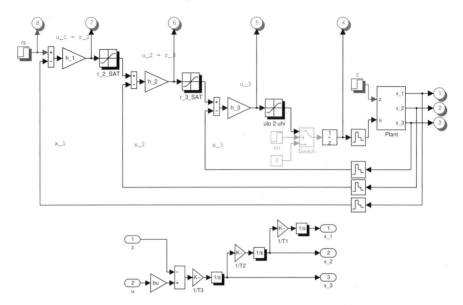

Fig. 6.36. The "cascade limiter" control system with the third-order plant

It is built around the $N = 2$ structure as the inner subsystem, where indices in the innermost loop have been incremented by one:

$$x_2 \rightarrow x_3, \ x_1 \rightarrow x_2, r_{2_{SAT}} \rightarrow r_{3_{SAT}}$$

To this is added a third integrator in the plant structure for x_1, and an additional proportional feedback with gain h_1 and output u_1, and a symmetrical setpoint saturation on it, to $r_{2_{SAT}}$.

6.13.3 Transient Response

This is shown in Fig. 6.37, where $r_{3_{SAT}} = 0.5$ has been taken from Chapter 4, and with two values of $r_{2_{SAT}}$ to illustrate the effect of its tuning.

Clearly, the design parameter $r_{2_{SAT}}$ should be made as large as possible to shorten the run-up transient, but from Fig. 6.37 there is an upper threshold to avoid overshooting in the "finals" approach phase. The parameter $r_{3_{SAT}}$ is of minor influence, but if it is significantly reduced (such as to avoid the overshoot on x_2 in the initial phase of the run-up), then the movement of u gets more constrained, and $r_{2_{SAT}}$ must be reduced as well.
Suitable numerical values would be $r_{3_{SAT}} = 0.30 \ \rightarrow \ r_{2_{SAT}} = 0.24$.

In Fig. 6.37 this produces a run-up settling time to $|u(t)| < 0.05 \ \forall \ t \geq t_E$ of $t_E - t_0 = 4.8$ s, where the minimum-time would be 3.175 s, *i.e.* an increase to approximately 150%.

Note also that this is caused by an extended phase in the run-up transient, where u is not at its limits, but at zero. And this is due to the approach speed constraint $r_{2_{SAT}}$ being constant, and set to what is determined by the finals approach. Such "coasting" on the trajectory will lengthen the run-up time linearly with increasing size of r_1.

This is clearly suboptimal, and less than what is produced by the TG with awf design. But there is the usual tradeoff with robustness.

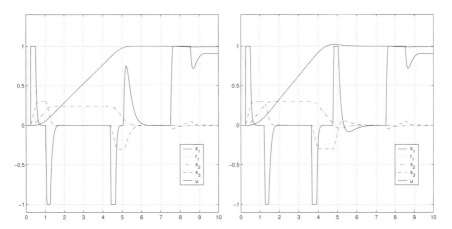

Fig. 6.37. Run-up and load swing responses of the "cascade limiter" control system with the plant $1/(sT)^3$:   (left) with $r_{2_{SAT}} = 0.24$, (right) with $r_{2_{SAT}} = 0.30$

6.13.4 Stability Analysis

As mentioned above, the system now has n_p saturation nonlinearities. A work-around using one saturation element has been presented in Chapter 5 for actuator slew and stroke limitations. It shall be extended here.

Consider first the innermost loop for x_3 in Fig. 6.36. There, the control saturations u_{SAT} are acting as slew limitations on x_3. And its setpoint limitations $r_{3_{SAT}}$ are acting as stroke saturations. From the transients on u in the nonlinear phases in Fig. 6.37, one infers that $r_3(t)$ will change no faster than stepwise. Then the innermost loop may be conservatively replaced by a linear one with

$$G_3 = \frac{x_3}{r_3} = \frac{1}{s\tau_3 + 1} \; ; \; \text{where from Chapter 5} \quad \tau_3 = T\frac{r_{3_{SAT}}}{u_{SAT}} \qquad (6.87)$$

Extending this one step outward to the x_2 loop yields

$$G_2 = \frac{x_2}{r_2} = \frac{1}{s\tau_2(s\tau_3 + 1) + 1} \quad \text{with} \quad \tau_2 = T\frac{r_{2_{SAT}}}{r_{3_{SAT}}} \qquad (6.88)$$

Considering finally the x_1 loop, the saturation r_{2SAT} is now the only nonlinearity in the "conservative approximation", with index ca; see Fig. 6.38

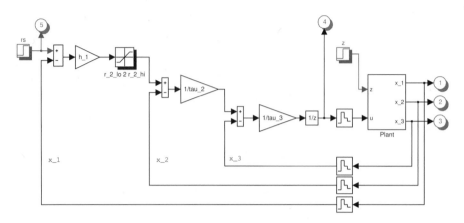

Fig. 6.38. The "conservative approximation" system

For the *stability test* the main results of Chapter 2 apply

$$F_{ca} + 1 = \frac{sT[s\tau_2(s\tau_3 + 1) + 1]}{s^3 T\tau_2\tau_3 + s^2 T\tau_2 + sT + h_1}$$

$$= \frac{(sT)(s^2\tau_2\tau_3 + s\tau_2 + 1)}{h_1\left(s^3 \frac{T\tau_2\tau_3}{h_1} + s^2 \frac{T\tau_2}{h_1} + s\frac{T}{h_1} + 1\right)}$$

$$\text{and} \quad (\Delta_{max_{ca}} + 1) = \frac{r_{2SAT}}{r_1\,h_1} \tag{6.89}$$

Compared with the case without cascade limiters, the shape of the Nyquist contour will be deformed significantly in the favorable direction: for $\omega \to 0$ it converges to zero as sT does, *i.e.* along the positive imaginary axis. And this shows stability for "almost all" r_1 step sizes, which covers the practical needs. Note that the stability of the linear system with the closed-loop characteristic polynomial $s^3 T\tau_2\tau_3 + s^2 T\tau_2 + sT + h_1$ must be checked as well. This leads to restrictions on τ_3 and τ_2, *i.e.* on r_{3SAT} and r_{2SAT} (they may not be made arbitrarily small).

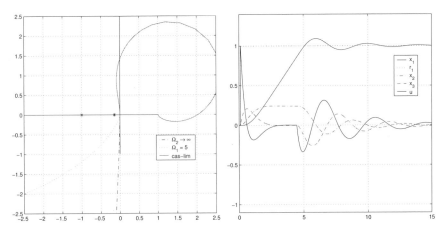

Fig. 6.39. Stability test with Eq. 6.89 and run-up response of the system in Fig. 6.38 with $r_{2_{SAT}} = 0.24$, $r_{3_{SAT}} = 0.30$

6.14 Selection for Approach Speed

The main drawback of the cascade limiter method is that the additional constraints are constants, and must be set such that the "finals" trajectory has no perceptible overshoot. This is too time-consuming for the run-up transient as a whole. It would be convenient to have variable constraints. And those can be implemented very easily, if the selectors method is used instead.[2]

The method shall be introduced with the double-open-integrator chain plant, where it can be visualized in the phase plane. Then it shall be generalized to the $n_p > 2$ case.

6.14.1 The Design Procedure

It consists of

1. replacing the cascade limiter structure in Fig. 6.36 by the nearly equivalent "cascade of R_j plus selectors" with $j = 2, \cdots n_p$, as shown in Fig. 6.41,
2. where initially the setpoints of the additional constraint controller pairs are constants, and determined as in the cascade limiter method. The transition to the "finals" approach phase yields the value for $r_j|_0$.
3. Then those setpoints are replaced by functions of the distance to the origin.
4. A typical setup is a linear combination of the state errors

$$\mathbf{r}(\mathbf{x}) = -Q(\mathbf{x} - \overline{\mathbf{x}}) + \mathbf{r}|_0 \qquad (6.90)$$

---

[2] As an **exercise**, try to find an equivalent structure based on the cascade limiter concept.

with the coefficients matrix Q of appropriate dimensions. Note the sign convention.

The design method for $\mathbf{r}(\mathbf{x})$ of Eq. 6.90 is used in "sliding mode" systems, *e.g.* [10]. There, the functions must be selected such that u does not saturate during a substantial time interval, when the system state moves along the "sliding mode" trajectory. Note that, in such sliding mode systems, the control gains are set very high, and "high-frequency chattering" appears on u. This is not the case here, where the control is linear on $u_{lo} \cdots u_{hi}$. In other words, the switching planes of the sliding mode system turn into corridors of finite width along both the "sliding surface" $u_{2_{hi}} = 0$ and $u_{2_{lo}} = 0$.

From this, a last item is generated for the design procedure:

5 Select the coefficients in Q such that, along the "sliding" trajectory, $u(t)$ is not forced into its saturations u_{SAT}.

From Fig. 6.37 this is feasible at least for all coefficients in Q being zero, and $\mathbf{r}\big|_0$ small enough (such that the finals trajectory is linear).

6.14.2 The Second-order Case $G(s) = 1/(sT)^2$

- Concerning item 3 of the design process, a possible setup for the functions $\mathbf{r}(\mathbf{x})$ would be parabolas, such that u will not saturate while the trajectory runs along them.
- Applying item 4 produces here

$$r_{2_{hi}} = -q(x_1 - r_1) + r_{2_{hi}}\big|_0 \quad \text{and} \quad r_{2_{lo}} = -q(x_1 - r_1) + r_{2_{lo}}\big|_0 \quad \text{with} \quad q \geq 0 \tag{6.91}$$

- From item 5, q should be selected such that $u(t)$ does not saturate during an excessive time interval. In other words, the line $u_{2_{hi}} = 0$ in the phase plane must stay to the lower left of the "finals" parabola generated for u_{lo}; see Fig. 6.40. Thus, the value of q is determined by the maximum initial conditions arising in the specific application.

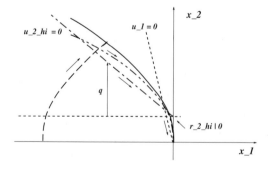

Fig. 6.40. Visualization of the design process for $G = 1/(sT)^2$ in the phase plane

The Control Structure

The control structure is shown in Fig. 6.41.

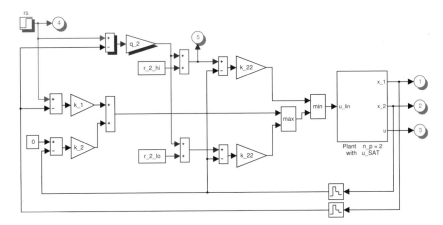

Fig. 6.41. The control structure for $G = 1/(sT)^2$

The linear controllers R_1, and $R_{2_{hi}} = R_{2_{lo}}$ are designed to

$$\Omega_1 := 10 \quad \text{and} \quad \Omega_2 := 20 \quad \text{with} \quad T_s = 0.01 \qquad (6.92)$$

Transient Response

The simulation results are given in Fig. 6.42. There, $r_{2_{hi}}|_0 := 0.24$ has been selected, such that $u(t)$ does not run into the saturations during the finals trajectory (see chapter 4). This is for improved robustness.

Then three values of q_2 are used. The lower limit case $q_2 = 0$ yields constant setpoints and corresponds to the cascade limiter method, although with lower values for $r_{2_{hi}}$, as one would use there. The upper limit case is to select $q_2 = q_2^*$, such that also during the second part of the run-up transient u saturates nearly everywhere, and thereby generates the minimum-time approach parabolas in the phase plane; see Fig. 6.42 (bottom left). For this benchmark case, and for instantaneous transition of u between saturations, the numerical value is q_2^{max}. And then taking into account the delayed transition in the linear corridors, tuning yields q_2^*.

$$q_2^{max} = 2\left(1 - r_{2_{hi}}|_0\right) \quad \rightarrow \quad q_2^* \approx 0.95\, q_2^{max} \qquad (6.93)$$

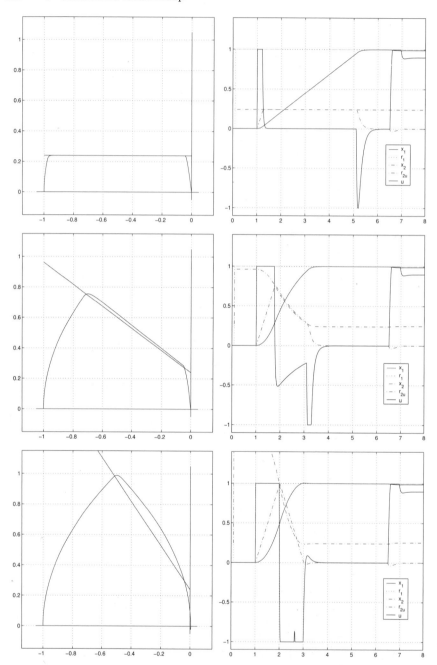

Fig. 6.42. Responses of the control structure for $G = 1/(sT)^2$, with $r_{2_{hi}}|_0 := 0.24$: (left) phase portrait; (right) time response; (top) $q_2 = 0$; (center) $q_2 = 0.5\ q_2^*$; (bottom) $q_2 = q_2^*$

Consider now the run-up time as a measure of performance. From the simulations, and taking the end time t_E such that $|u(t)| \leq 0.05 \ \forall \ t \geq t_E$,

q_2/q_2^*	0	0.5	1.0	Min. time
$t_E - t_0$ [s]	4.7	2.65	2.20	2.0

There is a substantial gain by increasing q_2 from zero while still at small values, but it diminishes for q_2 approaching q_2^*. In other words, as a design rule-of-thumb, not much will be lost in run-up time for $q_2/q_2^* > 0.5$, but robustness will improve.

Stability Test

The procedure of partitioning the phase plane from Chapter 4 is used. The "finals" approach under R_1-control is linear (by choosing $r_{2_{hi}}\big|_0$ appropriately low). Then, for initial conditions close enough to the "sliding surface" $u_{2_{hi}} = 0$, the results of Chapter 3 apply.

For the *linear subsystem*:

$$F + 1 = \frac{1 + R_2 G_2}{1 + R_1 G_1} \ ; \ \text{where}$$

$$1 + R_2 G_2 = 1 + k_{22}\frac{1}{sT} + q_2 k_{22}\frac{1}{(sT)^2} = \frac{(sT)^2 + k_{22}(sT) + q_2 k_{22}}{(sT)^2}$$

i.e.

$$F + 1 = \frac{(sT)^2}{(sT)^2 + k_2 sT + k_1} \ \frac{(sT)^2 + k_{22}sT + q_2 k_{22}}{(sT)^2}$$

$$= \frac{(sT)^2 + k_{22}sT + q_2 k_{22}}{(sT)^2 + k_2 sT + k_1}$$

As $q_2 \ll k_{22}$, the numerator can be approximated by

$$(sT)^2 + k_{22}(sT) + q_2 k_{22} \approx (sT)^2 + (k_{22} + q_2)(sT) + q_2 k_{22}$$

$$= (sT + k_{22})\,(sT + q_2)$$

$$\text{that is} \quad F + 1 \approx \frac{\Omega_2 T}{(\Omega_1 T)^2}\ \frac{\frac{s}{\Omega_2} + 1}{\frac{s}{\Omega_1} + 1}\ \frac{sT + q_2}{\frac{s}{\Omega_1} + 1} \tag{6.94}$$

Note that

$$\lim_{\omega \to 0}\left(|F(j\omega) + 1|\right) = q_2 \frac{\Omega_2 T}{(\Omega_1 T)^2} \ \geq \ 0 \tag{6.95}$$

as from the structure $\quad q_2 \geq 0$.

That is, the Nyquist contour evolves in the right half plane, except at $\omega \to 0$ and for $q_2 = 0$. In other words, the stability properties improve with increasing q_2 values. The low limit is for $q_2 = 0$, *i.e.* for the cascade limiter method. This is intuitively clear, as the constraints on $u(t)$ get weaker.

And for the *nonlinear subsystem*, again from Chapter 3:

$$\Delta_{max} + 1 := \frac{r_{2_{hi}} k_{22}}{r_1 k_1} = \frac{r_{2_{hi}}}{r_1} \cdot \frac{\Omega_2 T}{(\Omega_1 T)^2} \tag{6.96}$$

That is, the position of the straight line for the test is independent of q_2, and the same as for $q_2 = 0$.

Then, by extending the initial conditions laterally away from the sliding surfaces, the active system part is the R_2-SAT control, with

$$F + 1 = \frac{1}{1 + R_2 G_2} = \frac{(sT)^2}{(sT)^2 + k_{22}(sT) + q_2 k_{22}} \approx \frac{sT}{sT + k_{22}} \frac{sT}{sT + q_2}$$

$$\text{and with} \quad q_2 \ll k_{22} \quad \rightarrow \quad F + 1 \approx \frac{sT}{sT + k_{22}} \tag{6.97}$$

and for the nonlinear part

$$\Delta_{max} + 1 = \frac{u_{SAT}}{r_{2_{hi}}(r_1) \, k_{22}} = \frac{u_{SAT}}{(q_2 r_1 + r_{2_{hi}}|_0) \, k_{22}} \tag{6.98}$$

The Nyquist contour starts from the origin as $(sT)^2$ does, which indicates local stability only. However, as $q_2 \ll k_{22}$, the Nyquist contour will not proceed far into the left-hand plane, such that the radius of attraction will be sufficiently large, and covers what is needed for most applications.

Fig. 6.43 shows the Nyquist contours for the three cases of Fig. 6.42.

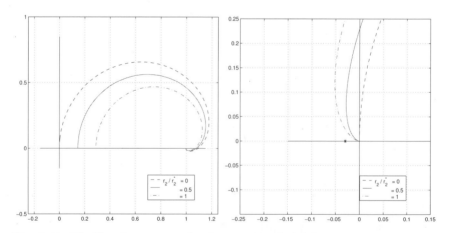

Fig. 6.43. The Nyquist contours for the partitioned stability test:
(left) for trajectories along the sliding surface to the origin; (right) for trajectories from the initial conditions to the sliding surface; where * indicates $\Delta_{max} + 1$ for $r_1 = 1.0$

To *summarize*: for $q_2/q_2^* \to 1$ the stability properties of the trajectories from the initial conditions to the sliding surface turn out to be more critical than for the rest, from near to the sliding surface up to the final equilibrium. Note that this property is also due to the particular choice for the value of $r_{2_{hi}}\big|_0$.

6.14.3 The Third-order Case $\quad G(s) = 1/(sT)^3$

First, the approach from Sect. 6.14.2 will be used here.
The initial step is again to select the value of $r_{2_{hi}}\big|_0$, such that the "finals" trajectory will be linear, *i.e.* $u(t)$ does not meet its saturation. This again will provide nice stability properties for all trajectories originating near the sliding surfaces, as for $0 < \omega/\Omega_1 \ll 1$ the Nyquist contour will have the same typical shape as in Fig. 6.43 (left).

> **Exercise**
> Show this.

The second step is to look at the lateral approach to the sliding surface. It now has to be done with a double-open-integrator chain plant, in combination with the saturation u_{SAT}. As demonstrated in previous chapters, this tends to overshooting for large Ω_2 entries (which one needs to produce an acceptable tracking quality along the sliding surface), and a large initial control error $e_2(0)$ (that is, a large lateral offset from the sliding surface).
Starting from the initial equilibrium at $x_1 = -r_1$, $x_2 = 0$, the initial control error is

$$e_2(0) = r_{2_{hi}}(0) - x_2(0) = \left(q_2\, r_1 + r_{2_{hi}}\big|_0\right) - 0 \qquad (6.99)$$

In other words, avoiding overshoots of $x_2(t)$ relative to the sliding surface by q_2 alone will result in a quite restrictive upper bound on q_2. But this will slow down the run-up unduly, *i.e.* result in poor performance.
As has been shown above, the overshoot may also be reduced by constraining the approach speed to the sliding surface, *i.e.* by introducing an additional constraint pair on x_3. The reference values $r_{3_{SAT}}$ can be set constant, see Fig. 6.44, or optionally increasing with $|e_2|$.

The Control Structure

The control structure is shown in Fig. 6.44 (top).
The linear controllers R_1, and $R_{2_{hi}} = R_{2_{lo}}$, $R_{3_{hi}} = R_{3_{lo}}$ are designed to

$$\Omega_1 := 5 \quad \Omega_2 := 10 \quad \text{and} \quad \Omega_3 := 20 \qquad (6.100)$$

Transient Responses

Consider Fig. 6.44 (center).
First $r_{2_{hi}}\big|_0$ is tuned manually to $:= 0.15$. This produces a linear "finals" trajectory.

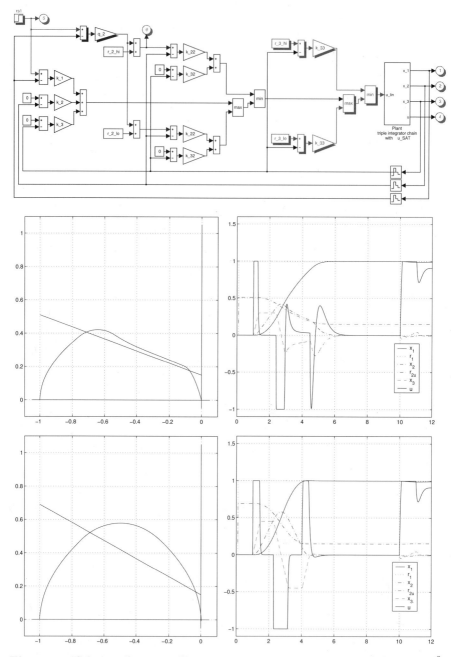

Fig. 6.44. "Selection for approach speed". The control structure for $G = 1/(sT)^3$ and its responses, with $r_{2_{hi}}|_0 := 0.15$:
(center) $q_2 = 0.36$, $r_{3_{SAT}} = 0.30$; (bottom) $q_2 = 0.54$, $r_{3_{SAT}} = 0.45$

Then q_2 is increased from zero to shorten the run-up time as much as possible, and at the same time $r_{3_{SAT}}$ is lowered to avoid excessive overshoot on $x_2(t)$ at the sliding surface. A reasonable combination is $q_2 = 0.36$, $r_{3_{SAT}} = 0.30$. A second combination, $q_2 = 0.54$, $r_{3_{SAT}} = 0.45$, is used in Fig. 6.44 (bottom), to shorten the run-up time further. But $u(t)$ saturates much longer, and robustness is reduced.

Run-up times to $|u| \leq 0.10$ are summarized in the following table.

q_2	0	0.36	0.54	Min. time
$t_E - t_0$ [s]	7.65	4.75	3.60	3.175

Concerning stability, the structure now has three nonlinear additive elements in series. Therefore, the test will be more involved, and shall not be followed up here.

Exercise
Investigate the stability properties.

To *summarize*: The structural complexity is much higher than for the second-order case. This makes tuning and stability testing more involved. The performance is acceptable from a qualitative point of view, but note that the main quantitative measure of performance, which is the relative increase of settling time versus the minimum-time, is higher that for the second-order case.

6.15 Selection for Lower-bandwidth R_1-Control

Looking at the phase plane plots in Sect. 6.14 and comparing them to what has been discussed in Chapter 4 suggests a new approach to design. The basic idea is to reinterpret the inclined sliding surface $u_{2_{hi}}(e_1)$ as being generated by a *second R_1 controller*, with less aggressive (lower-bandwidth) tuning.

6.15.1 The Design Procedure

- The "second" R_1 controller is denoted by $R_{12_{hi}}$.
- It shall have a *lower* bandwidth $\Omega_{12_{hi}} < \Omega_1$, such that the line $u_{12_{hi}} = 0$ is more inclined to the left
  (note that from Chapter 4 its slope is given by $(\Omega T)/n_p$),
- and an additional *vertical offset* $u_{12_{hi}}|_0 > 0$, such that it does not cross the origin, as the line $u_1 = 0$ does
  (this is to avoid interference of the $R_{12_{hi}}$ control with the linear R_1 control for small deviations).
- The same setpoint $r_1(t)$ shall be applied to both R_1 and $R_{12_{hi}}$ controllers.
- The same approach is applied in the opposite direction with $R_{12_{lo}}$, with $u_{12_{lo}}|_0$ and $\Omega_{12_{lo}}$.
  This is visualized in Fig. 6.45.

- The *offsets* in both directions are design parameters. They are set to the control saturation values, *i.e.* $u_{12_{hi}}|_0 := u_{hi}$ and $u_{12_{lo}}|_0 := u_{lo}$.
  This is an intuitive compromise to have low interference with the R_1-loop (in the linear regime), and also to provide reasonably bounded transfer conditions on x_2, x_1 from the R_{12} regime to the R_1 regime.
- And the design parameter *slope* is derived via Ω from the de-tuning method; see below. The slopes for up and down movement may be designed to different values.

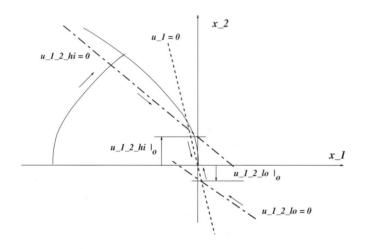

Fig. 6.45. Visualization of the design process in the phase plane

6.15.2 Stability Properties

The last item on the design procedure list connects to the stability analysis. Again the partitioning approach is used.

In the first phase of the run-up transient, the $R_{12_{hi}}$ controller will be selected, as in this time window $u_{12_{hi}}(t) < u_1(t)$, see Fig. 6.45 (this is mainly due to $\Omega_{12_{hi}} < \Omega_1$). And then the control saturation u_{lo}, u_{hi} is applied. In other words the active subsystem is the $R_{12_{hi}}$-SAT control loop.

Consider now the selector to be blocked, such that no transfer can occur. Then the basic stability analysis from Sect. 6.4 can be performed directly. The bandwidth Ω_{12} is a design parameter (and not fixed from the linear specifications as Ω_1 is). Thus, the de-tuning method from Sect. 6.7 applies, and yields Ω_{12}.

But there is the offset $u_{12_{hi}}|_0 := u_{hi}$. For the final steady state it has to be compensated by the nonzero controller output $\bar{u}_{12_{hi}} = -u_{hi}$, which requires

a (small) steady state control error $\bar{e}_{12_{hi}} = u_{hi}/k_{12} = u_{hi}/(\Omega_{12}T)^3$.
Thus the $R_{12_{hi}}$-SAT control loop (while the selection for R_1 is still blocked)
will stabilize the run-up transient close to the origin (at $\bar{e}_{12_{hi}}$). Therefore,
while crossing over the (blocked) transfer conditions to R_1 at time t_T, the
values of the state vector $\mathbf{x}(t_T)$ are bounded. Thus, the associated $(u_{lin})_{12_{hi}}$
at t_T is bounded as well to some finite range above the offset u_{hi}.

Now the selector is de-blocked. Then the transfer condition at time t_T

$$(u_{lin})_1\big|_{t_T} < (u_{lin})_{12_{hi}}\big|_{t_T}$$

must be such that the state vector $\mathbf{x}(t_T)$ lies within the radius of attraction
for the R_1-SAT loop. This puts an upper bound on the ratio Ω_{12}/Ω_1.

So far, this is only a plausibility argument.

Exercise
Try to prove this.

6.15.3 Transient Response

The case of $G(s) = 1/(sT)^3$ is considered first; see Fig. 6.46.

The *structure* is much less complex than the previous one for "selection for
approach speed" in Fig. 6.44.

The *design* is as follows:

- The offsets are set to u_{hi} and u_{lo}, as per the design procedure.
- The bandwidth of the R_1 loop is set to $\Omega_1 := 5$, as above.
- For R_{12}, a first possible design target is good stability properties. So Ω_{12}
  is taken from the de-tuning method, which indicates stability with the on-
  axis circle criterion for $\Omega = 1.575$ (see Fig. 6.14 (top row)), and therefore
  set $\Omega_{12} := 1.575$.
- A second possible design target for R_{12} is the shortest run-up settling time
  without overshoot. Then, the value for Ω_{12} has to be tuned manually with
  simulations, yielding $\Omega_{12} \sim 1.91$. In Fig. 6.14, this led to $\Omega \sim 1.975$.
  In other words, the de-tuning method seems to produce a good initial value
  for the second design target as well.

The *performance* (measured by the settling time t_E to $|u(t)| < 0.05$, and listed
below) is substantially improved, *e.g.* compared with the TG and awf method
in Sect. 6.10.

Ω_{12}	1.575	1.91	Min. time
$t_E - t_0$ [s]	3.85	3.25	3.175

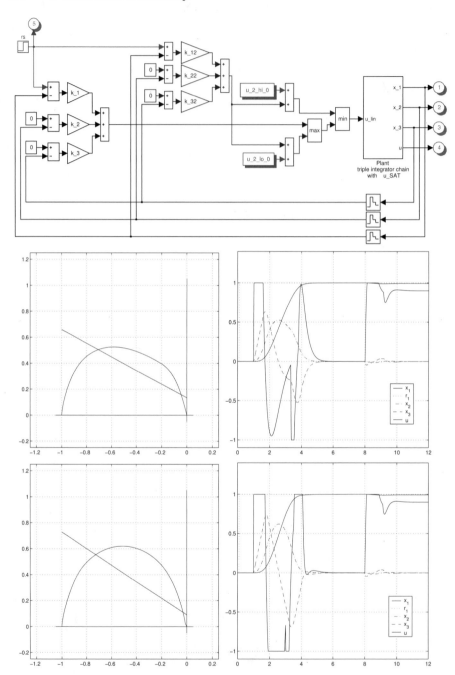

Fig. 6.46. "Selection for lower-bandwidth R_1 control".
The control structure for $G = 1/(sT)^3$ and its responses, with $\Omega_1 = 5.0$, and
(center) $\Omega_{1_2} = 1.575$; (bottom) $\Omega_{1_2} = 1.91$; see text

From the experiments, and for the first design target $\Omega_{1_2} := 1.575$, $u(t)$ will touch the saturations only for short time intervals, when the new loop switches in. So a reasonable robustness can be expected. Nevertheless the increase in run-up time versus the minimum-time trajectory is small (approximately 20 %).

Also from the experiments, and for the second design target $\Omega_{1_2} := 1.91$, the run-up time $t_E - t_0$ is quite sensitive to the choice of Ω_{1_2}. This is not surprising, as $u(t)$ runs along the saturations most of the time, and thus robustness must be low.

6.15.4 Checking the design procedure

The case of $G(s) = 1/(sT)^2$ shall be used, with $\Omega_1 := 10$. Then for stability by the on-axis circle criterion, $\Omega_{1_2} := 2.75$ (see Fig. 6.12).

The simulation results are given in Fig. 6.47. The settling time is $t_E - t_0 \approx 2.5$ [s] versus $t_E - t_0|_{min} = 2.0$ [s].

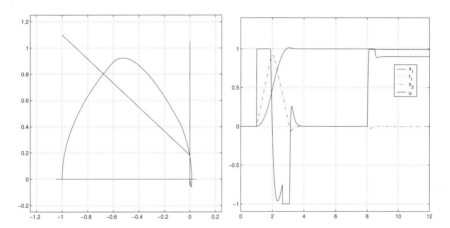

Fig. 6.47. Response with "Selection for lower-bandwidth R_1 control" for $G = 1/(sT)^2$,
$\Omega_1 = 10$, $\Omega_{1_2} = 2.75$

To *summarize*: the "selection for weak R_1-control" method scores high regarding such comparative characteristics as complexity of structure, number of tuning parameters, stability properties, and performance of both run-up and load swing transients.

6.16 Links to Model Predictive Control

6.16.1 Introduction

MPC is one of the three fundamental approaches (see Chapter 1) to the design of control systems with input and output constraints. It is based on numerical optimization of system trajectories rather than the analytical approach (by the maximum principle) or the intuitive design (by antiwindup or overrides). There are excellent books on the subject, for instance see [57, 58]. This is also a very active research area, see the proceedings of the recent *Control and Decision Conferences* (CDC), and the *American Control Conferences* (ACC), *etc.* One of the research topics is on the relations of antiwindup and MPC.

In the following we shall focus on one particular approach ([59] and references therein, and [60]), which has been developed quite recently, and which nicely fits to what has been presented so far.

First, three benchmark problems are stated. Then the classic solution is applied, with numerical trajectory optimization repeated at each sampling instant, *i.e.* with moving horizon. Next the explicit form of MPC is discussed for one of the benchmarks, and the link to the "selection for lower-bandwidth control" method is established.

The support of Tobias Geyer and Francesco Borrelli from our laboratory is gratefully acknowledged.

6.16.2 The Benchmarks

In order to illustrate the relations to the designs resulting from the intuitive approach, the three main benchmark problems used so far will be investigated.

- The *plant* shall be modelled by:
  - Case *i*, for one integrator

$$G_u(s) = \frac{1}{sT} \tag{6.101}$$

  - Case *ii*, for a double integrator chain

$$G_u(s) = \frac{1}{(sT)^2} \tag{6.102}$$

  - Case *iii*, for a triple integrator chain

$$G_u(s) = \frac{1}{(sT)^3} \tag{6.103}$$

The time constants shall be equal for simplicity, and set to $T = 1$ s. They shall be known a priori with a sufficiently small tolerance, such that

robustness need not be investigated for the moment.
The small delay D used before shall be omitted here.
Also, there shall be no high-frequency measurement disturbance present.

- For the *test sequence*, the focus is on the run-up phase from standstill $(r_1 = 0)$ to the nominal setpoint value $(r_1 = 1.0)$. However, the equivalent set of initial conditions shall be applied

$$y_1(0) = -1.0, \text{ and where needed } \quad y_2(0) = y_3(0) = 0$$

and the transient to the origin shall be optimized.
No further small setpoint steps or load steps shall be applied.

Exercise
You may want to extend this.

- For the *constraints*, and consistent with the current chapter, only input saturations u_{lo}, u_{hi} shall be considered. They are set symmetrical to the final steady state value $\bar{u}$, which is from the specification of the test sequence

$$\bar{u} = 0 \quad \rightarrow \quad u_{lo} = -1.0; \; u_{hi} = +1.0 \tag{6.104}$$

Also, there shall be no constraints on slew rate.

- The *controller* for the small range shall be linear (to avoid chattering on $u(t)$ around steady state), and with time-invariant coefficients (for ease of implementation).

- Regarding *performance*, the resulting trajectory shall be by design optimal with respect to the given value function (see Sect. 6.16.3). In other words, the value function and its entries are now the design parameters to be selected in a suitable manner.
  In addition to being optimal in the above sense, the transient shall have at most an acceptably small overshoot ($\lesssim 2.5\%$) in the finals approach phase. This is in compliance with usual applications specifications. It must be achieved by appropriate selection of the design parameters in the value function.

6.16.3 Classic Model Predictive Control

The Optimization Problem

The system to be controlled (the plant) is

$$x(t+1) = Ax(t) + Bu(t)$$
$$\text{and} \quad y(t) = Cx(t) \tag{6.105}$$

which describes a linear time-invariant system, in discrete-time form at time $t = kT_s$ and $t + 1 := (k + 1)T_s$, where $x(t) \in \mathbb{R}^n$, $u(t) \in \mathbb{R}^m$, $y(t) \in \mathbb{R}^p$, and the pair (A, B) is stabilizable.

The constraints are

$$u_{lo} \leq u(t) \leq u_{hi} \text{ with } u_{lo} \leq u_{hi} \text{ as } m\text{-dimensional vector}$$

$$\text{and}\quad y_{lo} \leq y(t) \leq y_{hi} \text{ with } y_{lo} \leq y_{hi} \text{ as } p\text{-dimensional vector} \quad (6.106)$$

The optimization is specified as follows. It is assumed that all components of the state vector x at time t are measured. Then the value function is taken as the quadratic cost functional with weights Q on the sequence of sampled transient states x from t up to $t + N_y$, weights P on the terminal state at $t + N_y$, and weights R on the control input sequence u over the same time window as x:

$$J(U, x(t)) = x'_{t+N_y|t} P x_{t+N_y|t} + \sum_{k=0}^{N_y-1} \left[x'_{t+N_y|t} Q x_{t+N_y|t} + u'_{t+k} R u_{t+k} \right]$$

$$(6.107)$$

where $x_{t+k|t}$ denotes the predicted state vector at time $t+k$, which is produced by applying the input sequence

$$U = [u_t, \cdots, u_{t+k-1}] \quad (6.108)$$

to the plant model Eq. 6.105. For the weights it is assumed that

$$Q, P \quad \text{is symmetric and semi-definite or definite}$$
$$R \quad \text{is symmetric and definite}$$
$$\text{and}\quad (\sqrt{Q}, A) \quad \text{is detectable.} \quad (6.109)$$

The optimization problem then is to minimize the value of J by using the control inputs u_j in the input sequence U of Eq. 6.108 as N_u independent optimization variables, subject to the constraints

$$y_{lo} \leq y_{t+k|t} \leq y_{hi}, \quad k = 1, \cdots, N_c$$
$$u_{lo} \leq u_{t+k} \leq u_{hi}, \quad k = 0, \cdots, N_c$$
$$x_{t|t} = x(t)$$
$$x_{t+k+1|t} = A x_{t+k} + B u_{t+k}, \quad k \geq 0$$
$$y_{t+k|t} = C x_{t+k|t}, \quad k \geq 0$$

$$\text{and}\quad u_{t+k} = K x_{t+k|t}, \quad N_u \leq k < N_y \quad (6.110)$$

where N_y, N_u, N_c are the prediction horizons for the output y, the input u, and the constraints c, with $N_u \leq N_y$ and $N_c \leq N_y - 1$.

This is the full problem statement for both input and output constraints. In this section, only the input constraints shall be considered, and the high and low output constraints shall be set to $+\infty$ and $-\infty$ respectively.

The "Finals" Approach

In this setting, the control is specified to end up as a linear state feedback control law in the linear "finals" approach to the origin. Therefore the gain matrix K in Eq. 6.110 is set to the linear quadratic controller with infinite time horizon (to obtain the time-invariant gains specified above), using weights Q_L, R_L, and with no terminal weights, $P_L = 0$.

Set first $r = 0.01$.

Furthermore, to enable a direct comparison with the intuitive design from above, the weights in Q_L are selected such that the same feedback gains and the same closed-loop bandwidth Ω result. This leads to the following numerical values for the diagonal entries in Q_L (the off-diagonal elements are all set to zero):

case i: single integrator plant

Ω	10	5	2.5
q_{11}	1	0.25	0.0625

case ii: double integrator plant

Ω	10	7.071	5	3.80	3.40	2.75
q_{11}	100	25	6.25	2.08525	1.33625	0.5719
q_{22}	2	1	0.5	0.28875	0.23120	0.15125

case iii: triple integrator plant

Ω	5	3.25	2.93	2.5	2.0	1.575
q_{11}	156.25	11.785	6.25	2.4413	0.64	0.1525
q_{22}	18.75	3.345	2.2	1.175	0.48	0.1844
q_{33}	0.75	0.3175	0.25625	0.1875	0.12	0.0750

The Nonlinear Phase of the Transient

Now there is no longer a closed-form solution and the optimization has to be performed numerically. This is done by re-formulating the problem into a Quadratic Program (QP).

By inserting

$$x_{t+k|t} = A^k x(t) + \sum_{j=0}^{k-1} A^j B u_{t+k-1-j} \tag{6.111}$$

into Eq. 6.110

$$V(x(t)) = \frac{1}{2} x'(t) Y x(t) + \min_U \frac{1}{2} U' H U + x'(t) F U$$

$$\text{subject to} \quad GU \leq W + Ex(t) \tag{6.112}$$

where U in Eq. 6.108 is the optimization vector, H is symmetrical and positive definite, and H, F, Y, G, W, E follow from Q, R. And Y is usually removed, as the focus is on the dependence on U.

This QP has to be solved at each time step.

Again, the weights Q, R are to be selected. Intuitively set

$$Q := Q_L; \quad \text{and} \quad R := R_L \tag{6.113}$$

to have continuity in the energy of motion along the trajectory. In other words, the q_{ii} reported in the tables above are adopted for the benchmarks.
For the sampling time, as in [59, 60],

$$T_s := 1.0 \quad \text{and} \quad N \geq 2 \tag{6.114}$$

The prediction horizons N thus cover a sufficient part of the run-up response. This has the advantage of low dimensions of the matrices, and a fast optimization. However, the performance for disturbance suppression will be low, due to the large delay. Therefore, here

$$T_s = 0.050 \quad \text{and} \quad N = 50 \tag{6.115}$$

i.e. a prediction horizon of 2.5 s. The optimization now takes much longer, and may not be able to finish in real time.

Transient Responses

For case i the simulation results are shown in Fig. 6.48. This establishes a base line for the next cases, and is for comparison with the corresponding case in Chapter 2 with P control; Fig. 2.7.

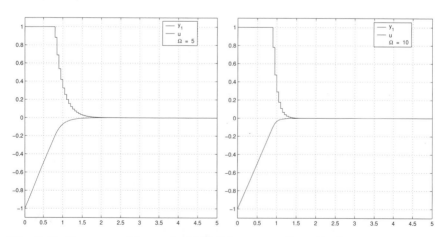

Fig. 6.48. Response with standard MPC for case i, $G = 1/(sT)$:
$\Omega = 5$ (left); $\Omega = 10$ (right)

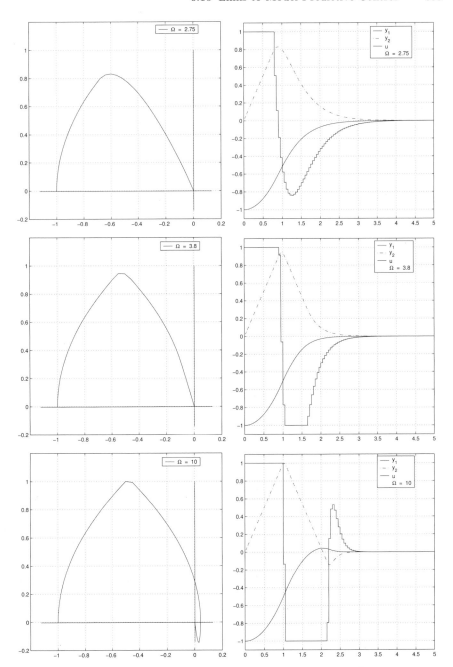

Fig. 6.49. Response with standard MPC for case ii, $G = 1/(sT)^2$:
$\Omega = 2.75$ (top); $\Omega = 3.80$ (center); $\Omega = 10$ (bottom)

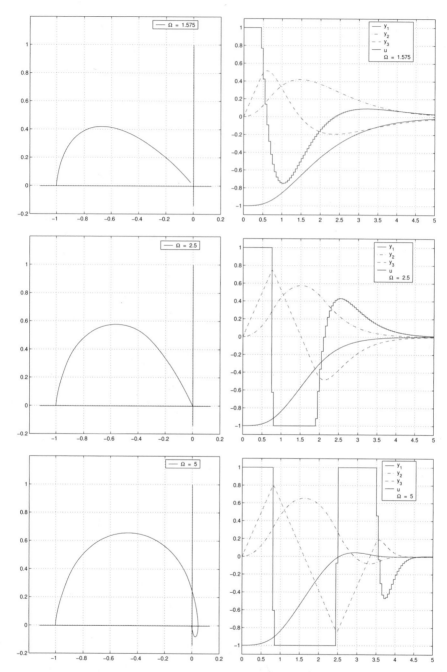

Fig. 6.50. Response with standard MPC for case *iii*, $G = 1/(sT)^3$:
$\Omega = 1.575$ (top); $\Omega = 2.50$ (center); $\Omega = 5.0$ (bottom)

For case *ii*, the simulation results in Fig. 6.49 (center) and (bottom) show in the first part of the response a "bang-bang" sequence on $u(t)$, similar to what the minimum-time optimal solution would require, followed by a linear "finals" approach phase. Qualitatively, this corresponds well with what has been found for the generalized awf design. There is also an overshoot in $y_1(t)$ for the design parameter $\Omega := 10$ and for the equivalent weights (q_{11}, q_{22}). Note that the response is still the optimal solution for those values.

To avoid such overshoot, intuitively the design parameter Ω must be selected below some upper limit. In this particular benchmark layout, this was found to be at $\Omega \approx 6.0$ (not shown). And the particular response turns out to be nearly minimum-time as well.

For case *iii*, Figure 6.50 shows the same trends. Again there is an upper limit on the value of the design parameter Ω, if overshoot is to be small enough. In this particular case simulations produce $\Omega \approx 3.25$ (not shown).

To *summarize*: the run-up response, produced by the standard MPC algorithm and with the design choices as described above, also turns out to be fitting well into the practically motivated specifications, as they were used before in the intuitive design approach.

Exercise
Explore this more broadly by
- considering plants with both negative and positive internal feedbacks,
- extending the test sequence to small setpoint steps and load swings,
- increasing the sampling time T_s,
- and reducing the control prediction horizon N_u.

6.16.4 Explicit Model Predictive Control

The time required for computing anew the optimal U^* at each sampling instant may exceed what is available from T_s.

One approach is to split up the computation into two steps. The first one is performed off-line, and consists of the optimization, *i.e.* determining the U^* sequence as a function of current state $x(t)$, and storing this (loosely speaking) in a look-up table. The second step consists of entering the look-up table, reading out the control U^* and implementing it. This is much less time consuming, and can now be performed on-line in real time.

This idea has been developed since the 1980'ies at least, mainly in the Operations Research field, and has found its current form as the "Explicit MPC approach" [59, 60].

The main result is, that U^* need not be computed and stored at length N_u. Its first move $u^*(t)$ is the only one to be implemented in the receding horizon technique, and it can be generated by a local linear affine control law:

$$u_j^* = -K_j\, x_j + q_j \tag{6.116}$$

where the index j points at the one of a finite number of finite-sized, compact regions in state space that contains the current $x(t) \to x_j$, and K_j is the corresponding gain matrix, and q_j is the offset for this region. This makes the control law *affine* to the standard state feedback law $u = Kx$. The output u_j is then put through the saturation element $u_{lo} \leq u(t) \leq u_{hi}$ and applied to the plant input.[3]

In this setting, region # 1 contains the origin, and contains all $x(t)$, from where on the control law is linear, *i.e.*

$$u_1^* = K_1 x_1 + q_1 \quad \text{with} \quad K_1 := K_L \quad \text{and} \quad q_1 := 0 \tag{6.117}$$

As has been discussed at length in previous sections, this region is of finite size, which strongly depends on linear closed-loop bandwidth Ω, and thus on the choice of Q, R. The adjacent regions are then determined by algorithms given in [60]. They are also of finite size. Their number N_{mpc} increases with the number N_u. Increasing N_u will add new areas adjacent to the existing ones.

Benchmark Case *ii*, with a Slow Sampling Rate

This is illustrated in Fig. 6.51, [60], for
- the double integrator plant with both time constants $T = 1.0$,
- saturations $-1.0 \leq u(t) \leq +1.0$,
- an area in the phase plane delimited by ± 5,
- weights $q_{11} = 1.0$, $q_{22} = 0$, and $r = 0.10$,
- and sampling time $T_s = 1.0$,
- using the finite-time version.

For $N_u = 2$, nine regions result from the "Explicit MPC" algorithm. Their delimits are shown in Fig. 6.51 (top right). The resulting control laws for the regions are (from [60])

$$
\begin{array}{ll}
\#\,1 & u^* = [-0.579, -1.546]\, x \\
\#\,4 & u^* = [-0.435, -1.425]\, x - 0.456 \\
\#\,5 & u^* = [-0.435, -1.425]\, x + 0.456 \\
\#\,2, 6, 8 & u^* = +1.0 \\
\#\,3, 7, 9 & u^* = -1.0
\end{array}
\tag{6.118}
$$

---

[3] Note that this general result is valid for both input and output constraints.

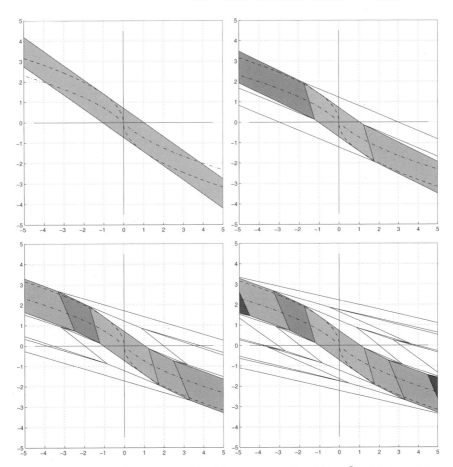

Fig. 6.51. Regions with Explicit MPC for case *ii*, $G = 1/(sT)^2$, with $T_s = 1.0$:
$N_u = 1$ (top, left), 3 regions; $N_u = 2$ (top, right), 9 regions;
$N_u = 3$ (bottom, left), 19 regions; $N_u = 4$ (bottom, right), 33 regions

Fig. 6.51 also indicates the "finals" trajectories (see Sect. 4.4), that would
result from instantaneous switching between the u saturations $(--)$, and by
including a look-ahead for the delay of one sampling time T_s $(-\cdot-)$, in order
to avoid a final overshoot.

Discussion[4]

- Note that the lower left delimit of region # 1 coincides with the line for
  $u = +1.0 = u_{hi}$ produced by the linear affine control law $u_1^* = -K_1 x$; and
  correspondingly for the upper right delimit with $u_1^* = u_{lo} = -1.0$.

---

[4] As an **exercise**, investigate further the properties stated here.

- The same holds for the lower left delimit of region # 4. It coincides with the line for $u = u_{hi}$ produced by the linear affine control law $u_4^* = -K_4 x + q_4$; and correspondingly for the upper right delimit with $u_4^* = u_{lo}$. And the same holds for region # 5.

- The control input u to the plant is continuous across the region delimits. In other words, there are no excessive bumps on u when transiting from one region to the next, but the time derivative of u may be discontinuous.

- Consider the phase plane plot for $N_u = 2$ first.
  By inspection, the regions # 1 to # 9 can be assembled into three "super-regions" A, B, C, i.e.
  - A consists of regions # 1, # 4, and # 5 and is a corridor, where u is in its linear range, $u_{lo} < u(t) < u_{hi}$,
  - B consists of regions # 2, # 8 and # 6, i.e. the lower left area, where $u = u_{hi} = +1.0$,
  - C consists of regions # 3, # 9 and # 7, i.e. the upper right area, where $u = u_{lo} = -1.0$.

- In super-region A, the regions # 1 and # 4 can be distinguished not only by checking the current x with the delimits from the table above, but equivalently by calculating both $u_1^* = K_1 x$ and $u_4^* = K_4 x + q_4$ first, and then performing a Max-selection between u_1^* and u_4^*. The same holds for regions # 1 and # 5, but then with a Min-selection between u_1^* and u_5^*. The result of this evaluation shall be denoted as u_A^*.

- The delimits between super-regions A and B can be implemented equivalently by performing a Min-selection between the result of the previous step u_A^* and the upper saturation limit $u_{hi} = +1.0$, and similarly for A and C by a Max-selection between u_A^* and $u_{lo} = -1.0$.

- Also, observe in Fig. 6.51 that the super-region A is situated with respect to the "finals" parabolas in such a way, that trajectories will enter first the linear corridor *before* meeting the "finals" parabola branch. This is favorable to avoid overshoot; see Sect. 4.4.

- Observe that this holds only in a restricted area of x around the origin. For trajectories evolving outside this area, the linear corridor would be entered *later* than the "finals" parabola being crossed over. Thus, a final overshoot will be unavoidable.

- This is avoided if N_u is increased. The optimization process adds two other regions to the linear corridor, which are more inclined, and again fit in *before* the "finals" parabolas. In other words, trajectories in a wider area of the phase plane will enter the extended super-region A before the "finals" parabola branch is encountered. And final overshoot may then be avoided for this larger area as well.

- Note that this argument applies not only for increasing N_u, but also for $N_u = 1$. There, the optimization process results in a linear corridor, which is region #1 from $N_u = 2$, but without the delimits to regions #4 and #5. Then the area in the phase plane is much smaller, where the linear corridor is entered *before* meeting the "finals" parabola. In other words, there will be a final overshoot except for small initial conditions.

Benchmark Case *ii*: A More Realistic Sampling Rate

So far a comparatively large sampling time ($T_s = 1.0$ s) has been used intentionally. Now it shall be lowered to $T_s = 0.250$, which is about halfway to what has been used with the standard MPC ($T_s = 0.050$).

In a first step this allows a higher linear closed-loop bandwidth Ω. Here, the weights are increased to

$$q_{11} = 3.0; \quad q_{22} = 0.3; \quad \text{and} \quad r = 0.10$$

Therefore, region # 1 will be more inclined, less wide, and its vertical size will shrink as well. Also, the adjacent regions in the linear corridor will decrease in size. But they will still be situated *ahead* of the corresponding "finals" parabola branch with zero delay. But it will be entered later, *i.e.* at x_2 values closer to the minimum-time trajectory.

Finally, N_u will have to increase substantially, as the overall prediction time $N_u T_s$ must not change significantly. An overall prediction time of 2.0 s leads to $N_u = 8$. Then the optimization process generates as the number of regions $n_R = 129$ (!); see Figure 6.52. For comparison, the solution for $N_u = 4$ with $n_R = 33$ is also shown.

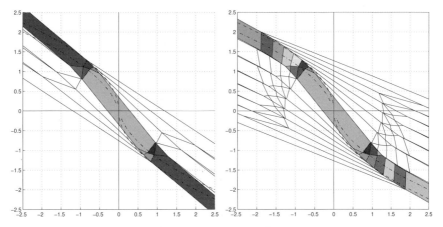

Fig. 6.52. Regions with Explicit MPC, $G = 1/s^2$, with $T_s = 0.25$: (left) $N_u = 4$, $n_R = 33$; (right) $N_u = 8$, $n_R = 129$

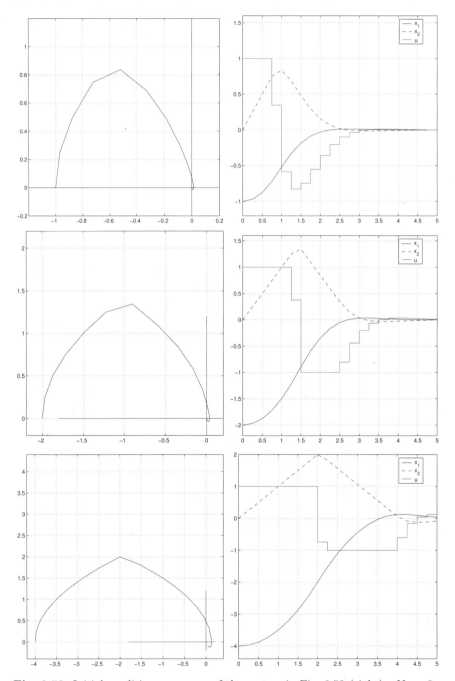

Fig. 6.53. Initial condition responses of the system in Fig. 6.52 (right), $N_u = 8$: $x_1(0) = -1.0$ (top); $x_1(0) = -2.0$ (center); $x_1(0) = -4.0$ (bottom); note the differents scalings of the plots

By further analysis of Fig. 6.52 (right), 108 regions have saturating first control moves of the type

$$u^*(t) = -(0x_1 + 0x_2) \pm u_{sat}$$

They form two large super-regions to the lower left with u_{hi}, and to the upper right with u_{lo}. And only 21 regions have a linear affine control law with non-vanishing gains. They form a linear corridor similar to what was found for $T_s = 1.0$, and predicted above. Its structure adjacent to region # 1 is slightly more complex, however.

The run-up response is shown by the equivalent initial conditions response in Fig. 6.53 for three different setpoint step sizes: $x_1(0) = -1.0, \ -2.0, \ -4.0$.

Direct *implementation* (as it was described for the slow sampling rate above) gets much more tedious. A very large number of region delimits have to be implemented and then sorted through in real time.
In a first step, intuitively, the regions from the linear corridor lead to 21 linear affine controllers, and u is evaluated from their outputs by successive Max-Min-selection. And then the Max-Min-selection to $u_{lo}, \ u_{hi}$ can eliminate checking the 108 regions.
This evaluation can be further simplified and speeded up by lumping adjacent regions in the linear corridor into larger ones, with constant gains and offsets in that region. Owing to this approximation the trajectory will be suboptimal, and more so if the lump size is increased.

This lumping approximation shall be illustrated on the double integrator plant with both time constants $T = 1.0$. Also let $T_s \rightarrow 0$ for simplicity. Then for the "finals" parabola with $u = u_{lo} = -1.0$

$$x_1 = -a_{lo} \, x_2^2 \quad \text{with} \quad a_{lo} = \frac{1}{2} \frac{1}{|u_{lo}|} \tag{6.119}$$

The tangent to this parabola at $(x_{1_t}, \ x_{2_t})$, using Eq. 6.119, is

$$x_2 = -\frac{x_{2_t}}{2x_{1_t}} x_1 + \frac{x_{2_t}}{2} = \frac{|u_{lo}|}{x_{2_t}} x_1 + \frac{x_{2_t}}{2} \tag{6.120}$$

The linear affine controller produces the straight line for $u = u_{lo}$ in the phase plane:

$$x_2 = -\frac{k_1}{k_2} x_1 + \frac{q - u_{lo}}{k_2} \tag{6.121}$$

where $k_1 := (\Omega T)^2$ and $k_2 := 2\zeta \Omega T$ with design parameters $\Omega, \ \zeta$.

Now let this border line for $u = u_{lo}$, Eq. 6.119, and the tangent to the parabola, Eq. 6.120, coincide. Then, for the slope coefficient

$$\frac{|u_{lo}|}{x_{2_t}} = \frac{k_1}{k_2} = \frac{\Omega T}{2\zeta} \quad \rightarrow \quad x_{2_t} = \frac{2\zeta}{\Omega T} |u_{lo}| \tag{6.122}$$

and for the offset coefficient q

$$\frac{x_{2_t}}{2} = \frac{q - u_{lo}}{k_2} \quad \rightarrow \quad q = u_{lo} + x_{2_t} 2k_2 = u_{lo} + 2\frac{\zeta}{\Omega T}|u_{lo}|\zeta\Omega T = u_{lo} + 2\zeta^2|u_{lo}|$$

(6.123)

Inserting numerical values yields

$$\zeta = 1; \quad u_{lo} = -1.0; \quad \rightarrow \quad q = +1.0 ; \quad i.e. \text{ further} \quad q := u_{hi} \qquad (6.124)$$

In other words, the offset q from Eq.6.123 does not depend on the choice of the bandwidth ΩT for the affine controller.

However, q depends on the design damping ratio ζ. Lower values for ζ produce a lower q, *i.e.* the linear corridor is lowered with respect to the "finals" parabola, and this increases settling time (all as one would intuitively expect).

And from Eq. 6.124, with the usual assumptions $\zeta = 1$ and symmetric saturations on u, produces an offset equal to the saturation value. This is easy to memorize as a design rule. But keep in mind that many assumptions have been made. So this design rule should not be overextended.

By the way, Eq. 6.124 also provides a background to the (intuitive) design choice of the offset in Sect. 6.15.

Finally, reasonably sized lumps for the affine control laws have to be aggregated, "reasonable" referring to the compromise between the reduction in number of affine control laws and the loss in optimality. This reduces to reasonable choices of their relative bandwidths Ω_j. The basis is given by the bandwidth Ω_1, as it is specified for linear operation around the final steady state, in region # 1. Intuitively then, from the shape of the "finals" parabola, a reasonable choice for the affine control laws in the adjacent lumped regions # 2^-, # 2^+ of super-region A (the linear corridor) would be

$$\Omega_{\#2-} := \Omega_{\#2+} := \frac{1}{2}\Omega_{\#1} \qquad (6.125)$$

And for the next lumped regions # 3^-, # 3^+ in the linear corridor set

$$\Omega_{\#3-} := \Omega_{\#\ 3+} := \frac{1}{2}\Omega_{\#2-} = \left(\frac{1}{2}\right)^2 \Omega_{\#1} \qquad (6.126)$$

Exercise
Investigate this design rule further.

6.16.5 Comparison with Previous Design Methods

All this is strikingly similar to what has been investigated in previous sections. So it is not surprising that the transient responses shown above are not that different.

For $T_s = 1$ and $N_u = 1$, the link to the "design by de-tuning Ω", Sect. 6.7, is evident. Both approaches lead to a linear state feedback $u_1^* = K_1 x$, with subsequent Max-Min-selection for the input saturations u_{lo}, u_{hi}. Previously, K_1 has been designed by pole assignment and stability considerations for the nonlinear case, whereas here it is designed as a Linear Quadratic Regulator using the weights Q_L, R_L, and by manipulating those as the design parameters suitably (to obtain the specified step response characteristics).

For $N_u \geq 2$, the link is obvious to Sect. 6.15, the "selection for lower-bandwidth control" approach. Again, the structures are the same, as far as there are linear state feedback laws with offsets generating the u_j^* for the regions other than $j = 1$. The parameters for the final linear controller in region # 1 are designed using Linear Quadratic-design instead of pole assignment, but resulting in the same state feedback gains, if the design parameters Q_L, R_L are adjusted accordingly.

The affine controller parameters (gains and offsets) for the regions other than # 1 have been determined above on a nonlinear stability basis, whereas here they are determined by optimization. Here, the u applied to the plant is selected by checking first in which region the current $x(t)$ is, and then computing the corresponding $u^*(t)$. The saturation is incorporated into the control law for each region. In contrast, in the override design, u is determined by calculating first the u_j, with j such that all regions in the linear corridor are pointed at, and subsequent Max-Min-selection, and a final Max-Min-selection for the input saturations u_{lo}, u_{hi}.

In the cases studied here, both approaches are "equivalent", as they yield the same result. The control input u to the plant is the same for a given $x(t)$ within its overall delimits for the optimization ($x_{i_{min}} \leq x_i \leq x_{i_{max}}$, $i = 1, 2, \cdots, n$). Note that this equivalence greatly reduces the implementation effort. And it seems plausible for the general case as well, but no proof is available yet.

The insight from the Explicit MPC solution also helps to improve the intuitive design. For the double integrator plant, the Explicit MPC with subsequent lumping for the regions in the linear corridor has led to the design recommendation for offsets being set to the saturation values and for bandwidths for the linear control laws being staged by a factor of ~ 2.

In the example shown in Fig. 6.47, based on design for stability (see Fig. 6.12), the first rule is adhered to, while the second rule is clearly violated: The staging factor is $\Omega_1/\Omega_2 = 10.0/2.75 \approx 3.64$, which explains the final overshoot.

And the MPC-based design also suggests a systematic improvement: insert an additional override controller pair, which is designed to $\Omega = 5.0$, and again with offsets u_{hi} and u_{lo} respectively.

Exercise
Check this last statement.

6.17 Summary and Outlook

In this chapter, plants have been considered with dominant order larger than one, *i.e.* cases not covered by the Ziegler–Nichols rules, and where PI(aw) controllers can no longer provide an acceptable closed-loop performance.
Therefore, state feedback controllers of P^+ and $I(aw)$-P^+ type are used. They are designed to a given closed-loop bandwidth Ω as the main design parameter, and damping $2\zeta := 2$. It is assumed that all dominant state variables are directly measurable, and no observer (with its additional dynamics) is needed. The non-modeled dynamics of the plant are accounted for by putting an appropriate upper limit on Ω.

If input constraints (actuator saturations) are inserted, and if advancing into the nonlinear range, then performance progressively deteriorates, and may end up with oscillatory unstable responses. This happens for the algebraic state feedback already, without an additional state variable in the controller from the integral action. The cause is that some plant internal state variables are pushed far away from their equilibrium values in the initial part of the trajectory, such that they cannot be brought back in time when approaching the new equilibrium. This is "plant windup".

Nonlinear stability properties may be assessed in a convenient way in the "stability charts". They show that plants with all internal negative feedback coefficients $a_j > 0$ (*i.e.* Hurwitz stable plants) are not very critical.[5] And the more open integrators there are in the plant, the more critical the situation gets. Finally, the situation is very critical for open-loop unstable plants.

Then several known design methods are investigated (note that such a selection will always be subjective). From a stability point of view, they all aim to increase the radius of attraction sufficiently for meeting practical design targets. Nonlinear stability analysis is the main systematic design tool here, while performance is checked qualitatively by benchmark simulations.
Recently, methods from "Robust Control" have received much attention in research, "robust" referring to the effect of saturation. They assume and imply that the system is already stable, and then the focus is on quantitative performance. However, to date, their application still needs extensive numerical calculations.
The last method in this list, the "selection for lower-bandwidth control", turns out to be very similar to what MPC in its Explicit form produces, *i.e.* affine linear control laws for the optimal first moves. Investigating this "similarity" further can help in mutually refining both design approaches. In this view, MPC generates the optimal trajectory as an upper bound for the intuitive (override) design. And the intuitive design helps to produce implementation-friendly control strategies, and also suboptimal ones, in a controlled manner.

---

[5] "Critical" refers to how sensitive the stability properties are to the maximum size of the overload on the saturation.

7

Generalized Override Control

7.1 Introduction

In Chapter 3, Override Control has been developed on the basis of a common but rather simple case, *i.e.* positioning control on a rigid point mass. Several structures have been investigated, and transient performance was found to be excellent. This has then be extended in Chapter 4 to situations where, in addition, there is a high and low saturation on the control input u to the plant.

This may be considered as too narrow a view by the design engineer, who is confronted with a large variety of control problems. From a more theoretical point of view, the above case implies a number of assumptions, which need to be relaxed.

The aim of this chapter is to discuss those assumptions in more detail. We shall show that the design elements from above are quite versatile and produce good solutions for a large field of practical applications. We shall also show some links to the design approach by MPC.

Sect. 7.2 considers situations where the plant dominant response is of first order (mainly of the open integrator type), where PI(aw) controllers are best suited. The selector/override idea shall be used in different novel ways.

Then, in Sect. 7.3, situations are investigated where the plant response is of higher order, and where PI(aw) control must be augmented with state feedback or related add-on features, similar to the ideas in Chapter 6 on generalized antiwindup.

In Sect. 7.4, the class of G_2 having a zero in or near the origin is discussed. This relates to an important and successful application in the power industry.

Up to now, the override action has been on the control variable. This paradigm is investigated next in Sect. 7.5, by shifting the override action to the reference of the main loop, which is similar to using "reference conditioning" instead of "control conditioning"; see Chapter 2. This has some structural advantages, but we shall also find significant drawbacks.

The second paradigm was that, during the output constrained parts of the trajectory, the focus of control is on running along the constraint as closely as possible ($y_2 \rightarrow r_{2_{hi}}$, $r_{2_{lo}}$), regardless of the transient excursions of the main controlled variable $y_1(t)$. In Sect. 7.6, a compromise design strategy is proposed and investigated.

Finally, links to MPC are established in Sect. 7.7.

7.2 Systems with Dominant First-order Plants

Consider a tank, which serves as a buffer between an upstream production unit delivering the inflow, and several downstream units with time-varying outflows from the tank, or *vice versa* several inflows and one outflow. There may be other parallel flows to the single inflow and single outflow, but this is not considered further here. The purpose of the buffer tank is to absorb transient mismatches of inflow and outflow, such as to have less load swings on the production unit.

A level control system shall be installed. It can act either on the inflow or on the outflow, depending on the specific overall situation. The opposite flows then act as the disturbance. If the inflow is controlled, then the total outflow is the disturbance, and *vice versa*.

The closed-loop bandwidth of the level control is set rather low, such that the controlled flow will match the disturbance flow in a longer term mean, but will not follow large short excursions. In other words, disturbance suppression is to be low at high frequencies and high at very low frequencies.

But then, on rare occasions, the flow mismatch may be both higher and longer than designed. Then the tank will overflow or fall dry. This must be avoided, even if the controlled flow unit has to be maneuvered more strongly. And this upset on the controlled flow should be eliminated as soon as possible, after the excessive flow mismatch returns to its normal size again.

The constrained variable y_2 and the main controlled variable y_1 are now the same, $y_2 := y_1$, and thus the plant transfer functions in both the main and the override paths are the same, $G_2 := G_1$. This is a special case of $N_2 = N_1$. Also, we have "loose" control for small level deviations and "tight" control of the same level along its constraint values.

In the following, we shall consider first the case of acting on the inflow only, then while acting on the outflow only, and also a mixed strategy. Next we shall consider the case where there is no main level control, as the controlled flow is determined from outside the buffer tank. Finally, we shall look at a more complex application with multiple control variables and overrides.

7.2.1 Actuation on the Inflow

The Control System

The control system Fig. 7.1 is put together from the elements of Chapter 3.

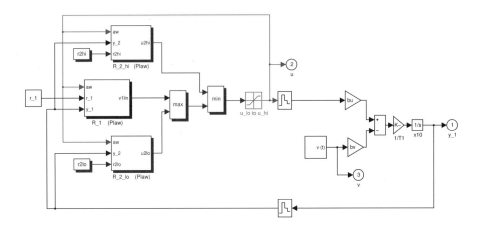

Fig. 7.1. Level control system on a buffer tank, actuation on the inflow

It consists of a PI(aw) controller R_1 for level y_1 acting on the inflow u_{in} with settings for low bandwidth: for the simulation, poles are assigned to $\Omega_1 T := 0.10$, where T is the nominal filling time of the tank, which is time-scaled to $:= 1.0$ for the simulation. The integral action sees to high disturbance suppression at low frequencies $\omega \ll \Omega_1$. The setpoint r_1 for the long-term mean of the level y_1 is set to 0.50. And the mean flow through the system is set to 0.50 units.

Excessive level excursions are avoided by two level override loops $R_{2_{hi}}$, $R_{2_{lo}}$, with setpoints at $r_{2_{hi}} := 0.99$; $r_{2_{lo}} := 0.01$, both acting on u_{in}. Both override control loops shall have the same gain and reset parameters. This implies that the tank has constant surface area with height. If not, the parameters can easily be adjusted to the local plant transfer function. For the simulation, their closed-loop bandwidths are set to $\Omega_2 T := 2.0$, *i.e.* by a factor of 20 higher than in the R_1 loop. Finally, the inflow is limited to its working range by the saturation block with limits at u_{lo}, u_{hi}.

All PI(aw) controllers are of form **D** from Chapter 2, with sampling time $T_s = 0.050$, and with compensating awf gains $k_{a_1} = k_{a_2} := 2$.

Transient Response

Fig. 7.2 shows the simulation results for three disturbance (outflow) profiles:

(a) A "normal" sequence of outflow variations:

+0.25 for 2 time units followed by −0.25 for 2 time units,

such that the level stays away from the constraint setpoints. Note that the inflow variations u are much smaller than the imposed outflow variations v. Note also the slow equilibration by the weak R_1 level control.

(b) A "rare" case of excessive length of flow mismatch:

+0.25 for 4 time units, followed by −0.125 for 8 time units.

Here, the level transiently runs into the upper level constraint and the override is activated. Note that, after the excessive outflow reduction v is finished, the inflow u_{in} returns approximately to the value before the override became active. Thus the corresponding functional specification stated above is met. The second phase of outflow mismatch now lets y_1 run into its opposite constraints. Finally, there is a linear slow equilibration transient, with u_{in} not straying far from its steady state value 0.50.

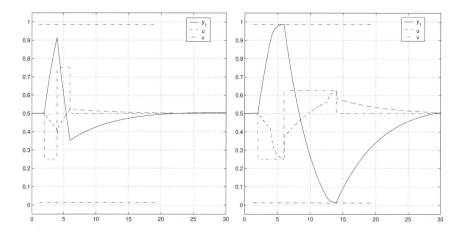

Fig. 7.2. Transient response of the level control system Fig. 7.1:
(left)   for the "normal" flow mismatch sequence (a);
(right) for the "rare" flow mismatch sequence (b), leading into the upper level constraint, and then into the lower one

Stability Analysis

The mismatch sequence (a) is such that the system stays within its linear operating range, and thus stability is as designed through R_1.

For case (b) this no longer holds. Note that now stability is of the input-output type, and not of the initial condition response type, which has been built upon so far. So, $\mathcal{L}_2$-stability (see [12], chapter 6) is to be investigated. This shall not be followed through in detail here, but only outlined.

Consider the disturbance input deviation from its steady state value $\Delta v = v(t) - \overline{v}$. Note that its "energy" $\int_0^\infty \Delta v^2(t)\,dt$ is bounded, as $v(t)$ is a finite length pulse returning to zero at its end. And the nonlinear control system is globally asymptotically stable (which shall be shown in an instant). In other words, it will dissipate this finite energy and thus $e_1 = r_1 - y_1$ will decay to zero.

We shall focus on the typical $v(t)$ shapes in Fig. 7.2. There two phases may be distinguished.

In the first phase, **A**, the system stabilizes on the constraint setpoint, which is taken as a new "final equilibrium". Then the equilibrium of the R_1 loop before applying the v step is seen as the initial condition with respect to this new "final equilibrium". The control system now has the R_2 loop as its new "main loop", while the R_1 loop functions as the override loop.

Then the initial conditions stability may be checked as in Chapter 3. Using the basic result and inserting compensating awf gains

$$F_{\mathbf{A}} + 1 = \frac{s}{s + \Omega_2}\,\frac{s + \Omega_1}{s} = \frac{s + \Omega_1}{s + \Omega_2} \tag{7.1}$$

The Nyquist contour evolves in the right half plane, and the circle test shows global asymptotic stability.

In the second phase, **B**, the system starts with initial conditions such that the override loop is switched in, and the "final equilibrium" is controlled by R_1 to $r_1 = 0.50$.

Then the flow mismatch step back to zero can be seen as a corresponding initial condition injection on the integral actions within the PI(aw) controllers. The basic result of Chapter 3 with compensating awf gains again yields

$$F_{\mathbf{B}} + 1 = \frac{s}{s + \Omega_1}\,\frac{s + \Omega_2}{s} = \frac{s + \Omega_2}{s + \Omega_1} \tag{7.2}$$

Its Nyquist contour again evolves in the right half plane, yielding global asymptotic stability also for the second phase.

A Different Control Structure

In practice, a control structure is often used, which is available in most commercial process control systems as "error progressive" or similarly.

It consists of one standard PI(aw) controller with a nonlinear characteristic on the control error $e = r - y$. The characteristic consists of a linear proportional element with very low gain placed in parallel to a deadspan element with high gain slopes, where the break points are adjustable to implement the level constraint setpoints. The low gain is to implement the low bandwidth control for the "normal" operation range, while the high gain is used for the high bandwidth limiting control.

This layout is certainly simpler than the override structure of Fig. 7.1. But it has at least two significant drawbacks.

One is that it will work well with pure P control in all loops (in fact it is easy to show that it is equivalent to the override structure with P controllers). But there are no direct means to re-tune the integral action as well. It stays fixed to the very slow setting for the normal operation, and therefore the response along the constraint will be excessively sluggish, and far below the performance from Fig. 7.2.

The second drawback is that there is no "look-ahead" effect on u while y_1 approaches the constraint limits, as the structure Fig. 7.1 has. To see this, replace the controller form **D** by the equivalent one of form **A**, which has a PD element upstream of the selection. The D part provides the "look-ahead" property. This is missing in the simpler form, where there are only P elements upstream of the selection. This may produce overshooting of the constraints. Both drawbacks will lead to shifting of the constraint setpoints inward from the physical level limits, and thus to losing the corresponding part of the buffer tank operational capacity.

7.2.2 Actuation on the Outflow: Direct vs. Reverse Control

The Control Structure

Consider now the design situation where the outflow is available for level control as u and the total inflow is the disturbance v. If the definition of the control error $e = r - y$ is maintained, then the negative level feedback of Fig. 7.1 turns into a positive feedback, *i.e.* the level loop will be unstable. This is due to u acting on the (positive) outflow, which enters with the *negative* sign in the mass balance. In other words, $G_u(s)$ now has a negative sign.

To cope with this, a block with gain -1 is inserted in the level controller error; see in Fig. 7.3. This feature is standard in commercial process control systems. The switch-selectable gain is either $+1$ or -1[1]. No other modifications are required. For instance, the initial values of the integral actions in the controllers are the same as in the previous case of inflow control, *i.e.* $u_1(0) := v(0)$ for initial steady state. Note also that u is positive and increases with increasing outflow. (Such simple design rules simplify the checking out in the real plant.)

Consider now the action of the $R_{2_{hi}}$ controller. In order to keep the level at the constraint setpoint $r_{2_{hi}}$, the outflow has to be *increased* with respect to u_1, whereas the inflow would have to be *decreased*. This calls for a Max-selector instead of the Min-selector on the inflow. And similarly the output of the $R_{2_{lo}}$ controller has to be connected to the Min-selector, whereas previously it was to the Max-selector.

In Fig. 7.3 this is implemented by "crossing over" the R_2 outputs, and conserving the graphic symbol for the Max-Min-selector block. (Again such a

---

[1] This is often denoted as "direct action" vs. "reverse action"

simple design rule helps to avoid mistakes and to interpret the signal flow graph.)

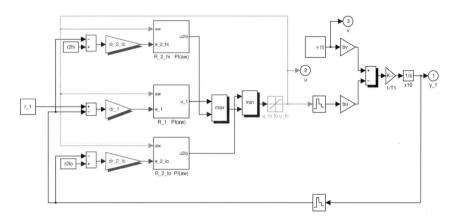

Fig. 7.3. Level control with actuation on the outflow ("reverse control")

Transient Responses

The correct functioning of this system is verified in Fig. 7.4, where the same disturbance sequence is now applied to the total inflow.

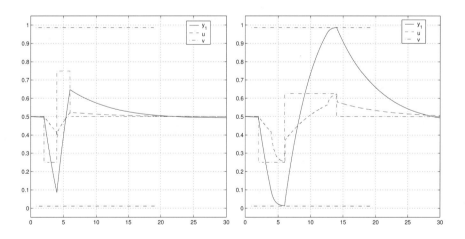

Fig. 7.4. Transient response of the level control system Fig. 7.3, ("reverse action"): (left) for the "normal" flow mismatch sequence (a);
(right) for the "rare" flow mismatch sequence (b), leading into both level constraints

Stability Analysis

There is no fundamental difference in the control structure apart from the signs, and this can be eliminated right away. Thus, no separate stability analysis is required.

7.2.3 Main Control on the Inflow and Overrides on the Outflow

The Control Structure

The next idea is to use a mixed strategy, where the inflow is controlled with the low bandwidth loop (to keep the fast outflow variations from the production unit). And the override action is now on the outflow, cutting off the disturbances such that the level does not exceed its constraints; see Fig. 7.5. Note that this requires some compact access to the total outflow signal. Here, there is a local flow-controller with output u_v, and its setpoint being the sum of all outflow demands v. In both the inflow and outflow control loops, there are "man-to-auto" mode transfer switches. They allow the awf loops to equilibrate before the loops are put into "auto"-mode, similar to what has been used previously.

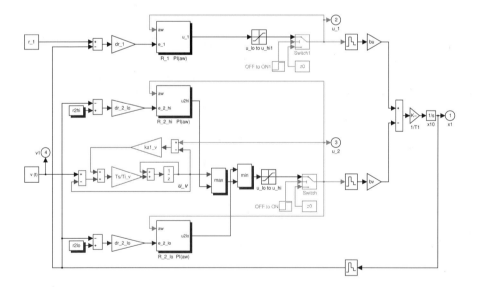

Fig. 7.5. Level control with actuation on the inflow ("direct control") for normal operation, and with level overrides acting on the outflow ("reverse control")

Transient Responses

From Fig. 7.6, the control structure performs as expected.

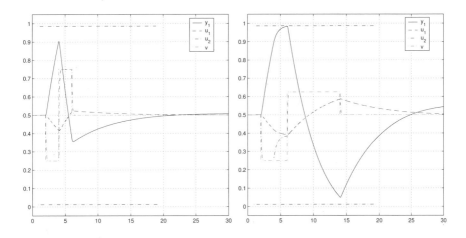

Fig. 7.6. Transient response of the level control system Fig. 7.5, with a mixed strategy; level main control on the inflow and overrides on the outflow:
(left)   for the "normal" flow mismatch sequence (a);
(right) for the "rare" flow mismatch sequence (b)

Stability Analysis

Again, there is nothing fundamentally different from the system Fig. 7.1, and thus the stability properties are the same.

7.2.4 Output Limitation with $N_2 > N_1$

The Basic Case

This situation arises when the control task from Sect. 7.2.3 is slightly modified: the main control variable (y_1) is no longer level, but inflow. Its setpoint shall be determined by other sources than the buffer tank level.

Then, obviously, the level can no longer be stabilized to a given setpoint as in the previous case, simply because there is no level control. Thus, the level will move up and down freely following the flow mismatch over time. This system is not asymptotically stable in the linear range. [2]

However, if the level constraints are met, then the override level control is activated and stabilizes the system at the constraint setpoint. This corresponds to phase **A** of Sect. 7.2.3, with R_2 playing the role of the main controller, and R_1 (the flow controller) being the override controller.

Denoting y_L as the measured level signal and y_F as the flow signal, and using

---

[2] You might consider this an ill-posed control problem. Nevertheless, this *is* a practical situation, as the application example below shows.

$$G_F(s) = \frac{y_F}{u} = \frac{1}{sT_F + 1}; \qquad G_L(s) = \frac{y_L}{u} = \frac{1}{sT_L}\frac{1}{sT_F + 1} \qquad (7.3)$$

where $T_F \ll T_L$. For the stability test, with compensating awf gains

$$F_{\mathbf{A}} + 1 \approx \frac{s\left(s + \frac{1}{T_F}\right)}{(s + \Omega_L)\left(s + \frac{1}{T_F}\right)}\frac{s + \Omega_F}{s + \frac{1}{T_F}} = \frac{s}{s + \frac{1}{T_F}}\frac{s + \Omega_F}{s + \Omega_L} \qquad (7.4)$$

where the Nyquist contour evolves in the right half plane, indicating global asymptotic stability.

For phase **B**, however:

$$F_{\mathbf{B}} + 1 \approx \frac{s + \frac{1}{T_F}}{s + \Omega_F}\frac{(s + \Omega_L)\left(s + \frac{1}{T_F}\right)}{s\left(s + \frac{1}{T_F}\right)} = \frac{s + \frac{1}{T_F}}{s}\frac{s + \Omega_L}{s + \Omega_F} \qquad (7.5)$$

which is not asymptotically stable (there is one pole at the origin). Thus, the basic assumption of the circle test for the linear subsystem is not fulfilled, and thus it may simply not be applied here.

An Application Case

A typical example of this situation is in hydropower control. The flow is generated by the power control loop of a power plant. During "normal" operation, the power setpoint, and thus flow, is generated by a supervising control system, such that the overall water content in the river system will stay about constant. In other words, the flow has to be approximately equal to the total upstream inflow to the hydrological system, and the power setpoint is adjusted accordingly. Therefore, the plant will deliver "base load" into the grid.

However if a large mismatch power flow occurs in the grid, large grid frequency deviations result. To avoid this, the "normal" power setting for the hydropower plant will be overridden such that the power station contributes to cover the load mismatch. Thus, the power station transfers from base load operation to load/frequency control mode, or to "primary" control mode.

But there are constraints. One is the finite working range of turbine flow. The second one comes from upsetting the water flow balances along the river. Reservoir levels will move away from their normal values. This can be tolerated up to limits which are dictated by navigation, regulatory agencies, or simply flooding. This leads to override loops from the adjacent reservoir level to the water flow loop. This may be the upstream level, or the downstream level, or even both. When the level constraints are met, they will move the actual flow back to near the equilibrium flow in the hydrological system, and thus shift power production back to the base load level. But this time interval may be sufficient to adjust production of thermal power plants along their typical load gradients of 5%/min, see Sect. 7.4.

Note that, after the grid mismatch has been driven to near zero by load shift on those other plants, the local reservoir level will stay at its constraint value as long as the local power setpoint is not moved in a way to establish a new equilibrium in the water system around the "normal" levels. In other words, the deficit or surplus in reservoir content which was accumulated during the phase of primary control must be compensated later by an equivalent water flow mismatch in the opposite direction. And this will upset all downstream reservoirs again.

This strongly suggests that such management of levels, flows, and power output is better performed by a central dispatcher for the river system. Any local action will upset neighboring subsystems in this strongly coupled system unduly.

In other words, such transients should be controlled by the central dispatcher, where for instance MPC is clearly a well suited design method. Note that this requires all information to be available at the dispatcher. In this control structure, the local overrides then evolve to additional local safety features, which will come into action only in abnormal situations.

This is a nice example of how simple override controls can cooperate with central optimizing controls in a complex multivariable system.

7.3 Systems with Dominant Higher-order Plants

So far the plant transfer functions G_1, G_2 have been of dominant first order, where PI(aw) controllers are best suited. As in Chapter 6, on generalized antiwindup, plants of dominant higher-order ($N_1, N_2 \geq 2$) shall now be considered. Again, state feedback is needed, possibly implemented in some form of multiple cascades. To simplify matters it is assumed that all relevant state variables are directly measured, and thus no observer is needed.

7.3.1 Open Integrator Chain Plants

Consider first the positioning of a rigid mass along one axis. The standard setup would be position control with speed constraint, as has been discussed at length in Chapters 3 and 4, i.e. $N_1 = 2$, $N_2 = 1$.

A different setup from practice is speed control as the main loop (see Chapter 2), but now with position constraints: the position is not relevant as long as it is off the end constraints. But it becomes dominant if the position approaches those end points, in order to avoid damaging the plant. Another case is elevator control, where the main control loop is for cabin speed, and position control becomes dominant when approaching the target floor position. In these cases $N_1 = 1$, $N_2 = 2$.

This is similar to what has been looked into with the buffer tank in Sect. 7.2.4. Thus the stability problem for the complete system can be considered as "ill posed", if speed is identified as the main controlled variable y_1 to r_1, and the position is identified as the secondary output y_2 constrained to $r_{2_{hi}}$, $r_{2_{lo}}$. Visibly, the complete system will not be at rest for $y_1 \to r_1 \neq 0$.
But it will be "well posed", if the control task is stated in the opposite way, by considering stability to the position setpoint $r_{2_{hi}}$, while the speed control y_1 to $r_1 > 0$ is considered as the override loop: it limits the speed to r_1 for this approach. This occurs symmetrically for y_2 down to $r_{2_{lo}}$, with negative speed $y_1 \to r_1 < 0$ being the constraint. And this is the basic problem from Chapter 3, although with a one-sided override only. But this does not change the sector bounds for the nonlinear stability test.

Generally speaking, testing stability for open integrator chains with $\Delta N = N_1 - N_2 < 0$ does not make sense. This will lead to $F + 1$ having $-\Delta N$ poles at the origin, and therefore not being Hurwitz as the sector criteria require. However, exchanging the role of the variables as described above leads to a well-posed stability problem for the equilibrium along the original constraint setpoint. Now $F + 1$ has $-\Delta N$ zeros at the origin, and F meets the requirements of the sector criteria.

The Case of $N_1 = 3$ and $N_2 = 2$

So far cases of $N_1 = 2$ have been considered. In positioning control this means that acceleration is proportional to the control input u, and not a distinct state variable. But this is not always adequate. Examples are elevators with "jerk" limitation, where the rate of change of acceleration is relevant. Then the control structure is extended by an additional open integrator, which is attributed to the plant; see Fig. 7.7.

The Control Structure

The *plant* is given as

$$G_1(s) = \frac{y_1}{u} = \frac{1}{sT_1} \frac{1}{sT_2} \frac{1}{sT_3} \quad \text{and} \quad G_2(s) = \frac{y_2}{u} = \frac{1}{sT_2} \frac{1}{sT_3} \qquad (7.6)$$

where integrators for T_2 and T_3 are physically the same. Numerical values are set to

$$T_1 = 10.0; \quad T_2 = 1.0; \quad T_3 = 1.0$$

Speed and actuator stroke constraints are

$$r_{2_{hi}} = +1.0; \quad r_{2_{lo}} = -1.0; \quad \text{and} \quad u_{hi} = +1.0; \quad u_{lo} = -1.0$$

The *controllers* are of the form **D** from Chapter 2 for the PI(aw) part with additional state feedbacks; see Fig. 7.7. The controller parameters are determined by pole assignment using

$$(s + \Omega_1)^4 = 0 \quad \text{and} \quad (s + \Omega_2)^3 = 0 \tag{7.7}$$

Furthermore, set $\Omega_2 := 2.0\Omega_1$, with Ω_1 as the key design parameter, and the sampling time at $T_s = 0.020$. The awf gains k_{a_1}, k_{a_2} are set to their compensating values: $k_{a_1} = 4.0$ and $k_{a_2} = 3.0$.

The *test sequence* starts at standstill at $x = 0$ with $u = 0$. Then the reference r_1 is stepped up to 1.0, while u is forced to zero by the standstill switch. One time unit later the run-up is enabled by this switch. Note that no small setpoint steps or load swings are applied later.

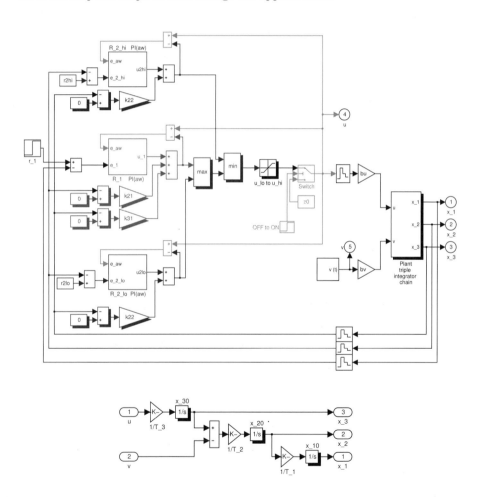

Fig. 7.7. Positioning control with output constraint on speed (integrator with T_2), and where acceleration is a significant state variable (integrator with T_3).
(top) the controller; (bottom) the plant

Transient Response

Fig. 7.8 shows the run-up responses for three sets of the design parameter Ω_1. For the first set (top figure), the u saturations are met only in the acceleration phase to the sliding equilibrium along the speed constraint. Thus, the stability analysis from Chapter 3 is feasible.

For the second set, Fig. 7.8 (center), u saturates in the "finals" phase from the R_1 controller. That is, the stability analysis of Chapter 4 applies. Finally, for the third set, the u saturations are dominant in the final approach phase of y_1 to r_1. From experiments not shown here, the y_1 loop is unstable for a small increase of Ω_1 to 3.333 (and Ω_2 to 5.0).

Stability Analysis

The first case, (a) in Fig. 7.9, is the output-constrained system without input constraints. For compensating awf gains:

$$(F+1)_a = \frac{s^3 \, (s+\Omega_1)}{(s+\Omega_1)^4} \frac{(s+\Omega_2)^3}{s^2 \, (s+\Omega_2)} = \frac{s}{s+\Omega_1} \frac{(s+\Omega_2)^2}{(s+\Omega_1)^2} \tag{7.8}$$

As $\Omega_2 > \Omega_1$, the Nyquist contour evolves into the right half plane, and shows asymptotic stability for almost all initial conditions $x_1 \neq 0, x_2 = 0, x_3 = 0$.

And the second case, (b) in Fig. 7.9, is the input-constrained R_2 loop, with

$$(F+1)_b = \frac{s^2 \, (s+\Omega_2)}{(s+\Omega_2)^3} = \frac{s^2}{(s+\Omega_2)^2} \tag{7.9}$$

which now evolves transiently into the left half plane, and thus indicates only local asymptotic stability, Note the small overshoot in the "finals" phase in Fig. 7.8 (bottom).

To complete the analysis, consider case (c) in Fig. 7.9 with the R_1 loop with input saturation, *i.e.*

$$(F+1)_c = \frac{s^3 \, (s+\Omega_1)}{(s+\Omega_1)^4} = \frac{s^3}{(s+\Omega_1)^3} \tag{7.10}$$

which crosses over the negative real axis. Thus, the describing function test indicates that unstable limit cycles appear, if the deviations are very large. But this is suppressed by the R_2 loops.

Exercise
Expand the system to allow linear range setpoint steps and load swings, as in Chapters 3 and 4, and check transient responses as well as stability.

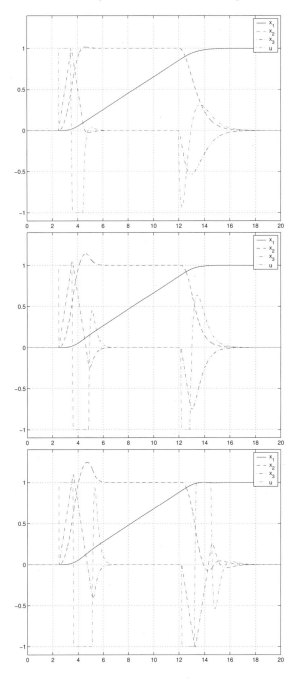

Fig. 7.8. Transient response of the control system in Fig. 7.7, with $\Omega_2 := 2.0\Omega_1$, (top) $\Omega_1 = 2.0$; (center) $\Omega_1 = 2.5$; (bottom) $\Omega_1 = 3.0$

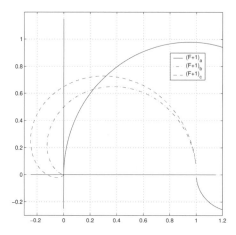

Fig. 7.9. Nyquist contours for the stability analysis of the positioning control system given in Fig. 7.7

The Case of $N_1 = 3$ and $N_2 = 1$

Now the acceleration shall be the constrained output in the positioning loop.

The Control Structure

Fig. 7.10 shows the override structure.

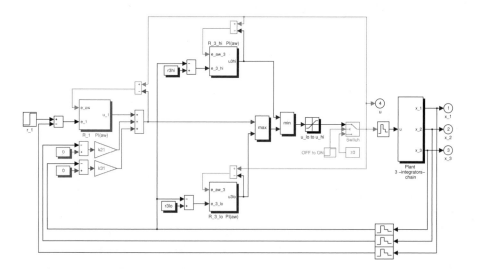

Fig. 7.10. The positioning control system, where acceleration is the constrained output

The structural elements and the parameters are the same as in the previous case, except

$$y_3 := x_3; \quad r_{3_{hi}} = +0.50; \quad r_{3_{lo}} = -0.50; \quad \text{and also} \quad \Omega_3 := 5.0\Omega_1 \quad (7.11)$$

again with Ω_1 as the key design parameter.

Transient Response

Fig. 7.11 shows the run-up responses.
With the exception of the set with "weak" R_1-control ($\Omega_1 = 1.0$), there is a distinct speed "windup" now, leading to an unacceptable overshoot of $y_1(t)$.

Stability Analysis

The three elements to consider are now (see also Fig. 7.12):

-   For case (a), that is for the output-constrained, but not input-constrained, "finals" approach, again with compensating awf gains

$$(F+1)_a = \frac{s^2}{(s+\Omega_1)^2} \frac{s+\Omega_2}{s+\Omega_1} \quad (7.12)$$

The Nyquist contour now evolves into the left half plane, thus indicating local stability.

-   For case (b), the input-constrained acceleration to the speed "sliding equilibrium"

$$(F+1)_b = \frac{s}{s+\Omega_2} \quad (7.13)$$

and the Nyquist contour is in the right half plane, indicating almost global stability.

-   For case (c), the input-constrained R_1 loop

$$(F+1)_c = \frac{s^3}{(s+\Omega_1)^3} \quad (7.14)$$

which, however, does not seem to be an important case in the transients shown in Fig. 7.11.

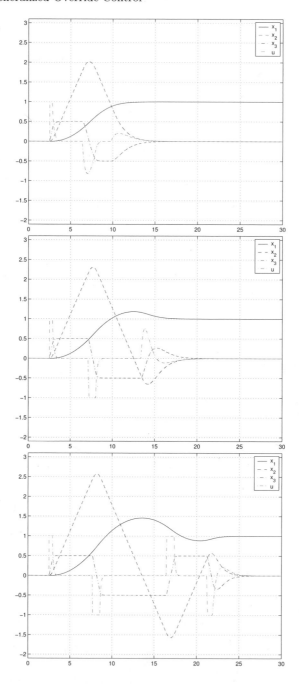

Fig. 7.11. Transient response of the control system in Fig. 7.10, with $\Omega_3 := 5.0\Omega_1$: (top) $\Omega_1 = 1.0$; (center) $\Omega_1 = \sqrt{2}$; (bottom) $\Omega_1 = 2.0$

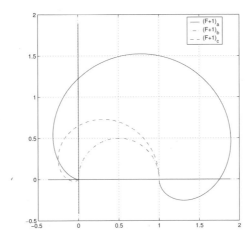

Fig. 7.12. Nyquist contours for the stability analysis of the positioning control system shown in Fig. 7.10

Using Multiple Output Constraints

In order to meet the general specifications on system response, the overshoots of $y_1(t)$ and of $y_2(t)$ must be suppressed.

This can be obtained in the previous case ($N_1 = 3, N_2 = 1$), Fig. 7.11, by inserting extra speed overrides to counter speed "windup" observed in the run-up transient. In the other case ($N_1 = 3, N_2 = 2$), extra acceleration overrides are inserted against the "windup" of x_3 in Fig. 7.8.

The *control structure* for the case $N_1 = 3, N_2 = 1$ is shown in Fig. 7.13. Set

$$\Omega_2 := 1.5\Omega_1; \qquad \Omega_3 := 5.0\Omega_1; \tag{7.15}$$

with Ω_1 as the key design parameter. And, as in the previous case, of acceleration constraint:

$$r_{3_{hi}} := +0.5; \qquad r_{3_{lo}} := -0.5; \tag{7.16}$$

The extra speed limits are taken from the case of speed constraint ($N_1 = 3,\ N_2 = 2$) above, but note that this is in fact a second design parameter, because the control problem does not assume a physical constraint on speed. Also, $T_s := 0.020$.

Comparing the *transient responses* shown in Fig. 7.14 with those in Fig. 7.11 documents the beneficial effect of the extra speed constraint. Note that this allows the key design parameter Ω_1 to be increased by a factor of $\sim$ 2.5, thus increasing k_{p_1} by a factor of $\sim$ 15, and improving the disturbance rejection accordingly.

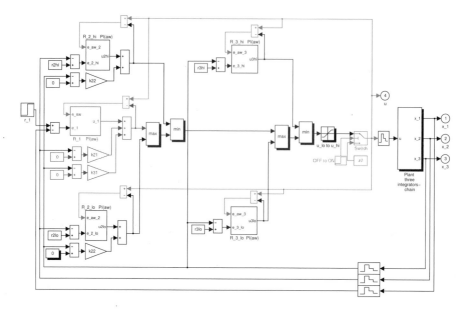

Fig. 7.13. Positioning control with output constraint on acceleration (x_3), and an extra (nonphysical) constraint on speed (x_2)

This also shows that there is a design compromise between thr increased bandwidth Ω_1 of the R_1 loop, and the extra speed constraint setpoint r_2, which must be lowered to avoid overshooting of $y_1(t)$. And this will increase the overall settling time. One way out is to replace the constant extra speed setpoint by a better-suited function; see Sect. 6.14.

A Design Guideline

The aim here is to extract some criteria to help the designer determine at a glance whether the given design problem is non-critical or critical, *i.e.* whether an in-depth investigation is needed or not.

The *plant* $u \rightarrow y_1$ to be considered is a chain of integrators of order N_1. The *controller* R_1 is a PI(aw) controller of form **D** of Chapter 2 with additional state feedbacks.

If there are input saturations (on u) as discussed in the previous chapters, the case of $N_1 = 1$ is "not critical". The case $N_1 = 2$ is "sensitive", as there may be a considerable overshoot and a low rate of convergence. And the case $N_1 = 3$ is "critical", as unstable limit cycles may appear.

For the output constraint feedbacks, PI(aw) controllers with state feedback are used. The constraints on output y_2 shall be on some integrator output *upstream* of y_1, *i.e.* $G_2(s)$ is a part of $G_1(s)$.

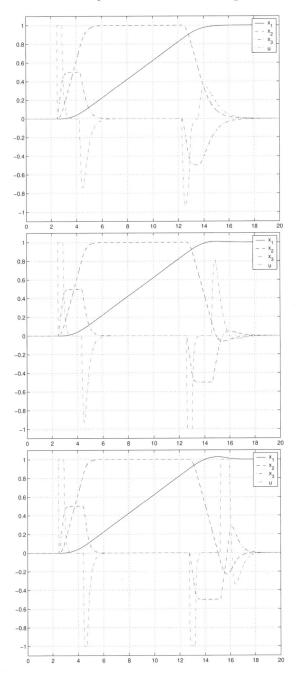

Fig. 7.14. Transient response of the control system in Fig. 7.13, with $\Omega_2 := 2.0\Omega_1$, and $\Omega_3 := 5.0\Omega_1$:
(top) $\Omega_1 = 2.0$; (center) $\Omega_1 = 2.5$; (bottom) $\Omega_1 = 3.0$

Its order is $N_2 \leq N_1$. This allows cancelation of zeros and poles in $F + 1$ without causing controllability and observability problems, and also yields the properties of $F + 1$ required by the sector stability tests.
Let

$$\Delta N := N_1 - N_2 \qquad (7.17)$$

Then for the cases with output constraints only, *i.e.* the $(F + 1)_a$-situations from above.

$$F + 1 = \left[\frac{s}{s + \Omega_1}\right]^{\Delta N} \left[\frac{s + \Omega_2}{s + \Omega_1}\right]^{N_2} \qquad (7.18)$$

The first factor shapes the Nyquist contour in the low-frequency range relevant for the nonlinear stability properties. The second factor is a lag-lead element. Its influence on the Nyquist shape in the relevant range is minor, if $\Omega_2 > \Omega_1$, which is a reasonable design assumption.
The first factor is what would appear for a plant G_1 with order ΔN and input saturations. And thus the stability charts (see Sect. 6.4.2) may be used for visualizing.
In other words, the stability analysis of $(F + 1)_a$ yields:

- The case $\Delta N = +1$ leads to favorable stability properties. It is therefore "non-critical". The same holds for $\Delta N = 0$, which is the design case of "loose-tight" y_1 control.
- $\Delta N = +2$ leads to a restricted area of attraction. It is therefore "sensitive", and will need a closer investigation, if very large inputs are to be expected.
- And the case $\Delta N = +3$ may be considered "critical", and requires an in-depth investigation.

In addition, input saturations are almost always present. Thus, the $(F + 1)_b$ and $(F + 1)_c$ situations from above must be discussed as well. However, they may be integrated into the same framework by using the result from Chapter 3, that input saturations are a special case of output saturations with $N_2 := 0$. This will generate additional cases of ΔN.
This leads to the general rule, that all the values of ΔN within the control system should not exceed $+1$. If this is not the case, then extra constraints should be inserted. This will improve dynamic properties considerably (see the multiple overrides example from above, where extra speed constraints have been inserted to have all the values of ΔN at $+1$).

These results from the stability analysis correlate well with the findings from the simulations. The control performance turns out to be very good from a practical design point of view. Note that this is not self-evident for the higher order plants.

Remarks

- For $\Delta N < 0$, the control problem is "ill posed", and must be turned around such that the equilibrium along the constraint setpoint is considered; see the case in Sect. 7.2.4.

- Only cases up to dominant order $N_1 = 3$ have been investigated by simulations. But from a practical point of view, cases with $N_1 > 3$ get difficult to control even in the linear range. Thus, it is highly recommended to re-design the control configuration, such as using better-placed actuators, *etc.*, to reduce N_1.

- Only run-up setpoint step responses have been considered, but no disturbance steps (load swings) have been applied.[3] Then, from the finding in Chapter 4, the extra constraints must be inserted *upstream* of the physical ones to avoid any violation of the physical constraints.

  Note that the structure in Fig. 7.13 complies with this rule, if the acceleration constraint is the physical one, and the speed constraint is the non-physical add-on. If this were the other way round, then the sequence of overrides on u must be exchanged as well.

- Consider inserting extra acceleration constraints on a positioning system with physical speed constraints, in order to reduce the corresponding ΔN, and thus to avoid overshoot in the run-up phase to the sliding equilibrium along the speed constraint. This may be counterproductive in the "finals" approach phase, as the braking capability is reduced by the extra acceleration constraint, and thus leads to a $y_1(t)$ overshoot (see Fig. 7.14 (bottom), at $t > 13$).

- This observation applies to plants with unstable poles and for disturbance responses, where any restriction on the control action may considerably reduce the radius of attraction.

The main restricting assumption so far has been on the plant being a chain of open integrators. This will be investigated next.

7.3.2 Output Constraint Control on a Flexible Transmission

This specific example has been selected because the plant contains an undamped resonance, and its non-transformed state space representation is not in control canonical form.

The aim is to clarify if the override design technique can cope as well with this type of control problem.

The Plant Model

This is a mechanical system of two (rigid) masses m_1, m_2 coupled by an elastic spring with zero mass and stiffness c_s. The actuator force F_u drives mass m_1, while a load force F_v acts on mass m_2. And y_1 is the speed w_2 of mass m_2.

Typically, such plant models appear in robot control and machine tools, where

---

[3] You may investigate this as an extended **exercise**.

the spring represents the gear elasticity, or in elevators with m_2 as the cabin mass, and the spring element would be the elastic cable.

Part of the plant may be rotative. Then replace forces by torques, m by Θ, and w by ω.

The state variables are
- momentum, *i.e.* speed w_1 of m_1,
- deformation of the spring $\Delta\ell_s$, and
- momentum, *i.e.* speed w_2 of m_2.

The driving force F_u shall not depend on the speed w_1, and similarly the load force F_v shall not depend on w_2. All friction forces are set to zero. And the constrained output y_2 shall be the coupling force $F_s = c_s\Delta\ell_s$, to avoid a plastic deformation.

Then

$$m_1\frac{d}{dt}w_1 = -F_s + F_u; \quad \text{with} \quad F_u = k_u u; \quad \text{where} \quad F_{u_{lo}} \leq F_u \leq F_{u_{hi}}$$

$$\frac{d}{dt}\Delta\ell_s = w_1 - w_2; \quad \text{and} \quad F_s = c_s\Delta\ell_s$$

$$m_2\frac{d}{dt}w_2 = +F_s - F_v$$

and as outputs:

$$y_1 = k_{y_1}w_2 \quad \text{and} \quad y_3 = k_{y_3}w_1$$
$$y_2 = k_{y_2}F_s \quad \text{where} \quad F_{s_{lo}} \leq F_s \leq F_{s_{hi}} \tag{7.19}$$

For the control problem to be "well posed", set

$$F_{s_{lo}} + \Delta F \;<\; F_v \;<\; F_{s_{hi}} - \Delta F$$
$$F_{u_{lo}} + \Delta F \;<\; F_s \;<\; F_{u_{hi}} - \Delta F \tag{7.20}$$

where ΔF is finitely small and positive, such as to allow stable linear control along the constraints within a small but finite interval.

All state variables shall be directly measurable, *i.e.* in addition to Eq. 7.19:

Next, the model is transformed into "per unit" form,[4] yielding

$$\tau_1\frac{d}{dt}x_1 = \quad 0x_1 + 1x_2 + 0x_3 - 1v$$

$$\tau_2\frac{d}{dt}x_2 = -1x_1 + 0x_2 + 1x_3$$

$$\tau_3\frac{d}{dt}x_3 = \quad 0x_1 - 1x_2 + 0x_3 + 1u$$

and as outputs $\quad y_1 = x_1; \quad y_2 = x_2; \quad \text{and} \quad y_3 = x_3 \tag{7.21}$

from which the state space representation in matrices T, A', B', C, D can be read directly (T stands for $diag(\tau_1, \tau_2, \tau_3)$).

The characteristic polynomial $|sT - A'| = d(s)$ then is

---

[4] Do this as an **exercise**, and obtain τ_1, τ_2, τ_3 as functions of the plant parameters.

$$d(s) = s\tau_1 \left[s^2\tau_2\tau_3 + \left(1 + \frac{\tau_3}{\tau_1}\right)\right] = s(\tau_1 + \tau_3)\left[\left(\frac{s}{\omega_r}\right)^2 + 1\right] \quad (7.22)$$

with
- the run-up time parameter $\tau := \tau_1 + \tau_3$ for the open integrator part, and
- the mechanical resonance frequency $\omega_r := \sqrt{(\tau_1 + \tau_3)/(\tau_1\tau_2\tau_3)}$, and
- with zero damping.

And thus for the transfer functions

$$G_1 = \frac{y_1}{u} = \frac{1}{d(s)}$$

$$G_2 = \frac{y_2}{u} = \frac{s\tau_1}{d(s)} = \frac{1}{s^2\tau_2\tau_3 + [1 + (\tau_3/\tau_1)]}$$

$$G_3 = \frac{y_3}{u} = \frac{s^2\tau_2\tau_3 + 1}{d(s)} \quad (7.23)$$

The Control System

Fig. 7.15 shows the plant model in Simulink form together with the controller structure. R_1 is an I(aw) controller with three additional state feedbacks. This special structure is needed to implement a negative coefficient k_{1_1} for pole assignments $\Omega_1 \ll \omega_r$; see Eq. 7.24.

The R_2 controllers are PI(aw) elements with additional state feedbacks, the same as in Fig. 7.7. The *controller parameters* are determined by pole assignment, as before, to Ω_1 for R_1, and to Ω_2 for both R_2 controllers, yielding

$$
\begin{aligned}
k_{0_1} &= \Omega_1^4 \tau_0 \tau_1 \tau_2 \tau_3 & k_{0_2} &= \Omega_2^3 \tau_0 \tau_2 \tau_3 \\
k_{1_1} &= 4\,\Omega_1^3 \tau_1 \tau_2 \tau_3 - 4\Omega_1 \tau_3 & k_{1_2} &= 3\,\Omega_2^2 \tau_2 \tau_3 - [1 + (\tau_3/\tau_1)] \\
k_{2_1} &= 6\,\Omega_1^2 \tau_2 \tau_3 - [1 + (\tau_3/\tau_1)] & k_{2_2} &= 3\,\Omega_2^2 \tau_3 \\
k_{2_1} &= 4\,\Omega_1 \tau_3 & k_{p_2} &= k_{1_2}; \quad T_{i_2} = k_{1_2}/k_{0_2} \\
k_{a_1} &= \Omega_1 \tau_0 \qquad\qquad (7.24) & k_{a_2} &= 3 \qquad\qquad\qquad (7.25)
\end{aligned}
$$

where the R_2 feedbacks have been designed using the denominator $d_2(s)$ of G_2 from above.

Note that, for the additional state feedback through k_{2_2}, the difference $x_3 - x_1$ is used in Fig. 7.15. With the standard state feedback of x_3 only, a steady state offset $\overline{e_2} \neq 0$ will appear during run-up.

Exercise
Check this statement.

The region of interest for Ω_1 shall be around the mechanical resonance value ω_r, while again $\Omega_2 := 2\Omega_1$.

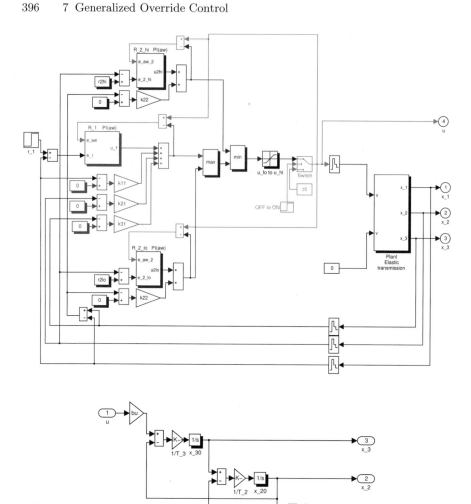

Fig. 7.15. The control system for speed control across the elastic coupling, and with constraints on the coupling force

For the simulation, set

$$\tau_1 = 1.0, \ \tau_2 = 0.02, \ \tau_3 = 1.0; \ \rightarrow \ \omega_r = 10.0 \quad \text{and} \quad T_s = 0.010 \qquad (7.26)$$

Furthermore, $F_v := 0$, and $r_{2_{hi}} := +0.5$, $r_{2_{lo}} := -0.5$.

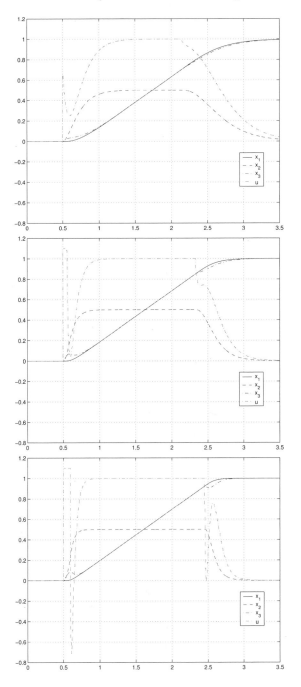

Fig. 7.16. Transient response of the control system in Fig. 7.15, with $\Omega_2 = 2\Omega_1$: (top) $\Omega_1 := 0.5\omega_r$; (center) $\Omega_1 := 1\omega_r$; (bottom) $\Omega_1 := 2\omega_r$

Note that the gain factor k_2 of G_2 is

$$k_2 = \frac{1}{1 + (\tau_3/\tau_1)}$$

Therefore, applying $r_{2_{hi}} := +0.5$ requires an input working range up to

$$u_{hi} > \frac{r_{2_{hi}}}{k_2} = r_{2_{hi}} \left[1 + (\tau_3/\tau_1)\right] = 1.0$$

which motivates setting $u_{hi} := +1.10$ and also $u_{lo} := -1.10$.

Transient Responses

The dynamic performance shown in Fig. 7.16 is again very acceptable. Notice in the R_1 loop the countermovement on $u(t)$, which is typical for plants with resonances, and which gets more pronounced for increasing bandwidth Ω_1.

Stability Properties

For the output-constrained (but not input-constrained) case, with compensating awf gains

$$(F+1)_a = \frac{s\left[s\left(s^2 + \frac{\tau_1 + \tau_3}{\tau_1 \tau_2 \tau_3}\right)\right]}{(s + \Omega_1)^3} \frac{(s + \Omega_2)^2}{s\left(s^2 + \frac{\tau_1 + \tau_3}{\tau_1 \tau_2 \tau_3}\right)} = \frac{s}{s + \Omega_1} \frac{(s + \Omega_2)^2}{(s + \Omega_1)^2} \quad (7.27)$$

In other words, the part in the plant transfer function from the resonance cancels in $(F+1)_a$, and thus is no longer of influence to the stability properties here.

But there is no such cancelation for the input-constrained cases (b) and (c), $i.e.$ in $(F+1)_b$ and $(F+1)_c$.

However, from the stability charts, case (b) has better properties than the double integrator chain, because the Nyquist contour starts in the right half plane.

And in case (c) the Nyquist contour starts vertically on the positive imaginary axis, which also indicates better stability properties than the double integrator chain.

This correlates well with the good performance observed in Fig. 7.16.

7.4 Load Gradient Control

This is a very successful application of advanced input and output constraint control with PID techniques. It has become a standard feature of control systems for thermal power units. It is also used in other situations, where $u(t)$ has to be moved in a ramping manner to run along the output constraints.

The term "Load Gradient" control[5] is often used in the power plant field. This may be misleading for an outsider. More precisely it would be "load ramping control". The gradient, which usually designates the *spatial* derivative, is not involved directly here, but rather the *time* derivative.

From a control point of view, this is a class of plants where the transfer function $G_2(s)$ in the general output constraint control structure of Chapter 3 has one zero at the origin (or very close to it). And the zero is not canceled by a corresponding pole, as has been the case up to now.

In other words, G_2 and G_1 shall now model *different* physical parts of the plant. Thus, the integrators in both transfer functions are not associated with the same physical balance equations, such that canceling zeros and poles from G_1 and G_2 may have secondary effects on observability and controllability. This difficulty did not arise in the applications investigated so far.

Considering implementation, its structure can be built from standard modules within industrial process control systems. This evidently contributes to its popularity.

We shall start with Henry's case as a motivating example. Deviating from our usual procedure, we shall use this directly as the current benchmark.

Then we shall perform the required modeling, design the linear controllers, and add the override and awf. This will require some additional elements to what has been used so far. On this structure, we shall investigate the transient response for performance, and check the stability properties.

7.4.1 Henry's Case

For many decades there was an iron rule in steam turbine operation about load changes: within a band of $\pm 10\%$ of rated load, fast load changes were allowed. For larger load swings a loading rate or "gradient" of $\pm 5\%$ / min was imposed. This was to constrain transient temperature differences in the thick-walled turbine casings, rotors, *etc.*, and thus to limit transient thermal stress and ultimately to avoid material fatigue, cracking and permanent deformations. This constraint had to be observed regardless of the thermal state of those thick-walled components, which is different after a long standstill (cold start), when temperatures are at ambient, from the state after a weekend shutdown (warm start), and from the state shortly after a unit trip (hot start).

---

[5] And other similar names, many of them proprietary.

From a control point of view, this is a feedforward output constraint strategy. It will have to be set to the most constraining situation, *i.e.* the cold start. It is conservative for all other situations; consequently, load-following to grid demand is slowed down unnecessarily.

In the early 1970s, analog electronic control equipment had replaced the previous generation of mechanic-hydraulic systems. It was much more flexible, and opened the way to a more refined strategy.

At that time Henry was a senior electronic control specialist with a major supplier of steam turbogenerator sets. His project was to design, implement and test an electronic control structure, which allows a better exploitation of the thermal state within the turbine.

In terms of Chapter 1, the idea was to replace the rigid feedforward scheme by an output constraint feedback control. More precisely (see chapter 3), the main (y_1) loop is for turbogenerator power output. The constrained variable (y_2) is an estimate of the most limiting thermal stress. It is obtained by measuring the temperature difference between the surface (with fast temperature response to heat input) and the interior of the thick turbine casing (with slow temperature response). And the control variable u is the turbine inlet valve stroke.

Henry designed such a control structure with an intuitive approach, using many of the ideas from the previous chapters. The first test was on a cold start and the loading was following the "iron rule" closely enough. Then the warm start was tested. Power was ramping up much faster as expected. This was so unexpected to the plant operations manager that he had to be restrained bodily from tripping the unit. But the test was successful. However, Henry detected some weakly damped transient behavior in the loading phase. He soldered in a larger capacitance on his printed circuit. On the unit loading test next day, the control performance was to his satisfaction.

Finally, this new scheme was such a success, that the supplier was not able to sell any new unit without this loading controller. It quickly expanded to steam generator control and to the whole power-generating industry. Since then it is a standard control feature.

As stated above, we shall use this case as the current benchmark, and start by developing a suitable model.

7.4.2 Modeling

Fig. 7.17 is a schematic of the relevant subsystems, the control valve with servomotor and input u, live steam pressure p_s and temperature ϑ_s, the turbine rotor and the generator rotor with its electric power output P_{el}, and the turbine casing modeled by two metal layers with temperatures ϑ_1, ϑ_2 at their respective centers of mass, and one isolation layer to ambient temperature ϑ_3. The steam temperature at the most critical location shall be ϑ_0.

PSfrag replacements

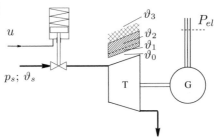

Fig. 7.17. The steam-turbine–generator subsystem with the thick-walled turbine casing, modeled by two metal layers and an isolation layer

Assume that steam pressure p_s and temperature ϑ_s are constant.

Then the *power output* of the turbine is proportional to mass flow, *i.e.* to u. Any small time constants from servomotors and other storage elements, as well as the generator dynamics, shall be replaced by a pure delay element e^{-sD_1}, *i.e.*

$$G_1(s) = \frac{y_1}{u} = \frac{P_{el_f}}{u_f} e^{-sD_1} \quad \text{with index } f \text{ for full scale}$$

and by scaling set $\quad \dfrac{P_{el_f}}{u_f} := 1$ $\qquad\qquad\qquad\qquad$ (7.28)

For the *metal temperature* model, it is assumed that heat conductivity within the layers tends to infinity. The finite conductivity is concentrated into thin interfaces. This leads to one first-order differential equation for the thermal energy content (heat balance equation) for each layer, and one algebraic equation for the heat flow $\overset{*}{I}_{k \to k+1}$ across each interface, which is driven by the temperature difference.

Also, the local steam temperature ϑ_0 along the length of the turbine will drop due to the expansion. This shall be proportional to mass flow, *i.e.* $\vartheta_0/\vartheta_s = \delta u$.

Then (see Fig. 7.18)

for layer 1 $\qquad m_1 c_1 \dfrac{d}{dt}\vartheta_1 = \overset{*}{I}_{0\to1} - \overset{*}{I}_{1\to2}$

$$\overset{*}{I}_{0\to1} = \alpha_f A\,(\vartheta_0 - \vartheta_1) \;=\; \alpha A\,(\vartheta_s \delta u - \vartheta_1)$$

$$\overset{*}{I}_{1\to2} = \beta A\,(\vartheta_1 - \vartheta_2) \qquad\qquad\qquad (7.29)$$

and for layer 2 $\qquad m_2 c_2 \dfrac{d}{dt}\vartheta_2 = \overset{*}{I}_{1\to2} - \overset{*}{I}_{2\to3}$

$$\overset{*}{I}_{2\to3} = \gamma A\,(\vartheta_2 - \vartheta_3) \qquad\qquad\qquad (7.30)$$

For brevity we use

$$x_1 = \begin{pmatrix} \vartheta_1 \\ \vartheta_s \end{pmatrix}; \quad x_2 = \begin{pmatrix} \vartheta_2 \\ \vartheta_s \end{pmatrix}; \quad \tau_1 = \frac{m_1 c_1}{\beta A}; \quad \tau_2 = \frac{m_2 c_2}{\beta A}; \quad k = \frac{\alpha_f}{\beta}; \quad \ell = \frac{\gamma}{\beta} \to 0 \tag{7.31}$$

which yields

$$\tau_1 \frac{d}{dt} x_1 = -(k+1) x_1 + x_2 + k\delta \cdot u$$

$$\tau_2 \frac{d}{dt} x_2 = \qquad\qquad + x_1 - x_2$$

and $\qquad\qquad y_2 = x_1 - x_2 \tag{7.32}$

This produces the transfer functions

$$\frac{x_1}{u} = k\delta \frac{s\tau_2 + 1}{s^2 \tau_1 \tau_2 + s\left[\tau_1 + \tau_2(k+1)\right] + k}$$

and $\qquad \dfrac{x_2}{u} = k\delta \dfrac{1}{s^2 \tau_1 \tau_2 + s\left[\tau_1 + \tau_2(k+1)\right] + k}$

i.e.

$$\frac{y_2}{u} = k\delta \frac{s\tau_2}{s^2 \tau_1 \tau_2 + s\left[\tau_1 + \tau_2(k+1)\right] + k} \tag{7.33}$$

which has the typical "zero at the origin" mentioned above.

PSfrag replacements

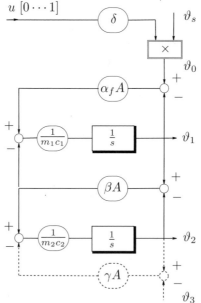

Fig. 7.18. Model structure for Fig. 7.17

However, if ℓ is not zero, but positive small, then the zero will be on the negative real axis, *i.e.* near but not at the origin.[6]
In order to take the non-modeled fast dynamics into account, a pure delay element e^{-sD_2} is appended.

The *numerical values* for the simulation are set to:

$$\alpha := \beta; \quad i.e. \quad k = 1; \quad \delta = 1.0; \quad \tau_2 = 60 \text{ s}; \quad \text{and} \quad \tau_1 = \tau_2/9 = 6.667 \text{ s}; \quad (7.34)$$

This is from setting $d_1 = d_2/3$, where d_j is the thickness of layer j.

With the material properties of ferritic steel, this leads to $d_1 \approx 10$ *mm*, *i.e.* $d_2 \approx 30$ mm, and to the overall wall thickness is $d_1 + d_2 \approx 40$ mm. Furthermore, set $D_1 = 3.0$ s; $D_2 = 0.5$ s

Defining τ as a new time constant $\tau := \frac{\tau_2}{3} := 20$ s

$$G_2 = e^{-sD_2} \frac{3(s\tau)}{(s\tau)^2 + [1/3 + (2 \cdot 3)](s\tau) + 1} \qquad (7.35)$$

It is convenient to do *time scaling* to τ, *i.e.* one time unit t' is now equivalent to 20 s, or $t' := t/20$. Consequently s is replaced by $s' = s/\tau$ in the transfer functions, where s' is in p.u. and no longer in s^{-1}.
Therefore, $D_1' := 0.15$; and $D_2' := 0.025$.

Notice that the values for the physical parameters are fairly realistic.

7.4.3 Controller Structures and Transient Response

Four control structures shall be discussed. Version **A** implements the "iron rule" strategy, and produces a base line for the subsequent designs. Then a simple version of override control (**B**), and two more complex ones (**C** and **D**) are discussed, where **D** will introduce a new design element, *i.e.* an extended form of awf.
The same investigation pattern shall be used, starting with the control structure, followed by the design of controller parameters, and transient responses. Stability analysis will be performed at the end for all four cases.

Version A: The "Iron Rule" Feedforward Approach

The *control structure* is implemented as one linear controller for y_1 and a subsequent nonlinear block on u, which shapes $u(t)$ according to the "iron rule" specifications in a feedforward manner; Fig. 7.19.

---

[6] Check this as a short **exercise**.

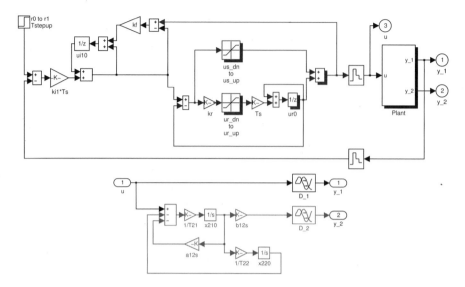

Fig. 7.19. Control structure for version **A**, the feedforward implementation

The parameters of the step-and-ramp generator are the instantaneous step size $u_{s_{up}}$, $u_{s_{dn}}$, and the ramp gradient set by the saturation in the integrator input at $u_{r_{up}}$, $u_{r_{dn}}$ with the feedback gain $k_r := 2$.

As to R_1, a PI controller with weak proportional action is best suited for the type of $G_1(s)$. Here, the proportional path is set to zero for simplicity, *i.e.* an integral controller is used. The awf loop to R_1 is taken over from a non-dynamic saturation.

In the *linear control design* phase, only R_1 need be determined. It is designed by pole assignment as previously in the time-scaled system:

$$0 = 1 + \frac{k'_{i_1}}{s'} 1 \rightarrow 0 = s' + \Omega'_1 \qquad i.e. \quad k'_{i_1} = \Omega'_1 \qquad (7.36)$$

with an upper limit Ω'_{1_u} (u stands for "ultimate")

$$\Omega'_{1_u} D'_1 \leq \frac{\pi}{2} \quad \text{and in numeric values} \quad \Omega'_{1_u} \approx 45 \qquad (7.37)$$

The other phase shift of $\pi/2$ up to the critical one (π) is contributed by the integrator. Note that here the delay D'_1 has been augmented by two times the sampling time T'_s, which is set to $T'_s = 0.005$ (*i.e.* $T_s = 0.10$ s in real time). The numerical value for the bandwidth shall be $\Omega'_1 = 12.0$, having a sufficient margin factor of ~ 4 to Ω'_{1_u}. And the awf gain k_f is set to the compensating value k^*_f.

Considering the awf gain, note that the I(aw) controller in Fig. 7.19 is in the output conditioning form (as shown in Fig. 2.11), and not in the generalized form used in Chapters 2 and 3 for the stability test. However, to obtain the same awf loop gains:

$$k_a \overset{!}{=} T_i \frac{k_f}{T_s} \quad \text{yields} \quad k_f = \frac{k_a}{T_i} \, T_s$$

and with the definition $\quad \dfrac{k_a^*}{T_i} := \Omega \ \rightarrow \ k_f^* = \Omega T_s \qquad (7.38)$

Also note that this result is independent of the dominant order of the closed loop.

As to the *transient responses*, all the following simulations are for a "cold start". At time zero the unit is at the spinning-no load condition: it is connected to the grid, the power control loop is on automatic and the power setpoint is at near zero. The turbine is at thermal equilibrium for this operating point, *i.e.* at low metal temperatures.
Then at $t = 1$ a power setpoint step is applied to 100 %, and both the transients for power y_1 and the constrained temperature difference y_2 are plotted.[7] On the left side the full transient is shown, and the right side zooms in on the first minute after the setpoint step $r_1 = 0 \rightarrow 1$ is applied to show more details about the dynamic performance of the y_1- and y_2-loops.

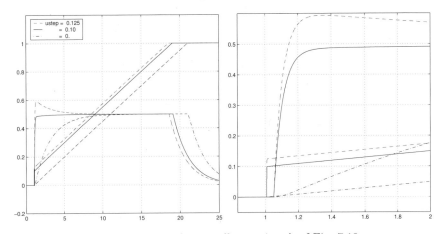

Fig. 7.20. Loading transient with controller version **A** of Fig. 7.19: (left) full run-up transient; and its initial phase (right)

Fig. 7.20 shows the responses for the feedforward strategy to constraining y_2. Three step sizes are investigated.

---

[7] Note that time units are in minutes in the following graphs of Sect. 7.4.

If no such step is applied ($u_{st} = 0$), then $y_2(t)$ slowly rises to its equilibrium. The power ramp-up is delayed by 2 time units with respect to the $u_{st} = 10\,\%$ step response. There, $y_2(t)$ moves quickly close below its constraint setpoint and then slowly rises to its final value. However, if a step size of $u_{st} = 12.5\,\%$ were applied, $y_2(t)$ overshoots and takes about 5 time units to return to the constraint level.

To *summarize*: the "iron-rule" step size of $u_{st} = 10\,\%$ reproduces the expected response quite well. This gives an indication about the validity of the model.

Version B: Standard Overrides

The basic idea is to go on using R_1 as designed for version **A**, and use PI(aw) controllers for the R_2.

The *control structure* **B** is shown in Fig. 7.21.

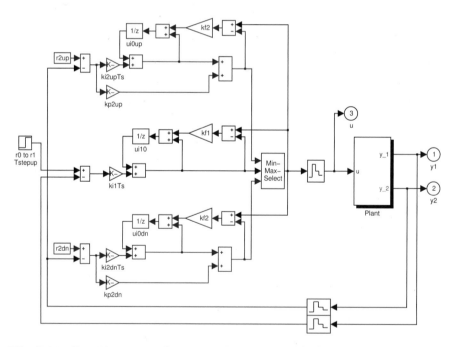

Fig. 7.21. Control structure for version **B**, with the PI(aw) feedback implementation for R_2

Note that the R_2 are in parallel form, such as to accommodate any negative k_p resulting from pole assignment to low Ω_2' entries.

The notation $_{hi}$, $_{lo}$ for the override controllers has been changed to $_{up}$, $_{dn}$, as this fits better here.

In the *linear control design* phase, the main controller R_1 is taken over from Version **A** without any changes.

And for R_2, the characteristic equation is (without the delay D_2):

$$0 = 1 + \frac{k_{p_2}'s' + k_{i_2}'}{s'} \frac{3s'}{s'^2 + (19/3)s' + 1} = s'^2 + s'\left(19/3 + 3k_{p_2}'\right) + \left(1 + 3k_{i_2}'\right)$$
$$(7.39)$$

that is

$$3k_{i_2}' = \Omega_2'^2 - 1; \quad \text{and} \quad 3k_{p_2}' = 2\Omega_2' - 19/3 \qquad (7.40)$$

with the upper limit for $\Omega_{2_u}' > 10$ (from simulations: $\Omega_{2_u}' \approx 15$ is suitable). And there is a lower limit as well, because negative controller parameters should be avoided. This leads to

$$\Omega_2' \geq \frac{1}{2}\frac{19}{3} = 3.1667$$

The awf gain is set to the compensating value ($k_{f_2} := \Omega_2' T_s$).

Numerical values for the bandwidth are set to $\Omega_2' = [3.1667, 4.0, 5.0]$, these having margins to Ω_{2_u}' of $\gtrsim [5, 4, 3]$ respectively.

Note that in Eq. 7.39 the pole at the origin from the PI controller is cancelled by the zero at the origin in $G_2(s)$ of the plant. In other words, one has to expect that the control error e_2 does not converge to zero for nonzero constant setpoints $r_{2_{lo}}$, $r_{2_{hi}}$. To keep this offset small, Ω_2' must be made as high as possible, *i.e.* not far below the limit Ω_{2_u}'.

The *transient responses* in Fig. 7.22 show how this structure performs. There is some overshoot of y_2 in the first minute. Then y_2 stabilizes to some value below $r_{2_{hi}}$. This steady state control error diminishes for higher entries of Ω_2', but the overshoot increases. The effect on power (y_1) is a slower run-up: with $\Omega_2' = 3.1667$, about 2 min are lost at the 100 % power endpoint. With $\Omega_2' = 4$, approximately 1 min is lost, and about 0.5 min with $\Omega_2' = 5$.

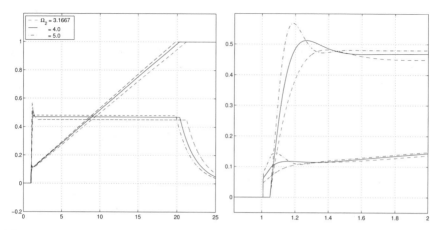

Fig. 7.22. Loading transient with controller version **B** of Fig. 7.21, with compensating awf gains $k_{f_1}^*$, $k_{f_2}^*$: (left) full run-up transient; and its initial phase (right)

Version C

In order to suppress this offset, an additional integrator is inserted at the controller output; see Fig. 7.23. Its function is to cancel the zero in G_2.

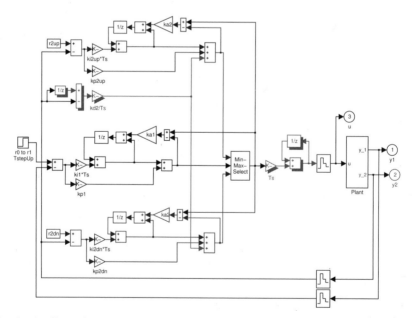

Fig. 7.23. Control structure for version **C**, with R_1 as PI(aw), and R_2 as "PI(aw) and D", followed by an integrator block after the selection

The integrator is inserted at the controller output downstream of the selector block. Note that this is one possible layout. There are several alternatives around in practice. The individual controller structures are PI(aw) for R_1, as $G_1(s)$ is now of dominant first order, and "PI(aw) and D" for R_2, as the denominator of $G_2(s)$ is of dominant second order.

The *linear control design* for the main controller R_1, setting $D_1 := 0$

$$0 = 1 + \frac{k_{p_1}s' + k_{i_1}}{s'} \frac{1}{s'} 1 \rightarrow s'^2 + s'k_{p_2} + k_{i_2} = 0 \qquad (7.41)$$

and thus

$$k_{i_1} = \Omega_1'^2 \quad \text{and} \quad k_{p_1} = 2\Omega_1' \qquad (7.42)$$

with an upper limit Ω_{1_u}' at ≈ 45, and no lower limit to avoid negative feedback parameters. Therefore, as above set $\Omega_1' := 12$.

For the override controller R_2, with $D_2 := 0$

$$0 = 1 + \frac{k_{d_2}s'^2 + k_{p_2}s' + k_{i_2}}{s'} \frac{1}{s'} \frac{3s'}{s'^2 + 19/3s' + 1}$$

$$= s'^3 + s'^2 (19/3 + 3k_{d_2}) + s' (1 + 3k_{p_2}) + 3k_{i_2} \qquad (7.43)$$

i.e.

$$3k_{i_2} = \Omega_2'^3; \quad 3k_{p_2} = 3\Omega_2'^2 - 19/3; \quad \text{and} \quad 3k_{d_2} = 3\Omega_2' - 1 \qquad (7.44)$$

with Ω_{2_u}' from simulations at ~ 16, and for the lower limit $\Omega_2' \geq \sqrt{\frac{19}{9}} = 1.45$. Therefore, set $\Omega_2' := [2.0,\ 3.0,\ 4.0]$, yielding a margin to $\Omega_{2_u}' \approx [8,\ 5,\ 4]$.

The *transient response* in Fig. 7.24 shows that this performs as expected.

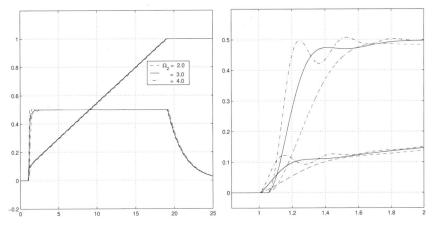

Fig. 7.24. Loading transient with controller version **C** of Fig. 7.23, with compensating awf gains: (left) full run-up transient; and its initial phase (right)

The steady state error on y_2 now tends to zero with all Ω_2' entries, and the run-up time to $y_1 := 100\ \%$ is increased only marginally (~ 0.08 min for $\Omega_2' = 3.0$, and ~ 0.2 min for $\Omega_2' = 2.0$). Considering the response in the first minute, $\Omega_2' = 3.0$ seems a good compromise. Then $u(t)$ rises to the 10 % level in about 0.2 min (compare this with version **A**). And there are no weakly damped modes in the $y_2(t)$ response.

Note that the size of the actual delay D_2 is the key factor for sizing Ω_2' and obtaining a good performance.

Version D

The structure **C** has two weak points, which must be addressed now.

One is the "PI(aw) and D" form of R_2. This does not perform well with high-frequency disturbances, and therefore should be avoided. The second weakness is the awf tracking signal being the output of the selector block. This implicitly assumes that the additional integrator output is always within the control saturations. Otherwise it will wind up.

The basic idea is to replace the structure of R_2 in **C**, which consists of a PID structure in series with an I element, by a "P+I+I^2" structure.

Correspondingly, the R_1 -controller changes to "I+I^2". Then the selection takes place on the control signal u (and not on its time derivative as in version **C**). Therefore, saturations on u can be implemented directly; e.g. see Fig. 7.1. Note that the "P+I+I^2" or "I+I^2" structures are not standard in all industrial process control systems. And this will also need a modification in the awf loop; see below.

The *control structure* **D** is shown in Fig. 7.25. The integrator which has been inserted downstream of the selector block is now moved upstream into all three controller outputs, as in Chapter 3, forms **H** to **K**.

Then the basic idea is to extend the awf concept of Chapter 2, such that the tracking feedback now acts on *both* integrators to keep them from winding up. In Fig. 7.25, this is implemented in the same way as in a standard observer arrangement, where the observer output $\hat{y}$ tracks the plant output y by proportional feedback with gains g_i of the observer error $e = y - \hat{y}$ to each of the i observer states (integrators). Here, the awf error $e_a = u(t) - u_2(t)$ provides proportional feedback to all controller integrators, with gains k_{f_i}.

For the *linear controller design* let

$$R_2 = \left(k_{p_2} + \frac{k_{i_2}}{s} + sk_{d_2}\right)\frac{1}{s} = k_{d_2} + \frac{k_{p_2}}{s} + \frac{k_{i_2}}{s^2} \qquad (7.45)$$

and similarly

$$R_1 = \left(k_{p_1} + \frac{k_{i_2}}{s}\right)\frac{1}{s} = \frac{k_{p_2}}{s} + \frac{k_{i_2}}{s^2} \qquad (7.46)$$

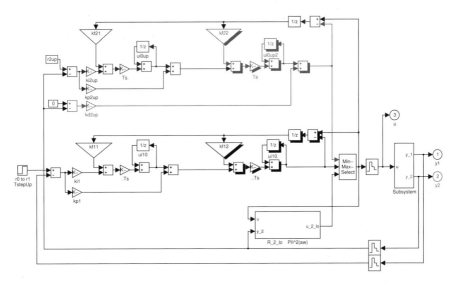

Fig. 7.25. Control structure for version **D**,
with R_1 as "I+I^2(aw)", and R_2 as "P+I+I^2(aw)"

The characteristic equations then are the same as for version **C**, and thus the controller parameters for R_1 and R_2 shall be taken over without change.

Note that the awf in Fig. 7.25 has been slightly modified from before: the summing point of the awf signal to the integrator input signal $e_1 k_{i_1}$ is moved upstream across the factor T_s, such that the $R_{1_2}(z) := T_s/1 - z^{-1}$ block now is equivalent to the continuous integrator $R_{2_1}(s) := 1/s$. This is equivalent if for the compensating awf gain $k_f^* := \Omega_2'$ instead of $k_f^* := \Omega_2' T_s$. This makes the design for the R_2 awf more transparent.

Now the awf gains k_{f_i} are designed. The "compensating awf" idea from the stability analysis in Chapters 2 and 3 and the noise effects discussion in Sect. 5.3 suggests using pole assignment for this awf feedback system as well, which is now of order higher than one. In this particular case, a first choice would obviously be (see Fig. 7.25 for definition of $k_{f_{21}}$ and $k_{f_{22}}$):

$$0 = 1 + k_{f_{22}} \frac{1}{s'} + k_{f_{21}} \frac{1}{s'^2} \,\, \hat{=} \,\, (s' + \Omega_2')^2$$

$$i.e. \quad k_{f_{21}} = \Omega_2'^2 \quad \text{and} \quad k_{f_{22}} = 2\,\Omega_2' \tag{7.47}$$

And for the R_1 controller:

$$0 = 1 + k_{f_{11}} \frac{1}{s'} + k_{f_{12}} \frac{1}{s'^2} \,\, \hat{=} \,\, (s' + \Omega_1')^2$$

$$i.e. \quad k_{f_{12}} = \Omega_1'^2 \quad \text{and} \quad k_{f_{11}} = 2\,\Omega_1' \tag{7.48}$$

An alternative to assigning the awf loop poles in the R_2 controller is

$$0 = 1 + k_{f22} \frac{1}{s'} + k_{f21} \frac{1}{s'^2} \; \hat{=} \; (s' + \Omega_2')(s' + \Omega_1')$$

i.e. $\qquad k_{f21} = \Omega_2' \Omega_1' \; \text{and} \; k_{f22} = \Omega_2' + \Omega_1' \qquad\qquad (7.49)$

The second version is used for the simulation. The reason will become clearer in the stability analysis below.

Transient responses

Again, this performs as expected; Fig. 7.26. There is no perceptible difference to the responses with version **C** in Fig. 7.24. Again, this will become clearer in the stability analysis.

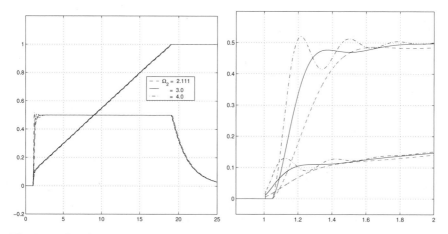

Fig. 7.26. Loading transient with controller version **D** of Fig. 7.25: (left) full run-up transient; and its initial phase (right)

Discussion

The only weakness now seems to be the small size of the "initial instantaneous" step of delivered power y_1. However, this can be tolerated, as the time interval for moving up u to 10 % of $\sim$ 10 s is acceptable as seen from dynamic power grid requirements.

Note also that the two-compartment model for the turbine casing is only valid for low frequencies. In fact, applying a step in u to a multi-layer model produces a sharp but short peak in temperature differences, *i.e.* thermal stress, which will accelerate surface ageing. Therefore, one is interested in reducing step size and step slope on $u(t)$. And a similar argument holds for thermal stresses in the steam generator. In other words, the shape of $u(t)$ produced by versions **C** and **D** is even better suited than the "step and ramp" shape from the base line version **A**. By the way: this is a typical discussion in thermal power plant operation on the tradeoff between faster grid load-following and longer life expectation for equipment.

For brevity, no small-sized power setpoint step responses are shown here, and also no cases other than "cold" loading. You may investigate this and other questions in the proposals for case studies in Sect. 7.4.5.

7.4.4 Stability properties

Version A

The relevant part of the system is shown in Fig. 7.27 (a). The nonlinear block consists of two nonlinear elements in parallel, and therefore does not comply directly with the assumptions on the sector nonlinearity.

But inspection of Fig. 7.27 (a) and its expected step response suggests a (conservative) approximation, which has the required properties, see Fig. 7.27 (b): the direct path is suppressed, and only the ramp part is considered.

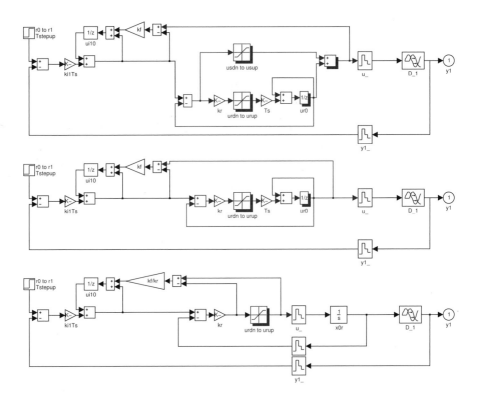

Fig. 7.27. Stability test for version **A**:   (top) original structure (a);   (center) conservative approximation (b);   (bottom) with shifted inputs to the awf (c)

This may be redrawn into Fig. 7.27 (c), by shifting the awf inputs to the input and output of the ramp saturation. Then, this is the system we have investigated in Chapter 2 and shown to be asymptotically stable in the large.

Exercise

Check the statement about Fig. 7.27 (b) being a conservative approximation, and check for which assumptions (c) is equivalent to (b).

Version B

The nonlinear system in Fig. 7.21 is the standard one from Chapter 3, and the main result from there can be applied directly.

$$
\begin{aligned}
(F+1)_B &= \frac{1}{1+R_1G_1}\ (1+k_{f_1}G_{a_1}R_{1_2})\ (1+R_2G_2)\ \frac{1}{1+k_{f_2}G_{a_2}R_{2_2}} \\
&= \frac{s'}{s'+\Omega'_1}\ \frac{s'+k_{f_1}}{s'}\ \frac{(s'+\Omega'_2)^2}{s'^2+a\,s'+1}\ \frac{s'}{s'+k_{f_2}} \quad \text{and with} \quad k_{f_1}=k_{f_2} \\
&= \frac{s'}{s'+\Omega'_1}\ \frac{(s'+\Omega'_2)^2}{s'^2+as'+1} \quad\quad\quad\quad\quad\quad\quad (7.50)
\end{aligned}
$$

where $a = 19/3$ from the $1:3$ thickness relation of wall layers.
Note that this is independent of the value inserted for $k_f = k_{f_1} = k_{f_2}$.

Fig. 7.28 shows the Nyquist plots for $\Omega'_1 = 12$ and $\Omega'_2 = [3.166,\ 4,\ 5]$. The Nyquist contour of $[F(j\omega)+1]_B$ for the stability test starts at $(-1;\ j0)$ in the vertical direction. This indicates good stability properties.

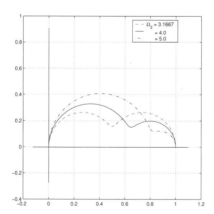

Fig. 7.28. Nyquist contour with controller version **B** of Fig. 7.21

Version C

Again, the structure Fig. 7.23 fits into the framework of Chapter 3.

$$(F+1)_C = \frac{s'^2}{(s'+\Omega_1')^2} \frac{s'+k_{f_1}}{s'} \frac{(s'+\Omega_2')^3}{s'(s'^2+as'+1)} \frac{s'}{s'+k_{f_2}}$$

and with $\quad k_{f_1} = k_{f_2}$

$$(F+1)_{C_a} = \frac{s'}{s'+\Omega_1'} \frac{s'+\Omega_2'}{s'+\Omega_1'} \frac{(s'+\Omega_2')^2}{s'^2+as'+1} \tag{7.51}$$

with the same remarks as for Eq. 7.50.

Fig. 7.29 shows the Nyquist plots for $\Omega_1' = 12$ and $\Omega_2' = [2.111, 3, 4]$.

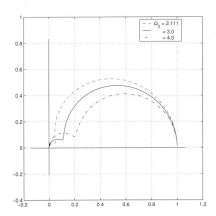

Fig. 7.29. Nyquist contour with controller version **C_a** of Fig. 7.23

The Nyquist contour starts from the origin in the vertical direction, and indicates asymptotic stability for almost all initial conditions.
Alternatively, let $\quad k_{f_1} = \Omega_1'; \quad k_{f_2} = \Omega_2'$. Then, from Eq. 7.51

$$(F+1)_{C_b} = \frac{s'}{s'+\Omega_1'} \frac{(s'+\Omega_2')^2}{s'^2+as'+1} = (F+1)_B \tag{7.52}$$

Version D

The main result of Chapter 3 applies again.

$$(F+1)_D = \frac{1}{1+R_1G_1} \left(1+k_{f_1}G_{a_1}R_{1_2}\right) \left(1+R_2G_2\right) \frac{1}{1+k_{f_2}G_{a_2}R_{2_2}}$$

$$= \frac{s'^2}{(s'+\Omega_1')^2} \frac{(s'+k_{f_1})(s'+k_{f_1})}{s'^2}$$

$$\times \frac{(s'+\Omega_2')^3}{s'(s'^2+a\,s'+1)} \frac{s'^2}{(s'+k_{f_1})(s'+k_{f_2})}$$

where $\quad k_{f_1} = \Omega_1'; \quad k_{f_2} = \Omega_2' \tag{7.53}$

Note the particular choice of the characteristic equation for the second-order awf loops.
Then

$$(F+1)_D = \frac{(s'+\Omega_2')^2}{s'^2+as'+1}\frac{s'}{s'+\Omega_1'} = \frac{(s'+\Omega_2')s'}{s'^2+as'+1}\frac{s'+\Omega_2'}{s'+\Omega_1'}$$

$$\overset{!}{=} (F+1)_{C_b} \tag{7.54}$$

The stability properties are the same. Therefore, the finding from the simulations is not surprising that the transient responses do not differ. Note that this is connected to the particular design of the awf gains.

7.4.5 Some Suggestions for Case Studies

The aim here is to extend the investigations such as to approach real design situations more closely. You may select items you are particularly interested in from the following menu. They are on structures and transient response, *i.e.* on simulations, and should be extended into stability analysis.

- Check the performance of the linear y_1 loop by small enough Δr_1.

- What is the effect of a nonzero heat loss $0 < \gamma \ll \beta$ (see Fig. 7.18)?

- Investigate D_2' up to 0.05, as the current value seems rather short for typical sensors.

- Investigate the effect of a high-frequency disturbance on measured temperature ϑ_1, with a peak-to-peak level of 0.002 of span. In other words, is the "PI(aw) and D" structure for R_2 a realistic option? How much is version **D** performing better?

- The input u may also saturate. This may arise if the y_1 loop is operated at very low or very high setpoints r_1, *i.e.* near to fully closed or fully open steam valves (see Fig. 7.17). Design suitable add-on awf loops to versions **A** through **D**.

- Replace the current casing model of Fig. 7.18 by a three-layer one, where layer 1 and 2 are as before, but there is a layer 3 attached to the outside with thickness $t_3 = 3t_2 := 90$ mm, but still with $y_2 = \vartheta_1 - \vartheta_2$ as the constrained output. This is more realistic for large, high-pressure and high-temperature units. Use versions **A** through **D**.

- Alternatively use a three layer model to catch better the surface thermal transients and thus get an indication of thermal ageing, using $t_1 = 3\ mm$, $t_2 = 9\ mm, t_3 = 27\ mm$, with $y_2 = \vartheta_2 - \vartheta_3$ for constraint control, and observing $y_3 = \vartheta_1 - \vartheta_2$, but not controlling it. Again use versions **A** through **D**.

- Investigate the "warm" and "hot" loading cases. For this, modify the model in Fig. 7.18 by having the heat transfer coefficient α not constant, but proportional to steam mass flow, *i.e.* to u.

  Hint: the y_2 model will be nonlinear. One possible design approach would be a linearized model having floating coefficients and robust controllers with fixed coefficients.

- And finally, if you are familiar with the variable pressure mode for power control of thermal units ....

7.5 Override Action on the Reference Input

7.5.1 Introduction

One of the basic assumptions in Chapter 3 has been that the additional feedback from the constrained output y_2 is acting on the control signal u, as in the "control conditioning" approach to antiwindup.

This may not be always easy to accept in practice. Consider, for instance, the attitude control system of an aircraft, where such a "breaking of the control loop" by the override and the minimum selector is difficult for the pilot to accept for safety reasons. He would prefer that the basic loop remains intact, and the override acts on his command inputs, such that he can switch out this additional function if necessary, and supply it in manual mode.

A second argument is that the aircraft will react quite differently to the pilot's stick inputs when the override takes over. In other words, there is no compromise between the original control and the constrained one. This second point will be looked into in Sect. 7.6.

A similar topic comes up with multivariable control systems and constraints. There, the design of the linear control system is much more involved and the linear structure of R_1 is more complex, because the interactions or cross couplings between plant inputs and outputs have to be accounted for. So it would seem unreasonable or even dangerous to disrupt this finely tuned structure by letting some override loop take over any plant input. But it would be much more transparent to have such overrides acting on the references to R_1, and thus on the next higher hierarchical level.

This motivates investigation of the consequences of overriding the setpoint instead of the control variable. This would be similar to the "reference conditioning" paradigm in antiwindup.

The single input loop shall be considered here (as this has been the case so far). A benchmark is specified first. Then three versions for the controller shall be looked into.[8] The standardized methodology from above is used. It starts with a structure (based on intuition). Then the linear feedbacks are designed. The transient performance on the benchmark is shown by simulations, and illustrates the performance achieved. Finally, the stability analysis is performed.

7.5.2 The Benchmark

This shall be the same as in Chapter 3, to allow a direct comparison. Consider the positioning of a rigid mass with an actuator force u against an unknown load force v. No friction and no spring forces shall be present. The main output to be controlled is the position y_1. And there is a constraint on speed y_2. A small delay D shall approximate the dynamics of actuator and sensors. That is

$$G_1 = \frac{y_1}{u} = e^{-sD}\frac{1}{s^2 T_1 T_2}; \quad \text{and} \quad G_2 = \frac{y_2}{u} = G_1 s T_1 = e^{-sD}\frac{1}{s T_2} \quad (7.55)$$

with numerical values

$$T_1 = 5.0 \ s; \quad T_2 = 1.0 \ s; \quad D = 0.025 \ s; \quad (7.56)$$

The test sequence shall be starting from equilibrium at $r_1 = 0$. A large setpoint step to $r_1 = 0.98$ is applied, which will lead into the speed constraints. During the run-up a load step $v = 0.90$ is applied to check the performance of the speed constraint feedback. After stabilizing at $r_1 = 0.98$ a small setpoint step is applied to check the linear response of the y_1 loop. Finally a load swing to $v = -0.90$ is injected to check the disturbance suppression.
The constraints are set to

$$u_{lo} = -1.20; \quad u_{hi} = +1.20; \quad r_{2_{lo}} = -0.50; \quad r_{2_{hi}} = +0.50 \quad (7.57)$$

The controllers shall be in discrete-time form, with sampling time $T_s = 0.010$.

7.5.3 The Design Targets

The final design targets shall be

- no significant overshoot of y_2 on $r_{2_{hi}}$ during run-up,
- no steady state offset on e_2, (which would slow down the run-up),
- no steady state offset on e_1,
- high closed-loop bandwidth for both linear loops (*i.e.* no significant bandwidth restrictions due to the nonlinear operation),
- acceptable closed-loop damping in both linear loops.

---

[8] Note that there are many other possible structures.

7.5.4 Form A: Nonlinear y_2-Feedback

This approach has been proposed, for example, for vehicle anti-roll-over [50] and aircraft attitude control [55].

Structure

From Fig. 7.30, this consists of a P-P cascade controller for R_1 with gains k_{s_1} for speed feedback and k_{p_1} for position control. And the nonlinear y_2 feedback consists of a P element with gain k_{p_2} and a deadspan element with breakpoints at

$$u_{2_{lo}} = k_{p_2} r_{2_{lo}} \quad \text{and} \quad u_{2_{hi}} = k_{p_2} r_{2_{hi}} \tag{7.58}$$

The override is additive on the main setpoint $r_1(t)$, thereby modifying it into the actual setpoint $w(t)$ for the y_1 loop.

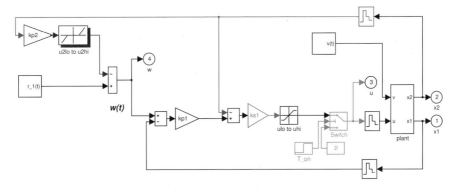

Fig. 7.30. Controller structure version **A**

It has also been suggested to replace the P element with gain k_{p_2} in Fig. 7.30 by a first-order lag with gain k_{p_2} and time constant τ_2 after the nonlinear element, where τ_2 is an additional design parameter. The idea is to smooth the override action further, and thus to reduce the effect felt by the pilot. On the other hand, this obviously compromises the performance of y_2 along $r_{2_{hi}}$ even further.

Exercise
- Investigate this alternative using the standardized methodology.
- Are there other suitable transfer functions R_2?

Linear feedback design

For small r_1 values, there is no override action, and the system response is linear. It is determined by R_1:

$$G_{lin} = \frac{R_1 G_1}{1 + R_1 G_1} \tag{7.59}$$

Neglecting the small delay D, *i.e.* designing for the dominant dynamics:

$$G_{lin_e} = \frac{k_{p_1} k_{s_1}}{s^2 T_1 T_2 + s T_1 k_{s_1} + k_{p_1} k_{s_1}} \overset{!}{=} \frac{\Omega_1^2}{(s + \Omega_1)^2} \tag{7.60}$$

For a given bandwidth Ω_1 then

$$k_{p_1} k_{s_1} = \Omega_1^2 T_1 T_2; \quad \text{and} \quad k_{s_1} = 2\Omega_1 T_2; \quad i.e. \quad k_{p_1} = (\Omega_1 T_1)/2 \tag{7.61}$$

where the bandwidth Ω_1 has an upper limit from the delay D.

For large inputs r_1, the nonlinear element produces an additional feedback of y_2. Consider the isolated y_2 loop (this will become clearer with version **B**). Its characteristic equation is

$$0 = 1 + k_{p_2} k_{p_1} \frac{k_{s_1}}{s T_2 + k_{s_1}} \rightarrow 0 = s + \frac{k_{s_1} + k_{p_2} k_{p_1} k_{s_1}}{T_2} \overset{!}{=} s + \Omega_2 \tag{7.62}$$

that is

$$k_{p_2} = \frac{1}{k_{p_1} k_{s_1}} (\Omega_2 T_2 - k_{s_1}) = \cdots = \frac{1}{\Omega_1 T_1} \left[\frac{\Omega_2}{\Omega_1} - 2 \right] \tag{7.63}$$

where the bandwidth Ω_2 has an upper limit from the delay D, and also a lower limit $\Omega_2/\Omega_1 > 2$, as k_{p_2} must be positive. And finally, if the design value for the bandwidth Ω_1 of the main loop is increased, then the override feedback gain k_{p_2} decreases, and therefore the tracking quality and disturbance suppression of the override loop decreases as well. In other words, there is a trade-off between the performance of the main loop (given by Ω_1, *i.e.* k_{p_1}, k_{s_1}), and the performance of the override feedback (given by k_{p_2} with fixed Ω_2 at its upper limit).

For numerical values on the benchmark:

$$\text{let} \quad D_e := D + T_s; \quad \text{and using} \quad \omega_u = \frac{2\pi}{4D_e} \quad \text{set} \quad \Omega_2 = \frac{\omega_u}{3} \rightarrow \Omega_2 := 15 \tag{7.64}$$

that is

Ω_1	0.75	1.50	3.00
k_{p_2}	4.80	1.0667	0.20

which clearly shows this trade-off.

Transient Response

The responses shown in Fig. 7.31 illustrate the results from above. With the lowest entry for Ω_1, there is a constraining effect on $y_2(t)$, but not up to specifications. However, the steady state control error e_1 is largely insufficient. For $\Omega_1 = 3.0$ this is much improved, but it is still not up to specifications, whereas the constraining effect on y_2 is not visible any more.

A better solution must be sought.

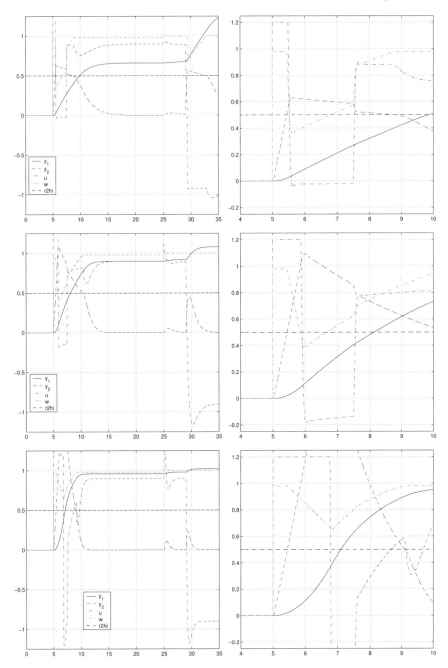

Fig. 7.31. Transients with controller version **A** of Fig. 7.30, with $\Omega_2 = 15$:
(top) $\Omega_1 = 0.75$;   (center) $\Omega_1 = 1.5$;   (bottom) $\Omega_1 = 3.0$;
(left) full transient, and initial phase (right)

Stability Analysis

At least this configuration has good stability properties. The structure Fig. 7.30 is in canonical form for the stability test. The nonlinear subsystem is the deadspan element with breakpoints from Eq. 7.58 and unity slope outside. And the linear subsystem has the transfer function

$$L = \frac{sT_1 k_{p_2}}{\left((s/\Omega_1) + 1\right)^2} \tag{7.65}$$

which has the properties required by the circle test. Its Nyquist contour $L(j\omega)$ evolves from the origin into the right half plane, and thus stability can be deduced for almost all initial conditions.

7.5.5 Form B: Override Structure with P Controllers

The results of form **A** are far from satisfactory. They are similar to what was found for the corresponding structure in Chapter 3. But a much better performing solution was developed there: the linear signal u_1 from the main loop is added to the input of the deadspan nonlinearity, which compensates its effect on u during nonlinear operation.

Structure

This idea is not directly applicable here, as u_1 was the output of a controller R_1 and not a reference input r, as it is the case here. However, there is a simple way to link the basic system from Chapter 3 to this case; see Fig. 7.32.

Fig. 7.32 (a) is a form of the generic structure of Chapter 3, where the nonlinear feedback acts within the controllers $R_j = R_{j_1} R_{j_2}$, $j = 1, 2$, such that R_{j_1} is upstream and R_{j_2} is downstream of the "nonlinear additive" element, which is highlighted in Fig. 7.32 (a). That is (variables are defined in the figures):

$$R_{1_1} = 1.0; \; R_{1_2} = 1.0; \quad \text{and } G_1 = \frac{y_1}{e} = k_{p_1} \frac{k_{s_1}}{sT_2 + k_{s_1}} \frac{1}{sT_1}$$

$$R_{2_1} = k_{p_2}; \; R_{2_2} = 1.0; \quad \text{and } G_2 = \frac{y_2}{e} = sT_1 G_1 = k_{p_1} \frac{k_{s_1}}{sT_2 + k_{s_1}} \tag{7.66}$$

From Chapter 3 this is equivalent to the structure with a Max-Min-selectors block placed on the control error e_1 of the main loop and having as override inputs $u_{2_{hi}} = k_{p_2} (r_{2_{hi}} - y_2)$ on the Min-selector and $u_{2_{lo}} = k_{p_2} (r_{2_{lo}} - y_2)$ on the Max-selector. From the simulation results in Chapter 3 this structure can be expected to perform much better, although there may be a tendency to overshooting on both $y_2(t)$ and $y_1(t)$. This structure also allows the "prefabricated" stability analysis of Chapter 3 to be applied directly. So this seems a reasonable base to start from.

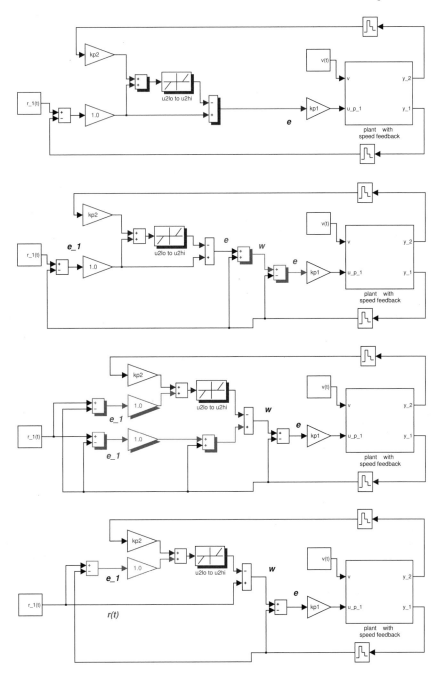

Fig. 7.32. Developing the basic override configuration (a), (top) into Controller structure version **B** (d), (bottom)

Then, in Fig. 7.32 (b), $y_1(t)$ is added and then subtracted in sequence downstream of the "nonlinear additive" block. This evidently has no effect on function. But it introduces the "setpoint minus measured variable" element upstream of the k_{p_1} block in the original structure in Fig. 7.30. In other words, Fig. 7.32 (b) now contains the complete R_1 loop of the original system.

Next, in Fig. 7.32 (c), the summation block inserted in Fig. 7.32 (b) is moved upstream along the linear path. And the $r_1 - y_1$ block at the input of R_{1_1} in Fig. 7.32 (a) is moved downstream.

Then, on the linear path, the consecutive subtraction and addition of $y_1(t)$ cancel, and the setpoint r_1 is left, as in Fig. 7.30; see Fig. 7.32 (d). But on the input to the deadspan element the $r_1 - y_1$ block is still there.
In other words, the structure in Fig. 7.32 (d) is equivalent to that in Fig. 7.32 (a). And Fig. 7.32 (d) is the original structure from Fig. 7.30, with one modification: the addition of the control error e_1 to the input to the deadspan element.

So far, the well-known block diagram manipulation rules have been applied. An alternative is the algebraic approach as follows. Start again with the basic structure from Chapter 3; Fig. 7.32 (a). There, with dsp denoting the deadspan function:

$$u = u_1 - \text{dsp}(w_2 + u_1) \quad \text{where} \quad u_1 = 1(r_1 - y_1)$$
$$i.e. \quad u = (r_1 - y_1) - \text{dsp}\left[w_2 + (r_1 - y_1)\right]$$
$$\text{re-written as} \quad u = \left[r_1 - \text{dsp}(w_2 + e_1)\right] - y_1 \; = \; w - y_1 \quad (7.67)$$

where w is the modified setpoint to be applied to the closed loop. And this is the structure version **B**; see Fig. 7.32 (d).

Linear feedback design

The controller R_1 for linear range operation is the same as in version **A**, and thus k_{p_1}, k_{s_1} are carried over with the same numerical values.

During nonlinear operation, the effect of $y_1(t)$ is canceled by the addition of $e_1(t)$ to the input of the deadspan element. And the effect of $y_2(t)$ is carried through without modification. Considering the Max-Min-selector implementation of Fig. 7.32 (a), then the closed loop transfer function (along the upper constraint) is

$$\frac{y_2(s)}{r_{2_{hi}}(s)} = \frac{k_{p_2} k_{p_1} \frac{k_{s_1}}{sT_2 + k_{s_1}}}{1 + k_{p_2} k_{p_1} \frac{k_{s_1}}{sT_2 + k_{s_1}}} = \frac{k_{p_2} k_{p_1} k_{s_1}}{sT_2 + k_{s_1}(1 + k_{p_1} k_{s_1})} \quad (7.68)$$

with the characteristic polynomial used in Eq. (7.62) for pole assignment in the R_2 loop in version **A**. Thus k_{p_2} is taken over from there.

Note that the gain of the transfer function in Eq. (7.68) is < 1. This predicts a steady state offset on e_2 (but no ramp-like behavior as in Fig. 7.31). Note also that the trade-off between gains $k_{p_1} k_{s_1}$ and k_{p_2} continues to hold. In other words, increasing Ω_1 will decrease the steady state errors e_1, but increase the offset e_2 during run-up.

This will have to be addressed next; see version **C**.

Transient Response

The simulations in Fig. 7.33 confirm the predictions. Only the case $\Omega_1 = 1.5$ is shown for brevity.

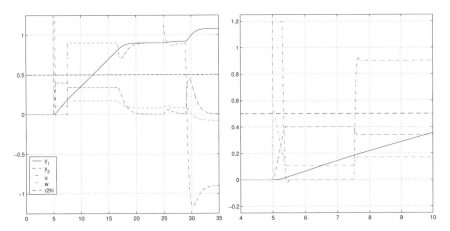

Fig. 7.33. Response with controller version **B**, Fig. 7.32 (d), with $\Omega_2 = 15$ and $\Omega_1 = 1.5$; (left) full transient, and initial phase (right)

Stability Analysis

As mentioned above, the equivalent structure in Fig. 7.32 (a) is used. Then

$$(F+1)_B = \frac{sT_1(sT_2 + k_{s_1})}{s^2 T_1 T_2 + sT_1 k_{s_1} + k_{p_1} k_{s_1}} \frac{1}{1} \frac{sT_2 + k_{s_1}(1 + k_{p_1} k_{p_2})}{sT_2 + k_{s_1}} \frac{1}{1}$$

$$= \frac{s(s + \Omega_2)}{(s + \Omega_1)^2} = \frac{s}{s + \Omega_1} \frac{s + \Omega_2}{s + \Omega_1} \tag{7.69}$$

As $\Omega_2 \gg \Omega_1$, the Nyquist contour evolves from the origin into the right half plane, thus indicating good stability properties.

7.5.6 Form C: Override Control with Integral Action

Although the controller structure **B** with P feedback is reasonably simple, the performance is not sufficient for most practical applications. The steady state control errors are too large, compounded by the trade-off effect discussed above. One approach would be to increase Ω_2 and thus increase k_{p_2}. But this requires not only a shorter sampling time T_s, *i.e.* a more powerful microprocessor, but also reducing the delay D. This means more powerful actuators and faster sensors, which may be not feasible technically or be very expensive. The usual approach is to add integral action. This shall be tried here as well. It is done in the R_1 control algorithm first, to drive $e_1 \rightarrow 0$. Then the structure of R_2 has to be selected for driving $e_2 \rightarrow 0$.

Structure

The first step is to insert the integral action into R_1. Here, this is done by an additional cascaded feedback through a discrete integrator; see Fig. 7.34 (center area). For its gain T_s/T_{i_1} the equivalent form $k_{i_1} T_s/T_0$ will be used. This integral action requires an awf with gain k_{a_1} from the saturations on u.

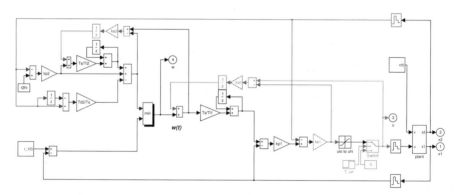

Fig. 7.34. Control structure version **C**

From the linear design (see below) this in fact drives $e_1 \rightarrow 0$, but produces a zero at the origin in G_2. Therefore, an additional integral action is required in R_2 to drive $e_2 \rightarrow 0$ during run-up; Fig. 7.34 (top left area). Its integral action again requires an awf. The awf reference signal is picked up downstream of the addition point for the awf from the u saturation. This interleaved awf loop structure is taken from the "slew and stroke" constrained actuator system (see Sect. 5.1 for more details).

The output constraint acts on the setpoint r_1 in Fig. 7.34 (lower left area), using the same structure as in Fig. 7.32.

There are other possible structures. For instance, the output constraint may act on the output of the integral action instead. This does not comply with the specification that the override acts on the overall setpoint, but it acts on the setpoint for the P cascaded solution from Fig. 7.32. In other words the integral action in the R_1 loop is thought to be part of the next hierarchical level.

Exercise
Develop this idea as version **D**, design the controller, and investigate both transient response and stability properties.

Linear Feedback Design

The y_1-feedback is designed first. Read from Fig. 7.34:

$$\frac{y_1(s)}{w(s)} = \frac{k_{i_1} k_{p_1} k_{s_1}}{s^3 T_0 T_1 T_2 + s^2 T_0 T_1 k_{s_1} + s T_0 k_{p_1} k_{s_1} + k_{i_1} k_{p_1} k_{s_1}}$$

$$\overset{!}{=} \frac{\Omega_1^3}{(s + \Omega_1)^3} := G_w(s) \tag{7.70}$$

that is for the R_1 controller parameters:

$$k_{s_1} = 3\Omega_1 T_2; \quad k_{p_1} = \frac{1}{k_{s_1}} \Omega_1^2 T_2 T_1; \quad k_{i_1} = \frac{1}{k_{p_1} k_{s_1}} \Omega_1^3 T_2 T_1 T_0; \quad \text{and } T_{i_1} := T_0/k_{i_1} \tag{7.71}$$

and for the compensating awf gain: $k_{a_1} = 3/(k_{p_1} k_{s_1})$.

Then the y_2-feedback loop is designed. Its characteristic equation is

$$0 = 1 + R_2(s) G_w(s) s T_1 \tag{7.72}$$

that is, R_2 must be of PID type to enable the usual pole assignment procedure. Let

$$R_2(s) = \frac{k_{d_2}(s T_0)^2 + k_{p_2}(s T_0) + k_{i_2}}{s T_0} \tag{7.73}$$

into the characteristic equation yields

$$0 = 1 + \left[k_{d_2}(s T_0)^2 + k_{p_2}(s T_0) + k_{i_2}\right] \frac{s T_1}{s T_0}$$

$$\times \frac{k_{i_1} k_{p_1} k_{s_1}}{s^3 T_0 T_1 T_2 + s^2 T_0 T_1 k_{s_1} + s T_0 k_{p_1} k_{s_1} + k_{i_1} k_{p_1} k_{s_1}} \tag{7.74}$$

set $K := k_{i_1} k_{p_1} k_{s_1}$

$$0 = s^3 T_0 T_1 T_2 + s^2 T_0 T_1 \left[k_{s_1} + k_{d_2} K\right] + s T_0 \left[k_{s_1} k_{p_1} + k_{p_2} \frac{T_1}{T_0} K\right]$$

$$+ \left[k_{s_1} k_{p_1} k_{i_1} + k_{i_2} \frac{T_1}{T_0} K\right] \tag{7.75}$$

Using $0 = (s + \Omega_2)^3$ yields

$$k_{d_2} = \frac{3}{(\Omega_1 T_0)^2} \frac{T_1}{T_0} \left[\left(\frac{\Omega_2}{\Omega_1} \right) - 1 \right]$$

$$k_{p_2} = \frac{3}{\Omega_1 T_0} \frac{T_1}{T_0} \left[\left(\frac{\Omega_2}{\Omega_1} \right)^2 - 1 \right]$$

$$k_{i_2} = \frac{T_1}{T_0} \left[\left(\frac{\Omega_2}{\Omega_1} \right)^3 - 1 \right] \tag{7.76}$$

In other words, a feasible solution demands $\Omega_2 > \Omega_1$. And with a given Ω_1, there is again an upper limit on Ω_2 from D and T_s from the innermost loop. Its gain is[9]

$$g_2 := k_{d_2} K = k_{d_2} \Omega_1^3 T_2 T_1 T_0 = 3\Omega_2 T_2 \frac{T_1}{T_0} \left[1 - \left(\frac{\Omega_1}{\Omega_2} \right) \right] = 3\, T_2\, \frac{T_1}{T_0}\, [\Omega_2 - \Omega_1] \tag{7.77}$$

and to have acceptable closed-loop damping, set

$$g_2 := 0.33\, g_{2_u} \quad \text{where} \quad g_{2_u} \approx \frac{\pi}{2} \frac{T_2}{D_e} = \frac{\pi}{2} \frac{1.0}{0.035} \approx 45 \quad i.e. \quad g_2 :\approx 15 \tag{7.78}$$

In numerical values for the benchmark, where

$$T_0 = T_2; \quad T_2 = 1.0; \quad \text{and} \quad T_1 = 5.0; \qquad \text{then} \quad \Omega_2 :\approx \Omega_1 + 1$$

Transient Response

The responses shown in Fig. 7.35 confirm the predictions from the linear design. The improvements compared with Fig. 7.33 are significant, and the specifications are now met. This design result may be considered as a good basis for further development.

Stability Analysis

The saturation on u shall not be considered here. You may investigate this case as an exercise.

With the output constraint loop, the standard situation of chapter 3 arises with

$$G_1 = k_{i_1} \frac{1}{sT_0} \frac{k_{p_1} k_{s_1}}{s^2 T_2 T_1 + sT_1 k_{s_1} + k_{p_1} k_{s_1}} \quad \text{and} \quad R_1 = 1;$$

$$G_2 = k_{i_1} \frac{T_1}{T_0} \frac{k_{p_1} k_{s_1}}{s^2 T_2 T_1 + sT_1 k_{s_1} + k_{p_1} k_{s_1}} \quad \text{and} \quad R_2 = \frac{k_{d_2}(sT_0)^2 + k_{p_2}(sT_0) + k_{i_2}}{sT_0}$$

---

[9] Check this as a small **exercise**.

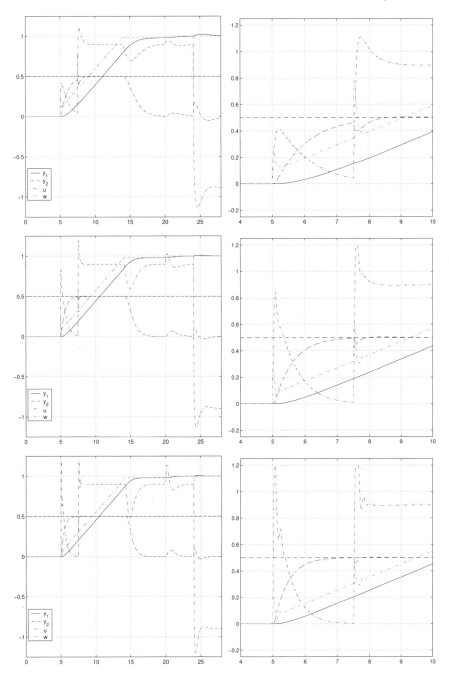

Fig. 7.35. Response with controller version **C**:   (top) $\Omega_1 = 1.50$; $\Omega_2 = 2.0$;
(center) $\Omega_1 = 2.25$; $\Omega_2 = 3.25$;   (bottom) $\Omega_1 = 3.0$; $\Omega_2 = 4.33$;
(left) full transient, and its initial phase (right)

and with $a_2 = \frac{k_{s_1}}{T_2}$ and $a_1 = \frac{k_{p_1} k_{s_1}}{T_2 T_1}$

$$(F+1)_C = \frac{s(s^2 + a_2 s + a_1)}{(s + \Omega_1)^3} \cdot \frac{1}{1} \cdot \frac{(s + \Omega_2)^3}{s(s^2 + a_2 s + a_1)} \cdot \frac{s}{s + \Omega_2}$$

$$= \frac{s}{s + \Omega_1} \cdot \left[\frac{s + \Omega_2}{s + \Omega_1} \right]^2 \tag{7.79}$$

As $\Omega_1 < \Omega_2$, the Nyquist contour starts from the origin and evolves into the right half plane, thus indicating good stability properties.

Note the relation of $(F+1)_C$ from Eq. 7.79 to $(F+1)_B$ from Eq. 7.69.

7.5.7 Summary

The practical motivation for having the override on the setpoint r_1 is to avoid "breaking the loop", and also to obtain a hierarchical (two-level) function structure.

It has been demonstrated here that having a nonlinear feedback (deadspan) in the override feedback will not produce an acceptable result, if practical gain restrictions due to the ever-present small delays are taken into account.

A feasible solution has been found by developing the controller structure from a special case of the basic one in Chapter 3. This has the benefit that the stability analysis can use the "pre-fab" procedure from Chapter 3.

It has also been demonstrated that this leads to a more complex R_2 controller: it has to be of the "PI(aw) and D" type, whereas in Chapter 3, a PI(aw) type was sufficient. Note that the additional D part is particularly disadvantageous in applications.

Also, the controller Fig. 7.34 contains two integral actions in series, because G_2 has a zero at the origin. This is due to the feedback loops in the "plant" for the override loop (with gains k_{p_1}, k_{p_2}). In Chapter 3 only one integrator was required.

Finally, the R_2 controller gains k_{i_2}, k_{p_2}, k_{d_2} being in series with the R_1-gains leads to severe restrictions on their admissible values, and thus to reduced disturbance rejection in the R_2 loop. This is similar to what was found in Chapter 2 for the "reference conditioning" approach. In numerical values for the benchmark, the upper limit was found $\Omega_2 \approx \Omega_1 + 1$, whereas in Chapter 3, $\Omega_2 = 3\Omega_1$ was feasible.

So, at this point, the "not breaking the loop" requirement leads to several distinct design drawbacks. Further research may show a way out.

7.6 A Compromise Between y_1- and y_2-Control

Another paradigm so far has been that if y_2 approaches its operational limit $r_{2_{hi}}$ or $r_{2_{lo}}$, then the control shall focus on driving y_2 along the respective limit regardless of how far the main variable y_1 may evolve from its setpoint r_1, as long as it ultimately can reach its setpoint. Otherwise, the control problem is considered to be not "well posed". In other words, there is a *"crisp"* decision between main control or constraint control, which keeps the design and the response relatively simple and transparent.

This may not always be the best strategy for a particular application. Some sort of compromise between main control and override control or "softening up the y_2 constraint" may fit the control specifications better.

This suggests using Fuzzy Logic Control techniques. Another approach would be a suitable extension of the previous techniques. Then, preferably, there should be no more than one design parameter governing the degree of compromise, in order to keep the design process well in hand.

The second approach shall be followed up here. The target application has already been mentioned in Sect. 7.5: the pilot would expect some (linear) compromise between y_2 and y_1 control in the "override" mode.
For other control problems,[10] another strategy would be to "override the overrides". The basic idea is to implement an additional upper and lower constraint on y_1 to keep it within a given range $r_{1_{lo}}$, $r_{1_{hi}}$ around r_1. For $y_1(t)$ within this range, the override from y_2 will be dominant, and will narrow down u. But near the constraints of y_1, this additional output constraint feedback will have to override the constraint action from y_2, and widen up the control u again. Thus, it will have to act on u downstream of the action from y_2. This strategy again leads to a "crisp" decision.

Structure

The basic idea is to *attenuate* the output w of the deadspan element by the new design parameter $1 - \beta$; see Fig. 7.36 (top).

$$u = u_1 - (1 - \beta)w \quad = u_1 - (1 - \beta)\mathrm{dsp}(w_2 + u_1) \quad \text{where} \quad 0 \le \beta \le 1 \quad (7.80)$$

and further

$$u = [1 - (1 - \beta)]u_1 - (1 - \beta)[u_1 - \mathrm{dsp}(w_2 + u_1)] \qquad (7.81)$$

see Fig. 7.36 (lower left), where the part $[u_1 - \mathrm{dsp}(w_2 + u_1)]$ is the expression for the "crisp decision" case.

---

[10] A typical example would be steam generator pressure control with fixed setpoint r_1, strong "load gradient" limitations, large load swings, and pressure limitations.

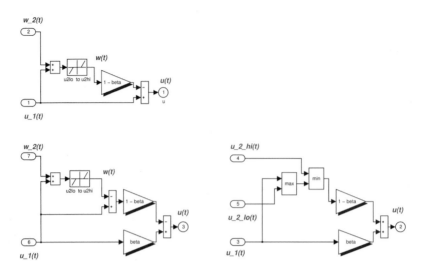

Fig. 7.36. Equivalent versions of the nonlinear element for implementing the compromise

- For $\beta = 0$, this is the crisp selection of either u_1 or w_2.
- For $\beta = 1$, $\Delta u = \Delta u_1$, and the R_1 loop is active everywhere.
- And for intermediate values of β, u is a linear combination of both cases.

For easier implementation on a current process control system, the nonlinear additive version is substituted by the Max-Min-selector combination; see Fig. 7.36 (lower right), with

$$u_{2_{hi}} = R_2(r_{2_{hi}} - y_2) \quad \text{and} \quad u_{2_{lo}} = R_2(r_{2_{lo}} - y_2)$$

$$u = \beta u_1 + (1 - \beta)\left[Min\left\{u_{2_{hi}}, Max\left(u_{2_{lo}}, u_1\right)\right\}\right] \tag{7.82}$$

In a next step, both the multiplication by β and the addition point of $(1-\beta)u_1$ are moved upstream across the Max-Min-selector block to its inputs

$$
\begin{aligned}
u &= \beta u_1 + Min\left[(1 - \beta)u_{2_{hi}}, \; Max\left\{(1 - \beta)u_{2_{lo}}, \; (1 - \beta)u_1\right\}\right] \\
&= Min\left[\beta u_1 + (1 - \beta)u_{2_{hi}}, \; Max\left\{\beta u_1 + (1 - \beta)u_{2_{lo}}, \; \beta u_1 + (1 - \beta)u_1\right\}\right] \\
&= Min\left[v_{2_{hi}}, \; Max\left\{v_{2_{lo}}, \; u_1\right\}\right]
\end{aligned}
$$

$$\text{where } v_{2_{hi}} = \beta u_1 + (1 - \beta)u_{2_{hi}} \quad \text{and} \quad v_{2_{lo}} = \beta u_1 + (1 - \beta)u_{2_{lo}} \tag{7.83}$$

In essence, an offset is added to the override controller outputs. This amounts to adding offsets to the inputs of the R_2 controllers. And the offsets are proportional to e_1. Note, that this is quite similar to what has been used in Sect. 6.14 as "selection for approach speed". So, one would predict a similar response, *i.e.* y_2 being proportional to the main control error e_1, and thus an exponential approach of y_1 to r_1.

Exercise

This is only a sketch. Substantiate it by considering the general case. Hint: start with a common integrator downstream of the selectors for both $e_1 \to 0$ and $e_2 \to 0$, and R_j controllers containing one proportional and all other feedback paths using derivatives.

The nonlinear element from Fig. 7.36 (top) is now inserted into the benchmark system from Chapter 3 (with override action on u); see Fig. 7.37.

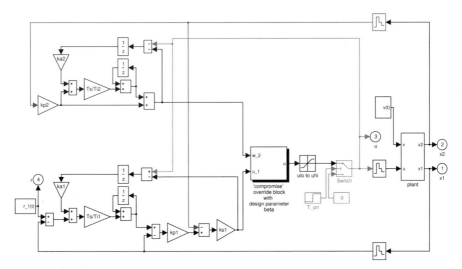

Fig. 7.37. Structure for position (y_1) control with speed (y_2) constraints, with a "compromise" override block from Figure 7.36 (top), with design parameter $0 \leq \beta \leq 1$

Transient response

The top left graph of Fig. 7.38 shows the response with the crisp decision $(\beta = 0)$ as a reference. Then β is increased in steps. The changes in response are characterized by two parameters:

- $y_{max}/r_{2_{hi}}$ shows how far the y_2 -constraint is overrun,
- and T_r shows the effect on performance of the main loop,

where the run-up time T_r is measured from the start signal to when R_1 takes over for the finals approach to r_1. From Fig. 7.38 read:

β	0	1/8	1/4	1/2	min. time
$y_{max}/r_{2_{hi}}$	1.0	2.0	2.9	4.0	5.0
T_r [s]	10	7.0	5.8	4.5	4.0

where the values for the minimum-time transient are appended in the last column for comparison.

Loosely speaking, the improvement in y_1 performance comes with a heavy cost increase on y_2.

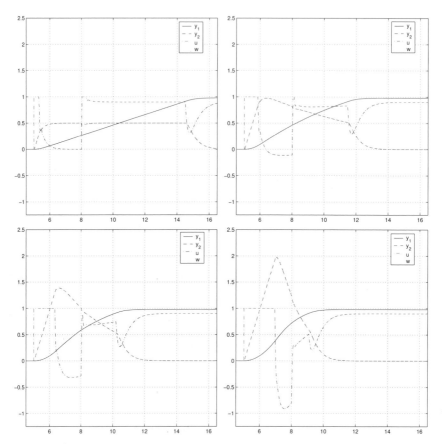

Fig. 7.38. Runup response for structure in Fig. 7.37, with $\Omega_1 = 3.0$; $\Omega_2 = 3\Omega_1$:
(top, left) $\beta = 0$;   (top, right) $\beta = 1/8$;
(bottom, left) $\beta = 1/4$;   (bottom, right) $\beta = 1/2$

Stability Analysis

Consider the nonlinear element of Fig. 7.36 (left) and insert it into the basic output constraint control system, with no constraints on u.

The nonlinear subsystem is the deadspan and unity slope element with the same break points and input signals as in Chapter 3.

The linear subsystem is the one from Chapter 3 again, but from Fig. 7.36 (left) with the attenuation factor $1 - \beta$ in series.

$$(F + 1)_\beta = (1 - \beta)(F + 1)\big|_{\beta=0} \tag{7.84}$$

Consider now the graphic stability test with a given overload of the deadspan element, $i.e.$ a given position Δ of the straight line on the negative real axis, and a given original Nyquist contour for $\beta = 0$. Then increase $\beta > 0$. The Nyquist contour will shrink, but its shape will not be deformed. And the straight line will not move. Therefore, the stability properties will continuously improve, with respect to the initial case of $\beta = 0$.

The next step is to include the saturations on u, as in Chapter 4.

Exercise
Compare this with the related case in Sect. 6.14.

7.7 Links to Model Predictive Control

7.7.1 Introduction

The MPC design method covers both input and output constraints, especially if output constraints are added to a control system with input constraints. Thus, it may serve as a reference for the intuitive design.
The Explicit MPC approach is used directly here. The main result from [59, 60] is again that the optimal controls which takes into account both the input and the output constraints turn out to be linear affine control laws where each one is valid for a finite part of the bounded state space area of interest.
The design is performed on the double- and triple-integrator plant benchmarks.

7.7.2 Case ii: The Double-integrator Plant

The benchmark of Sect. 6.16 shall be used, $i.e.$

$$Q = \begin{bmatrix} 1 & 0 \\ 0 & 0 \end{bmatrix}; \quad r = 0.10; \quad u_{lo} = -1.0; \quad u_{hi} = +1.0; \quad T_1 = T_2 = T = 1.0 \tag{7.85}$$

and with the output constraint

$$y_2 := x_2; \quad y_{2_{lo}} = -1.0; \quad y_{2_{hi}} = +1.0 \tag{7.86}$$

Design with Low Sampling Rate

The sampling time is set to $T_s = 1.0$.

The results of the optimization are reported in the phase plane representations in Fig. 7.39.

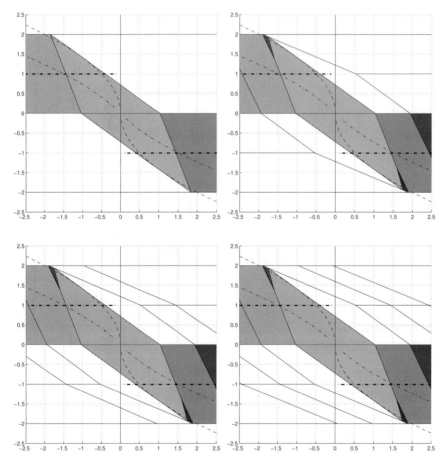

Fig. 7.39. Regions in the phase plane with linear affine control laws, for the double-integrator case, and for different control horizon entries, $T_s = 1.0$:
(top left) $N_u = 1$: 5 regions;   (top right) $N_u = 2$: 15 regions;
(low left) $N_u = 3$: 21 regions;   (low right) $N_u = 4$: 27 regions

Consider first the plot for $N_u = 1$. For region # 1 around the origin, the control law is

$$u_1^* = k_{1_1}x_1 + k_{2_1}x_2 + q_1 = -0.9653x_1 - 1.3895x_2 + 0 \qquad (7.87)$$

Then two horizontal regions are attached, with the affine control laws

$$u_{2,3}^* = k_{1_{2,3}}x_1 + k_{2_{2,3}}x_2 + q_{2,3} = -0.x_1 - 1.00x_2 \pm 1.00 \qquad (7.88)$$

The three regions are bounded by u_{lo}, u_{hi}.
Outside this linear corridor, to the lower left and the upper right respectively

$$u^* = 0x_1 + 0x_2 + 1.0 = u_{hi} \quad \text{and} \quad u^* = 0x_1 + 0x_1 - 1.0u_{lo} \qquad (7.89)$$

For $N_u = 2$, two small regions are added to the linear corridor in its upper left bend. In these regions the affine control laws are for the left-hand element and larger right-hand element respectively

$$u^* = 0x_1 - 1.0x_2 + 1.0 \quad \text{and} \quad u^* = -0.6154x_1 - 1.2870x_2 + 0.4156 \quad (7.90)$$

for the first elements of the optimal control sequence U^*, while the second elements are saturated. Two other regions are inserted symmetrically at the lower right bend.

For $N_u = 3$, additional regions are generated. Two are within the horizontal branches of the linear corridor. By inspection, the control laws are the same as in Eq. 7.88. All others are in the areas where either $u = u_{hi}$ or $u = u_{lo}$.

And for $N_u = 4$, again two additional regions appear in the linear corridor, where Eq. 7.88 holds, and all others are in the saturating region.

Design with a Higher Sampling Rate

The sampling time is set to $T_s = 0.25$, while all other parameters are the same as before. This generates Fig. 7.40.

For $N_u = 1$, the resulting control law for the central region # 1 is

$$u_1^* = -2.3155x_1 - 2.1505x_2 + 0.0 \qquad (7.91)$$

Attached to it are two regions with the resulting control laws

$$u_{2,3}^* = -0x_1 - 4.00x_2 \pm 4.00 \qquad (7.92)$$

The three regions are delimited to the lower left by $u = u_{hi}$ and to the upper right by $u = u_{lo}$.

For $N_u = 2$, two regions are added within the horizontal branches of the linear corridor, with the same control laws as in Eq. 7.92. Other larger regions are generated in the lower left area, where $u = u_{hi}$, and in the upper right area, where $u = u_{lo}$. Finally two small areas are generated in the upper left bend of the linear corridor, and symmetrically at the lower right bend. They are shown in more detail in Fig. 7.41 (left). For the small region to the upper left, the control law of Eq. 7.92 results, whereas for the larger region

$$u^* = -2.7132x_1 - 2.7804x_2 + 0.4445 \qquad (7.93)$$

as the first move of the optimal sequence, while the second one saturates at u_{lo}.

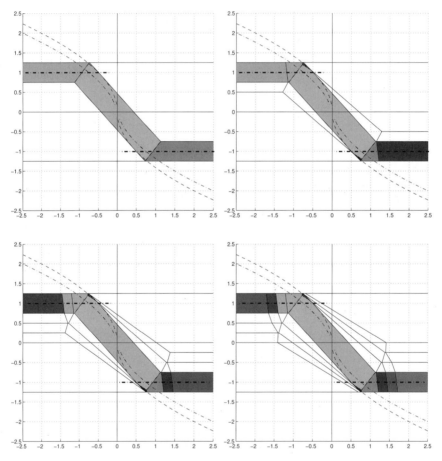

Fig. 7.40. Regions in the phase plane with linear affine control laws, for the double-integrator case, and for different control horizon entries, $T_s = 0.25$:
(top left) $N_u = 1$: 5 regions;   (top right) $N_u = 2$: 15 regions;
(low left) $N_u = 3$: 27 regions;   (low right) $N_u = 4$: 37 regions

For $N_u = 3$, again two regions are added in the horizontal branches of the linear corridor, and some others in the saturating areas with either $u = u_{hi}$ or $u = u_{lo}$. And another two small areas are added at the upper left bend; see details in Fig. 7.41 (right). For the smallest region in the horizontal branch the control law Eq. 7.92 is generated, while for the smallest region in the inclined part

$$u^* = -2.8164x_1 - 3.2503x_2 + 0.9399 \qquad (7.94)$$

for the first move in U^*, and

$$u^* = -0x_1 - 0x_2 - 1.0 \qquad (7.95)$$

for the following moves. For the larger region in the inclined part, the control law is the same as in Eq. 7.93.

And for $N_u = 4$, again two regions are added in the horizontal branches of the linear corridor, no further small regions appear at the bends, and all further regions are generated in the saturating areas.

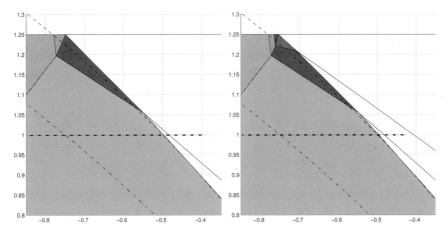

Fig. 7.41. Details of the regions in the phase plane of Fig. 7.40, $T_s = 0.25$: (left) $N_u = 2$, 15 regions;   (right) $N_u = 3$, 27 regions

Discussion

As described in Sect. 6.16, the control input $u^*(t)$ to the plant is evaluated at each time step by checking in which region the current $x(t)$ is, and then applying the corresponding affine state feedback control law. It is

either $u^* = -K_j x + q_j$ or $u^* = 0x + u_{lo}$ and $u^* = 0x + u_{hi}$ (7.96)

Now $u^*(t)$ may be generated equivalently for this case as follows:

- In the central region # 1 the linear state feedback without offset applies;

$$R_1 = -[k_{1_1}, \ k_{2_1}] \tag{7.97}$$

and with lateral saturations at u_{lo}, u_{hi}.
- In the horizontal branches, the control law $u^* = -k_{2_2}x_2 \pm q_2$ applies, which is independent of x_1, and laterally bounded by u_{lo}, u_{hi}. This may be rewritten as

$$u^* = k_{2_2} \left(\pm \frac{q_2}{k_{2_2}} - x_2 \right) k_{2_2} = (\pm r_2 - x_2) \tag{7.98}$$

that is, a P controller for R_2 with gain k_{2_2}, and setpoint $\pm r_2$, and subsequent saturations u_{lo}, u_{hi}.
- The small regions discussed in Fig. 7.41 are attributed either to region # 1 or the horizontal branch regions. This means that, for those attributed to region # 1, a different first control move will be generated, but the following moves will be saturated anyway. Thus, there will be a small loss in optimality for transients through these regions. Note that most transients will come in with $x_2 \rightarrow r_2$, and thus avoid these regions altogether.
- For all other regions, the optimal first control move is saturated.
- Thus, on the whole area of interest in the phase plane, the (slightly sub-) optimal control law can be generated by evaluating the outputs of the R_1 and the R_2 controllers first, subjecting them to a Max-Min-selection, and the result to a final saturation at u_{lo}, u_{hi}.

This is strikingly similar to what has been found for the intuitive design method.
The only difference is in the gain k_{2_2}. In the intuitive design, this has been a separate design parameter. Here, it is a consequence of the optimization process and cannot be selected at will. Note that in numerical values

$$r_2 = +1.0 = y_{2_{max}} \quad \text{or} \quad r_2 = -1.0 = y_{2_{min}} \quad \text{and} \quad k_{2_2} = 4.0 = \frac{1}{T_s} \tag{7.99}$$

In other words, this is a deadbeat controller along the state constraints. It would therefore degenerate to a very high gain controller for $T_s \rightarrow 0$, and this corresponds to what follows from the Maximum Principle, when it is applied to plants with state constraints.

Transient Responses

Finally, typical responses for initial conditions $x_1(0) = -1.75$, $x_2(0) = 0$ are shown in Fig. 7.42. There is a final overshoot on $x_1(t)$, which is due to the pole configuration resulting from the weights Q and r. From the simulations (not shown) no influence of the value of N_u is visible.
If the initial conditions are increased further (to $x_1(0) = -2.0$, $x_2(0) = 0$), an

"infeasible" trajectory is reported at $t = 3T_s$. This is due to the fact that u is not reduced sufficiently early from its saturation u_{hi}, such that an overshoot of $x_2(t)$ beyond its constraint value $x_{2_{hi}} = +1.0$ would result. And this is not admissible in the MPC context.

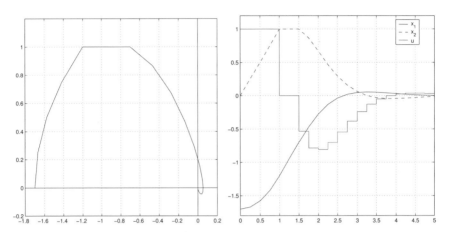

Fig. 7.42. Initial condition response of the system in Fig. 7.40, $T_s = 0.25$, $N_u = 1$

7.7.3 Case *iii*: The Triple-integrator Plant

The benchmark of Sect. 6.16 shall be used, *i.e.*

$$u_{lo} = -1.0; \quad u_{hi} = +1.0; \quad T_1 = T_2 = T_3 = T = 1.0$$

$$Q = \begin{bmatrix} 1 & 0 & 0 \\ 0 & 0.25 & 0 \\ 0 & 0 & 0.0625 \end{bmatrix}; \quad r = 0.10 \qquad (7.100)$$

with the output constraints

$$y_2 := x_2; \quad y_{2_{lo}} = -0.25; \quad y_{2_{hi}} = +0.25$$
$$\text{and} \quad y_3 := x_3; \quad y_{3_{lo}} = -0.25; \quad y_{3_{hi}} = +0.25 \qquad (7.101)$$

and with $T_s = 0.25$.

In the state space area of interest $-5 \le x_j(t) \le +5.0$, $j = 1, 2, 3$, the optimization yields the following numbers of regions n_R

N_u	1	2	3	4
n_R	7	33	93	181

The delimits in 3-D space of the seven regions for $N_u = 1$ are as follows, with $\underline{x} = [x_1, \; x_2, \; x_3]'$:

$$u = \begin{cases}
\begin{aligned}
&\text{region \# 1:}\\
&[-0.674,\ -1.504,\ -1.594]\,\underline{x} \quad \text{if} \quad \begin{bmatrix} -0.294 & -0.656 & -0.695 \\ +0.294 & +0.656 & +0.695 \\ -0.725 & +0.533 & +0.436 \\ -0.385 & -0.858 & -0.339 \\ +0.725 & -0.533 & -0.436 \\ +0.385 & +0.858 & +0.339 \end{bmatrix} \underline{x} \le \begin{bmatrix} +0.436 \\ +0.436 \\ +0.538 \\ +0.143 \\ +0.538 \\ +0.143 \end{bmatrix}
\end{aligned}\\[4ex]
\begin{aligned}
&\text{region \# 2:}\\
&+1.000 \quad\qquad\qquad \text{if} \quad \begin{bmatrix} -1.000 & 0 & 0 \\ 0 & +0.707 & +0.707 \\ 0 & 0 & +1.000 \\ 0 & -0.707 & -0.707 \\ 0 & 0 & -1.000 \\ +0.294 & +0.656 & +0.695 \end{bmatrix} \underline{x} \le \begin{bmatrix} +5.000 \\ -0.177 \\ -0.750 \\ +0.530 \\ +1.250 \\ -0.436 \end{bmatrix}
\end{aligned}\\[4ex]
\begin{aligned}
&\text{region \# 3:}\\
&-1.000 \quad\qquad\qquad \text{if} \quad \begin{bmatrix} +1.000 & 0 & 0 \\ 0 & +0.707 & +0.707 \\ 0 & 0 & +1.000 \\ 0 & -0.707 & -0.707 \\ 0 & 0 & -1.000 \\ -0.294 & -0.656 & -0.695 \end{bmatrix} \underline{x} \le \begin{bmatrix} +5.000 \\ +0.530 \\ +1.250 \\ -0.177 \\ -0.750 \\ -0.436 \end{bmatrix}
\end{aligned}\\[4ex]
\begin{aligned}
&\text{region \# 4:}\\
&[0,\ -2.000,\ -2.000]\,\underline{x} + 0.500 \ \text{if} \ \begin{bmatrix} -1.000 & 0 & 0 \\ 0 & -0.707 & -0.707 \\ 0 & +0.707 & +0.707 \\ 0 & -0.894 & -0.447 \\ 0 & +0.894 & +0.447 \\ +0.725 & -0.533 & -0.436 \end{bmatrix} \underline{x} \le \begin{bmatrix} +5.000 \\ +0.177 \\ +0.530 \\ -0.112 \\ +0.335 \\ -0.538 \end{bmatrix}
\end{aligned}\\[4ex]
\begin{aligned}
&\text{region \# 5:}\\
&[0,\ 0,\ -1.000]\,\underline{x} + 0.250 \quad \text{if} \quad \begin{bmatrix} -1.000 & 0 & 0 \\ 0 & 0 & -1.000 \\ 0 & 0 & +1.000 \\ 0 & +0.894 & +0.447 \\ 0 & -0.894 & -0.447 \\ +0.385 & +0.858 & +0.339 \end{bmatrix} \underline{x} \le \begin{bmatrix} +5.000 \\ +0.750 \\ +1.250 \\ +0.112 \\ +0.335 \\ -0.143 \end{bmatrix}
\end{aligned}\\[4ex]
\begin{aligned}
&\text{region \# 6:}\\
&[0,\ -2.000,\ -2.000]\,\underline{x} - 0.500 \ \text{if} \ \begin{bmatrix} +1.000 & 0 & 0 \\ 0 & -0.707 & -0.707 \\ 0 & +0.707 & +0.707 \\ 0 & -0.894 & -0.447 \\ 0 & +0.894 & +0.447 \\ -0.725 & +0.533 & +0.436 \end{bmatrix} \underline{x} \le \begin{bmatrix} +5.000 \\ +0.530 \\ +0.177 \\ +0.335 \\ -0.112 \\ -0.538 \end{bmatrix}
\end{aligned}\\[4ex]
\begin{aligned}
&\text{region \# 7:}\\
&[0,\ 0,\ -1.000]\,\underline{x} - 0.250 \quad \text{if} \quad \begin{bmatrix} +1.000 & 0 & 0 \\ 0 & 0 & -1.000 \\ 0 & 0 & +1.000 \\ 0 & +0.894 & +0.447 \\ 0 & -0.894 & -0.447 \\ -0.385 & -0.858 & -0.339 \end{bmatrix} \underline{x} \le \begin{bmatrix} +5.000 \\ +1.250 \\ +0.750 \\ +0.335 \\ +0.112 \\ -0.143 \end{bmatrix}
\end{aligned}
\end{cases}$$

Region # 1 is again the central one delimiting the linear state feedback with all three gains nonzero, and the offset being zero, as previously.

Regions # 2, # 3 are the saturating ones with all gains at zero, and the offsets at u_{hi}, u_{lo} respectively.

Then, in regions # 5, # 7, only the gain for x_3 is nonzero, at -1.0, and there is an offset $q_5 = +0.25; q_7 = -0.25$.

This may be rewritten as

$$u_5 = k_{3_5} \left(r_{3_5} - x_3 \right)$$

$$\text{with} \quad r_{3_5} = +0.25 = y_{3_{max}} \quad \text{and} \quad k_{3_5} = 1.0 = \frac{1}{4T_s} \quad (7.102)$$

and correspondingly for region # 7 with $r_{3_7} = y_{3_{min}} = -0.25$. The gain is at one fourth of the deadbeat value. This serves as limiting feedback for x_3 to r_{3_5} and r_{3_7}, similar to what has been found in case ii for the x_2 limitation there.

Finally, in regions # 4, # 6, the feedback law can be rewritten as

$$u_4 = k_{2_4} \left(r_{2_4} - x_2 \right) + k_{3_4} \left(r_{3_4} - x_3 \right)$$

$$\text{with} \quad r_{2_4} = +0.25 = y_{2_{max}}; \quad r_{3_4} = 0.0$$

$$\text{and} \quad k_{2_4} = 2.0 = \frac{1}{2T_s} = k_{3_4} \quad (7.103)$$

and correspondingly for region # 6 with $r_{2_6} = -0.25 = y_{2_{min}}$. This serves as limiting feedback for x_2 with the setpoints being r_{2_4} and r_{2_6}.

Exercise
Investigate the delimits listed above further.
Hint: note that many of the delimits do not depend on x_1. So drawing projections to the x_1, x_2 and x_2, x_3 planes may be helpful.

Note that the gains in regions # 4 to # 7 are produced by the optimization process and are not free design parameters anymore. In this case they turn out to be below the deadbeat values, but still make up a tightly tuned loop.

Finally, for increasing prediction horizons, $N_u \geq 2$, some regions are added laterally to the central region, where all three gains are nonzero, and the offsets as well are similar to what has appeared in case ii. Also, the corridors for limiting x_2 and x_3 are extended outward in the same fashion (the gains and offsets for the same type being the same), and there are more saturating regions.

Transient Response

A typical initial condition response is shown in Fig. 7.43 for $N_u = 1, 2, 3$.

The limiting of states $x_2(t)$ and $x_3(t)$ to their respective constraint values is clearly visible, although there seems to be a slow drift on $x_2(t)$ for $N_u = 2$. And for $N_u = 3$, a not well-damped mode is visible on $u(t)$ along this constraint. Further simulations show that this effect diminishes, however, with increasing control horizon N_u.

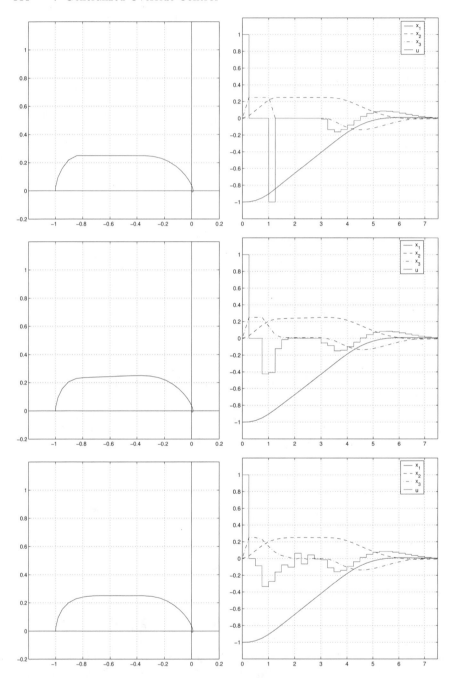

Fig. 7.43. Initial condition responses of the triple-integrator plant, Eq. 7.100 and
Eq. 7.101, $T_s = 0.25$:   (top) $N_u = 1$;   (center) $N_u = 2$;   (bottom) $N_u = 3$
(left) phase portraits;   (right) time responses

Again some problems with the feasibility of solutions have been experienced in the simulations, mainly in the initial phase of the trajectory. They increase while reducing N_u, increasing T_s, and increasing the weights in Q, or decreasing r.

The transient response is again very similar to what was obtained in Fig. 7.14 with the successive overrides structure from Fig. 7.13

7.7.4 Conclusions

The main findings from the benchmark problems investigated here are:

- The optimization process produces many regions, but a much smaller number of types of control law.
- The linear affine control laws have the same form as the output constraint feedback controllers R_2.
- The control laws along the state/output constraints are enforced by the optimization process, and are not free design parameters as in the override design.
- Those control laws are equivalent to tightly tuned controls, close to deadbeat. This makes them not very robust.
- The current u is selected in the Explicit MPC design by checking, in which region the current $x(t)$ is, and applying the corresponding control law, while in the override design the current x is applied to all control laws, producing a u_j for each of them, and finally determining the u to be applied to the plant by Max-Min-selection.
- However, note that the checking of regions in the MPC design is checking on which side of the hyperplanes delimiting the regions the current $x(t)$ is, and this may be seen as performing a Max- or Min-selection. This leads to the conjecture that both processes of evaluating u from $x(t)$ may be equivalent. Note that this is a plausibility argument only. A formal proof is not available yet, but would be very helpful for implementation purposes.
- Finally, the Explicit MPC design (at least in the form used here) may run into infeasible trajectories, and stop. This has not been found in the override design. This is because MPC considers state constraints in a much stricter way, and not as operational limits with some (small) tolerance band.

Such conclusions based on two benchmarks (although key ones, if seen from the previous investigations) cannot be more that tentative. Putting them on a more solid basis would be highly useful for industrial control design.

7.8 Summary

The override design technique is based on standard linear feedback controllers with awf combined with Max- and Min-selectors. It has been derived on a specific (but quite frequent) application, *e.g.* position control of a rigid mass with output constraints on speed and input constraints on actuator force. The technique has already been extended into generalized antiwindup.

Here, we have shown that this approach can be used in a much more general way. Several additional areas have been explored.

The first one is with plants of dominant first order and PI(aw) control. However the overrides are not generated from a secondary output but from the main control variable itself. Furthermore, the override is acting not on the inflow as before, but also on the outflow. The case of dominant plant order in the override loop being greater than in the main loop is also investigated.

Then the area is investigated where the dominant plant order in in the output constraint path is such that additional state feedback is needed in R_2 as well. It was found that the override control still performs well along the constraint. And for the overall system, the key parameter for stability and the radius of attraction was found to be the difference of dominant degrees in both paths. The situation is favorable, if the difference is 1. It deteriorates somewhat if the difference is 2, and markedly if ≥ 3. Then, inserting additional (non-physical) constraints may greatly improve the performance of the overall system.

Next, the case has been considered where the transfer function in the constraint path has a zero in or near to the origin. Then additional cascaded I actions are required. And the awf is extended accordingly.

Then, the basic paradigm of overrides acting on the control variable has been replaced by overrides acting on the main reference r_1, thus conserving its "loop integrity". Viable structures have been found, although they are more complex for comparable performance.

And the paradigm of a switch-over from control in the main loop to control along the constraint and back has been replaced by a "designed compromise".

Throughout, nonlinear stability analysis has turned out to be an efficient and effective tool for analysis and design. Performance has been judged in a more qualitative way by benchmark simulations. The Robust Control framework offers a quantitative step further.

Finally, the Explicit form of MPC has been applied. By inspection, some of the state space regions extend along the output constraints, and the optimal feedback laws in these regions are state feedbacks with offsets, which are directly related to the constraint values. In other words, the control structure produced by the Explicit MPC design is very similar to what the intuitive (override) design leads to. In general, the override design will have less regions. Thus, it will produce sub-optimal responses, though in a controllable way, because the optimal solution from MPC is so closely related.

8

Multivariable Control with Constraints

8.1 Introduction

From a very distant perspective, the control of multivariable systems may be designed either by a "top-down" or by a "bottom-up" approach.

The top-down approach is to consider an optimization of the trajectory from the current state of the plant to a new target state, while taking into account all the interactions and all the constraints on the control inputs and the state variables or measured outputs. This is where MPC comes up to its full potential. However there is the problem of infeasible trajectories, of model and disturbance uncertainties, *etc.*

The bottom-up approach starts from modeling the multivariable plant as non-coupled single-input-single-output (SISO) subplants, with suitable pairing of outputs to inputs, and designing the feedback controls with the methods for scalar u plus input and output constraints, such as developed in the previous chapters. For practical design purposes, this is admissible if the coupling transfer functions in the real plant are weak enough relative to the main transfer functions from the pairings mentioned above. If not, the control actions will interfere, and performance can be expected to degrade.

Nevertheless, we take this second approach here, as it is commonly used in design practice, and merits to be understood better.

In other words, the focus so far has been on control systems with one manipulated variable (*i.e.* $u(t)$ being a scalar), and one main controlled variable $y_1(t)$ (note that there may be additional measured variables in a cascade arrangement), and one constrained output $y_2(t)$. This covers input saturations as a special case.

We shall now step into more complex design situations, denoted for short as *areas*. First, a discussion in general terms shall provide an overview. Specific cases will then be investigated in the following sections.

8.2 An Overview

8.2.1 Multiple Overrides

One such area covers design situations with one manipulated plant input u, one main controlled variable y_1 for the linear operating range, but multiple constrained outputs $y_{2_j}, j = 1, 2, \ldots$. Recall that this has been discussed previously. In essence, only one of the constraints u_{2_j} can be "active" (or being "selected" or "switched through"), while the others are not. This follows from the override paradigm. Which one of the constraints is selected in the case of conflicting u_{2_j} values is defined by the sequence of the Max-Min-selectors along u_1: any particular selection masks all upstream selections, and is in turn masked by all downstream selections. This simple rule allows one to construct quite complex control strategies.

Note, however, that masked constraints may be violated. Avoiding this obviously needs extra actuators (see below). Also note that for physical reasons the most downstream selections must be for the actuator limitations, $i.e.$ the input saturations.

8.2.2 Multiple Actuators of the Same Type

The second area is again one controlled variable y_1, one linear controller R_1, but multiple actuators $u_{1_j}, j = 1, 2, \ldots$. They act jointly to produce a scalar input effect to the plant. Each of the actuators has its distinct working range limits. Together they build up the actuator assembly. The individual actuators shall move either in parallel ("joint control"), where each of them picks up its proportional share of the total input increment. Or they move in sequence ("split-range control"), $i.e.$ the next actuator will only move if all previous ones have attained their working range limits, and then it will pick up the whole input increment. In both cases it must be possible for individual actuators to be switched in or out without upsetting the main control loop unduly.

Typically, this is implemented by a cascade structure, where a control loop for total opening is inserted. Note that for such switching transients both the initial and the final equilibrium must be within the linear operating range. Otherwise the control problem will be "ill posed".

Some typical application examples are:

- Fossil-fuel fired steam generators with steam pressure control and with multiple burners as actuator subsystems, operated in split range mode at low load and in joint mode at high load.
- Hydropower stations with control for total electric power output, where individual turbine-generator units are the "actuator subsystems" within the actuator assembly, and usually are operated in joint control mode.
- Level control on run-of-river hydropower plants with multiple weirs or gates as actuator subsystems, which are usually operated in split-range mode.

8.2.3 Sequentially Acting Control Loops

A third area is closely related to the previous one. There is again one controlled variable y_1, and several actuators $u_{1_j}, j = 1, 2, \ldots$ are available. However, they act at different locations on the process (for instance on inflow and on outflow), and the plant response $u_{1_j} \to y_1, j = 1, 2, \ldots$ for each actuator is significantly different. This may be due to a different response of the actuator subsystem itself or of the plant. Also, the different actuators usually produce different side effects (i.e. have different constraints).

Then, typically, a separate controller R_{1_j} for each of the actuators is implemented. This provides the necessary parameters for individual tuning to the distinct plant responses, and for considering the individual constraints. All controllers use the same measured variable y_1 as input, but they have distinct setpoints r_{1_j}.

$$r_{1_{j+1}} := r_{1_j} + \Delta r_{1_{j,j+1}} \quad \text{with} \quad 0 < \Delta r_{1_{j,j+1}} \ll r_{1_j} \tag{8.1}$$

The setpoint increments must be sized such that a clear separation between control loops appears, *i.e.* one will be active while the others are in saturation mode, at least for steady state operation. Intuitively, the setpoint increment can be made small, such that y_1 will not run far off its nominal setpoint r_1, if integral action is used in all R_{1_j} controllers. Then obviously awf with appropriate gain will be needed to ensure a timely liftoff of actuator positions from their constraints.

A typical example is a steam-generator – steam-turbine – alternator unit. Its normal operating mode is power control. Then, in many cases, live steam pressure is used as the measured variable for adjusting the heat input to the steam generator and balancing it to the steam flow to the turbine and further on to electric power output of the alternator. And this is subject to multiple constraints on the load gradient both from the steam generator and steam turbine.

If now the generator is tripped from full load, the steam flow to the turbine must be reduced in less than 1 s to near zero to avoid rotor overspeed. But reducing the steam flow from the steam generator will take several minutes, and will stop at approximately 30 % (for steam generator operational reasons). The transient imbalance of steam flows results in rising live steam pressure, beyond its normal operating range. Then a second pressure control loop must come into action. It acts on a bypass valve subsystem to dump the excess steam flow directly to the condenser. Note that this is a waste of energy and thus should be kept to the minimum at all times. Intuitively a high performance pressure control loop with integral action will accomplish this.

The plants considered so far are modeled by one dominant compartment, where PI(aw) control is suitable. We shall now proceed to plants with more than one dominant compartment.

8.2.4 Parallel-acting Control Loops

The next area covers plants with two compartments C_1, C_2[1] coupled by a connecting line, which may be closed down by a switch or a valve; see Fig. 8.6 for an illustration.

Each of the compartments shall have its own outflow v_1, v_2, its own independently actuated flow u_1, u_2 and its own measured output y_1, y_2, which corresponds to its content (*i.e.* its own state variable x_1, x_2). Thus, for the dominant dynamics model with $q_{1 \to 2}$ denoting the crossflow in the connecting line:

$$\tau_1 \frac{d}{dt} x_1 = u_1 - v_1 - q_{1 \to 2} \quad \text{and} \quad \tau_2 \frac{d}{dt} x_2 = u_2 - v_2 + q_{1 \to 2} \qquad (8.2)$$

On this basis, two separate control loops can be installed, with controllers

$$R_j = \frac{u_j}{r_j - c_j x_j} \quad \text{where} \quad c_j = 1, \quad j = 1, 2 \qquad (8.3)$$

and with individual setpoints r_1, r_2. They will act as local control loops for each compartment, when the connecting line is shut down ($q_{1 \to 2} \equiv 0$).

Now the connecting line is opened. Then a crossflow will develop which is essentially driven by the difference of potentials $x_1 - x_2$. The most simple modeling description would be

$$q_{1 \to 2} := c(x_1 - x_2) \qquad (8.4)$$

Note that this sets to zero any inertia or storage effects in the connecting subsystem. And, by the way, hints at how to extend the description for more realistic models.

The crossflow $q_{1 \to 2}$ in Eq. 8.4 has a coupling effect on the local compartments; see Eq. 8.2. As long as the conductivity c of the connecting line is low, then the coupling effect is weak and the loops can be considered as separate systems. More interesting is the case of high conductivity, leading to strong crossflow effects. If now the setpoint r_1 is moved upward a small amount, then a strong flow from C_1 to C_2 results. To cover this in C_1 to keep $y_1 = c_1 x_1 \to r_1$, a high inflow u_1 is needed. And correspondingly the inflow to C_2 must be reduced by a large amount to keep $y_2 \to r_2$. In short, a small shift in setpoints r_1 vs. r_2 results in large countermovements on the manipulated variables u_1, u_2, and may easily drive them into opposite saturations. This property is known as "directionality". It is a major cause for windup.

Set further for the outflows v_1, v_2 in Eq. 8.2:

$$v_1 = -a_1 x_1; \quad v_2 = -a_2 x_2 \qquad (8.5)$$

---

[1] Note the change of meaning for the indices 1, 2.

Then

$$\begin{bmatrix} \tau_1 & 0 \\ 0 & \tau_2 \end{bmatrix} \frac{d}{dt} \begin{bmatrix} x_1 \\ x_2 \end{bmatrix} = \begin{bmatrix} -(a_1 + c) & +c \\ +c & -(a_2 + c) \end{bmatrix} \begin{bmatrix} x_1 \\ x_2 \end{bmatrix} + \begin{bmatrix} u_1 \\ u_2 \end{bmatrix} \quad (8.6)$$

and further for the common denominator of the four transfer functions $u_j \to y_k$ with $\tau_1 = \tau_2 = \tau$

$$\begin{aligned} 0 = |sT - A| &= [\tau s + (a_1 + c)] [\tau s + (a_2 + c)] - c^2 \\ &= (s\tau)^2 + (s\tau) [(a_1 + c) + (a_2 + c)] + [(a_1 + c)(a_2 + c) - c^2] \\ &= \left(s\frac{\tau}{c}\right)^2 + \left(s\frac{\tau}{c}\right) \left[\left(1 + \frac{a_1}{c}\right) + \left(1 + \frac{a_2}{c}\right)\right] + \left[\left(1 + \frac{a_1}{c}\right)\left(1 + \frac{a_2}{c}\right) - 1\right] \end{aligned} \quad (8.7)$$

Note the time scaling $\tau \to (\tau/c)$. If now the crossflow is comparatively strong, then

$$0 \le \frac{a_1}{c}, \frac{a_2}{c} \ll 1.0 \quad (8.8)$$

and the value of $det(A)$ of the open-loop system matrix A in Eq. 8.6 is small, or equivalently, the open-loop system is said to be "ill conditioned":

$$\left[\left(\frac{a_1}{c} + 1\right)\left(\frac{a_2}{c} + 1\right) - 1\right] = \left(\frac{a_1}{c} + \frac{a_2}{c}\right) + \left(\frac{a_1}{c}\frac{a_2}{c}\right) = \epsilon \quad \text{where} \quad 0 \le \epsilon \ll 1 \quad (8.9)$$

Thus, a small $\epsilon = |A|/c^2$ indicates strong directionality; see also Sect. 8.5.

For the special case $a_1 = a_2 = 0$, in Eq. 8.6 $det(A) = 0$ and

$$0 = \left(s\frac{\tau}{c}\right)^2 + 2\left(s\frac{\tau}{c}\right) = \left(s\frac{2\tau}{c}\right)\left(s\frac{\tau}{2c} + 1\right) \quad (8.10)$$

where $2\tau/c$ is the total filling time constant of the plant. One eigenvalue moves to the origin, yielding a slow open integrator response, while the other eigenvalue moves to the left, yielding a fast first-order lag response.
And if $a_1 = a_2 = a$ with $a > 0$, then

$$0 = \left(s\,2\frac{\tau}{c} + 2\frac{a}{c}\right)\left[s\,\frac{\tau}{2c} + \left(1 + \frac{a}{2c}\right)\right] \quad (8.11)$$

One eigenvalue is near the origin, resulting in a slow first-order mode, and the other is moved to the left again, producing a fast first-order response.
The slow mode is from the equilibration of the plant as a whole, and the fast mode from equilibration between both compartments.

8.2.5 The General Multi-input Multi-output Case

Many textbooks are available for the design and analysis of general multivariable control systems. They stress the linear range design [61], and investigate saturation effects as well [62].
Here, not more than a very brief introduction shall be given. The focus is on 2×2 systems; see Fig. 8.1.

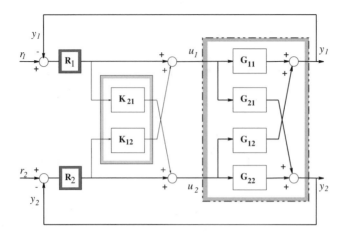

Fig. 8.1. The standard 2×2 multivariable system with its four plant transfer functions G_{11}, G_{21}, G_{12}, G_{22}, two controllers R_1, R_2, and two optional decouplers K_{12}, K_{21}

The usual representation of the plant as a transfer function matrix is used:

$$\begin{bmatrix} y_1(s) \\ y_2(s) \end{bmatrix} = \begin{bmatrix} G_{11}(s) & G_{12}(s) \\ G_{21}(s) & G_{22}(s) \end{bmatrix} \begin{bmatrix} u_1(s) \\ u_2(s) \end{bmatrix} \qquad (8.12)$$

with the feedbacks

$$\begin{bmatrix} u_1(s) \\ u_2(s) \end{bmatrix} = \begin{bmatrix} R_1(s) & 0 \\ 0 & R_2(s) \end{bmatrix} \begin{bmatrix} r_1(s) - y_1(s) \\ r_2(s) - y_2(s) \end{bmatrix} \qquad (8.13)$$

Note that in Eq. 8.13 the non-diagonal elements of the feedback matrix R are set to zero. Additional transfer functions may be inserted for decoupling; see Fig. 8.1 Closing the loops yields

$$\begin{bmatrix} 1 + R_1 G_{11} & R_2 G_{12} \\ R_1 G_{21} & 1 + R_2 G_{22} \end{bmatrix} \begin{bmatrix} y_1 \\ y_2 \end{bmatrix} = \begin{bmatrix} R_1 G_{11} & R_2 G_{12} \\ R_1 G_{21} & R_2 G_{22} \end{bmatrix} \begin{bmatrix} r_1 \\ r_2 \end{bmatrix} \qquad (8.14)$$

and further (using Cramer's rule) one of the four transfer functions $r_1 \rightarrow y_1$

$$\frac{y_1}{r_1} = \frac{1}{(1 + R_1 G_{11})(1 + R_2 G_{22}) - R_2 G_{12} R_1 G_{21}} (R_1 G_{11} - R_2 G_{12} R_1 G_{21})$$

$$(8.15)$$

The denominator in Eq. 8.15 is the same for all four transfer functions. It is rewritten as follows:

$$(1 + R_1 G_{11})(1 + R_2 G_{22}) - R_2 G_{12} R_1 G_{21}$$

$$= 1 + (R_1 G_{11} + R_2 G_{22}) + R_1 G_{11} R_2 G_{22} \left(1 - \frac{G_{12} G_{21}}{G_{11} G_{22}} \right)$$

$$= 1 + (R_1 G_{11} + R_2 G_{22}) + R_1 G_{11} R_2 G_{22} (1 - S)$$

$$\text{which defines} \quad S = \frac{G_{12} G_{21}}{G_{11} G_{22}} \tag{8.16}$$

where the transfer function S indicates the relative strength of the cross-coupling responses of the plant to the direct ones.

Four cases may be distinguished:

1. If either $G_{12} = 0$ or $G_{21} = 0$ (one-sided coupling) or both $G_{12} = G_{21} = 0$ (no coupling), then

$$S = 0 \tag{8.17}$$

and Eq. 8.15 degenerates to the standard result

$$y_1 = \frac{R_1 G_{11}}{1 + R_1 G_{11}} r_1 \tag{8.18}$$

2. If G_{12} and G_{21} are nonzero and have opposite signs, then

$$S < 0 \tag{8.19}$$

then the zeros of Eq. 8.15 tend to move to the right and may easily cross over into the right half plane. In other words, the system with both local loops closed will be oscillatory unstable, although it is stable if only one of the loops is closed. A typical such case is anti-surge and pressure control of compressors.

3. If G_{12} and G_{21} are nonzero and have the same signs, then if

$$0 < S < +1 \tag{8.20}$$

the zeros move as shown in Eq. 8.10 and in Eq. 8.11, and the directionality effect is strong.

4. And if

$$S > +1 \tag{8.21}$$

the cross-coupling responses are stronger than the direct ones, and the "input-output pairing" should be changed (the controller outputs should be connected the other way round, *i.e.* R_1 to u_2 and *vice versa*).

Saturations on the u_i

Consider first the case of $S < 0$. Generally speaking, if the saturation is encountered then this loop is opened. Thus only one loop stays closed permanently, and this intuitively should lead to better stability properties.
And for the case $S > 0$, the directionality leads to large opposite deviations in the controls, and thus may move the u_i into the saturations. Therefore, the further investigation shall focus on this case; see Sect. 8.6.

Decoupling

It is often proposed to stabilize (in case 2) or increase performance (in case 3) by "decoupling". From Fig. 8.1

$$K_{21}G_{22} \overset{!}{=} -G_{21} \quad \text{and} \quad K_{12}G_{11} \overset{!}{=} -G_{12} \tag{8.22}$$

such that the effect of the $G_{jk}, j \neq k$ is cancelled by the corresponding paths through the corresponding K_{jk}. Then the controller R_1 sees as its plant $\tilde{G}_1$:

$$\tilde{G}_1 = G_{11} + K_{21}G_{12} = G_{11} - \frac{G_{21}}{G_{22}}G_{12} = G_{11}(1 - S) \quad \text{and} \quad \tilde{G}_2 = G_{22}(1 - S)$$
$$\tag{8.23}$$

Thus the new plant response $\tilde{G}_j$ depends on the same key factor $(1 - S)$ as the denominator polynomial. It increases for opposite signs of the $G_{jk}, j \neq k$, and decreases if the signs are the same. It tends to zero for $S \to +1$, $i.e.$ for cases with strong directionality. So, to obtain a given closed-loop response, the controller gains must tend to infinity, $i.e.$ the controllers will not be implementable, and also the closed loops tend to be very sensitive to modeling uncertainty.
In other words, decoupling may be considered primarily for cases with $S < 0$, and not for cases with very strong directionality.

The design situations, discussed so far in general terms, shall be investigated now in more detail using benchmark cases, simulations, $etc.$

8.3 Parallel Actuators

8.3.1 Structure

The parallel actuators may be moved either in parallel, Fig. 8.2, or sequentially, Fig. 8.3. Only the nested loop for total opening (or an equivalent variable such as total flow, $etc.$) is given. Its setpoint r_2 is the output of the master controller, for instance the level controller, its outputs u_i are the n openings of the individual actuators within the actuator assembly. The u_i are then the inputs to the plant.

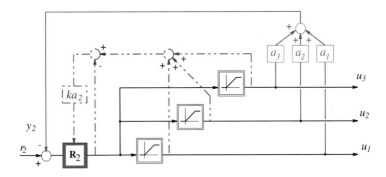

Fig. 8.2. The "nested loop" for three actuators moving in parallel

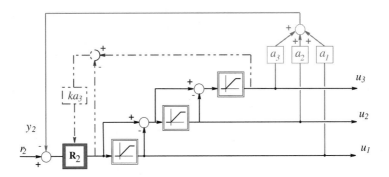

Fig. 8.3. The "nested loop" for three actuators moving in sequence

The saturation elements in Fig. 8.2 and 8.3 represent actuator subsystems having internally both slew and stroke limits, local position feedback, and a switch on their input to put them individually in automatic mode, manual mode, or shutdown. The controller R_2 is of I(aw) type, possibly with a weak P part to improve performance. The gains a_i represent the full-scale contribution percentage to total opening: if n actuators of equal size are installed, then $a_i = 1/n$.

8.3.2 Transient Response and Stability Analysis

This is left to the reader as an **exercise**:

-   Implement the "nested loop" in both forms with all details given above as a Simulink model.

- Investigate its response to setpoint steps, and to switch-out and switch-in of individual actuators.
- Add a level control loop, with the plant modeled by an integrator, and a PI level controller with suitably designed awf.
- Investigate its closed-loop response for a suitably designed test input sequence.
- Finally, investigate the stability properties.

8.4 Sequenced Control Loops

The example of a steam generator pressure control from Sect. 8.2.3 is carried further.

8.4.1 Structure

Fig. 8.4 shows the Simulink model. The main element in the plant is the mass balance of steam flows. The steam inflow is produced by vaporization. One outflow is to the turbine. It is switched from 100 % down to 0 % to simulate the effect of the speed controller following the load rejection. The second outflow is through the bypass valve(s). Heat input to vaporization is controlled by the steam pressure controller, of PI(aw) type. There is a lag to model the thermal storage capacities, a gradient limiter, and a typical stroke limit.

The *plant coefficients* are:

$$T_1 = 10 \text{ [s]}; \quad T_2 = 10 \text{ [s]}; \quad v_0 = 1.0; \quad v_1 = 0.0; \quad v_{up,dn} = \pm 1.0 \text{ [s}^{-1}] \quad (8.24)$$

- The load gradient limiter is set to ± 10 % of immediate step and ± 15 %/min of load ramp.
- The stroke limit on the heat input is set at $30 - 120$ %.
- The bypass shall have a working range of $0 - 120$ %.

The *pressure controller* (index $_1$) acting on the heat input is tuned to

$$R_1 : \quad k_{p_1} = 3.0; \quad k_{i_1} = 0.050; \quad ka_1 = 0.10 \quad (8.25)$$

with initial setpoint at 0.975, and then stepped up to 1.0 to show the linear closed-loop response.

The controller acting on the bypass valve opening (index $_2$) is tuned to

$$R_{2_{hi}} : \quad k_{p_{2_{hi}}} = 20; \quad k_{i_{2_{hi}}} = 25; \quad ka_2 = 1.0; \quad \text{with setpoint } r_{1_{hi}} = 1.05; \quad (8.26)$$

The sampling rate is set to $T_s = 0.10$.
Note that the bypass loop has to have the same high performance as the speed control loop, in order to keep the pressure overshoot low.

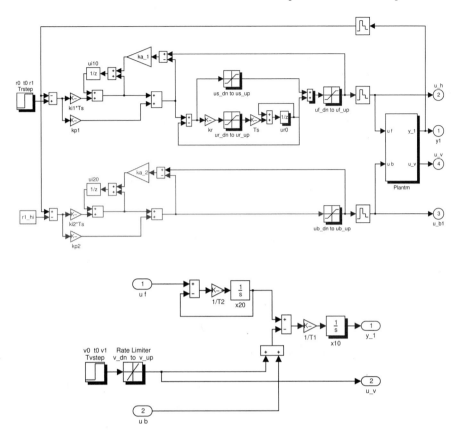

Fig. 8.4. Steam pressure control: the control structure
acting on the heat-input $(r_1 \rightarrow u_f)$;   and on the bypass valve opening $(r_{1_{hi}} \rightarrow u_b)$
and the plant model

8.4.2 Transient Response

Fig. 8.5 shows the simulation results.
The discussion is left to the reader as an **exercise**.

8.4.3 Stability Analysis

The basic effect of the bypass is that the total actuator working range is
enlarged on both slew and stroke, although only one-sided. Therefore, the
stability properties should improve.
Again, details are left to the reader as an **exercise**.

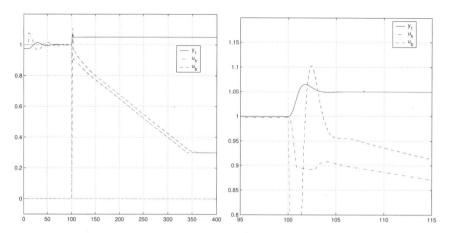

Fig. 8.5. Transient pressure response to the setpoint step at $t = 10$ s, and to the full load rejection at $t = 100$ s; with details in right-hand plot

8.5 Parallel Control Loops

The example of Sect. 8.2.4 with two compartments, local control and strong crossflow interconnection is developed further here.

8.5.1 The Plant Model

Fig. 8.6 shows the 2×2 plant model in its state space representation. This corresponds directly to Eq. (8.6) up to Eq. (8.9). Note that all fast dynamics from the sensors and actuators are suppressed, and that the crossflow has no inertia. Note also that the manipulated variable for compartment C_2 is the outflow.

The parameter entries for the simulation are

$$u_{1_{hi}} = +1.0; \quad u_{1_{lo}} = -1.0; \quad u_{2_{hi}} = +1.0; \quad u_{2_{lo}} = -1.0 \qquad (8.27)$$

and

$$\tau_1 = 1.0; \quad \tau_2 = 1.0; \quad a_1 = 0.5; \quad a_2 = 0.5; \quad c = 5.0 \qquad (8.28)$$

That is, the cross-coupling coefficient is a factor of ten larger than the local feedback coefficients.

Thus, from Eq. 8.11

$$0 = \left(s\frac{1}{5} + 0.10 \right) \left(s\frac{1}{5} + 2.10 \right) \quad \rightarrow \quad 0 = (s + 0.50)(s + 10.50) \qquad (8.29)$$

i.e. the eigenvalues or open-loop poles are spread by a factor of 21.

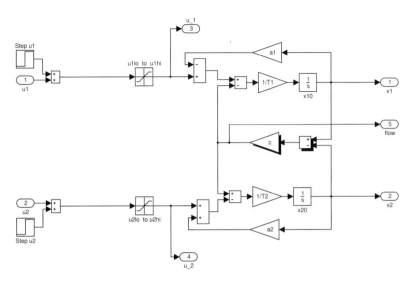

Fig. 8.6. The open-loop model of the 2×2 plant in its state space representation

This is visualized in the simulation in Fig. 8.7, where first a step is applied to u_1 and then to u_2. Note the time scaling.

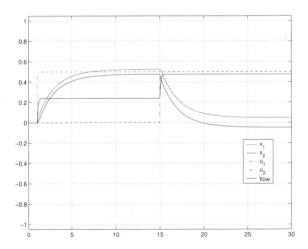

Fig. 8.7. Open-loop input step responses of the system in Fig. 8.6:
the fast mode on "flow" is due to the equilibration between both compartments,
and the slow mode is due to the equilibration of both compartments as a whole

8.5.2 State Feedback Control

The control structure in Fig. 8.8 consists of local state feedback with feedback
gains k_1, k_2 and with local feedforward gains kr_1 on the setpoint r_1, and kr_2
on r_2. This is equivalent to local P controllers with gains k_1, k_2, where the
setpoints are modified

$$r'_j = r_j + (kr_j/k_j), \quad j = 1, 2 \tag{8.30}$$

Note also the sign inversion in the local feedback on compartment C_2 to
conserve negative feedback.

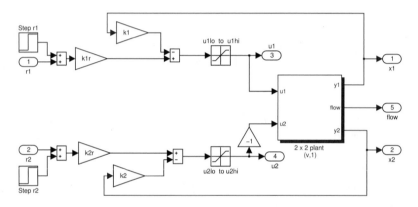

Fig. 8.8. The system from Fig. 8.6 with local P controllers

The state space representation is

$$\begin{bmatrix} \tau_1 & 0 \\ 0 & \tau_2 \end{bmatrix} \frac{d}{dt} \begin{bmatrix} x_1 \\ x_2 \end{bmatrix} = \begin{bmatrix} -[(a_1 + k_1) + c] & +c \\ +c & -[(a_2 + k_2) + c] \end{bmatrix} \begin{bmatrix} x_1 \\ x_2 \end{bmatrix} \\ + \begin{bmatrix} kr_1 & 0 \\ 0 & kr_2 \end{bmatrix} \begin{bmatrix} r_1 \\ r_2 \end{bmatrix} \tag{8.31}$$

The controller gains are set to

$$k_j = \Omega_j \tau_j - a_j, \quad j = 1, 2 \tag{8.32}$$

which applies the standard single-loop pole assignment approach to the isolated two compartments C_1, C_2.
As for this benchmark case

$$a_1 = a_2 = a \ \text{ and } \ \tau_1 = \tau_2 = \tau, \ \text{ set } \ \Omega_1 = \Omega_2 = \Omega \ \rightarrow \ k_1 = k_2 = k \tag{8.33}$$

The closed-loop denominator polynomial is, by replacing in Eq. 8.11 the local open-loop feedback coefficients a by the closed-loop ones $k + a$:

$$(s\tau + k + a)(s\tau + 2c + k + a) \tag{8.34}$$

In other words, the closed-loop poles of the interconnected system will be shifted significantly from the open-loop ones if

$$\frac{k + a}{c} > 1 \tag{8.35}$$

And the numerator polynomial of one of the four transfer functions, $r_1 \rightarrow y_1 = x_1$, is

$$kr_1 (s\tau + c + k + a) \tag{8.36}$$

This produces a steady state value of the step response

$$\frac{\overline{y_1}}{r_1} = \frac{kr_1}{2c + k + a} \tag{8.37}$$

Thus there will be a considerable steady state offset for finite gain values $k > 0$. To keep it small, again Eq. 8.35 applies.

For the simulations in Fig. 8.9:

$$\Omega\tau := 10; \quad \text{that is} \quad \frac{k + a}{c} = 2, \quad \text{and also} \quad kr := k + a = \Omega\tau \tag{8.38}$$

A setpoint step $r_1 = 0.15$ is applied first, while $r_2 = 0$, and then a setpoint step $r_2 = 0.15$ without changing r_1. Thus the crossflow variable must return to zero. Note the change in time scaling vs. Fig. 8.7.

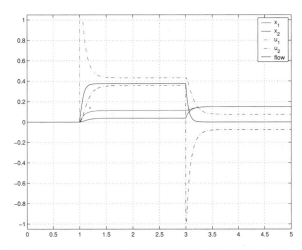

Fig. 8.9. Closed-loop response of the system in Fig. 8.8, with local P controllers

8.5.3 Additional Integral Action in the Local Controllers

This is to suppress the steady state errors. As the u_j may saturate, a standard awf is applied; see Fig. 8.10.

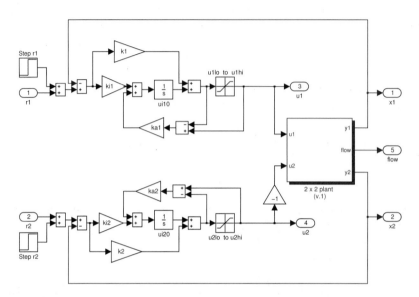

Fig. 8.10. Control structure for the plant in Fig. 8.6, with local PI(aw) controllers

The PI(aw) controller parameters are designed by pole assignment for isolated operation.

$$k_{p_j} = 2\Omega_j\tau_j - a_j; \quad k_{i_j} = (\Omega_j\tau_j)^2; \quad k_{a_j} = \Omega_j\tau_j \quad \text{for} \quad j = 1, 2 \quad (8.39)$$

Then the pole locations[2] for the interconnected system are at

$$-26.1803; \quad -10.0000; \quad -10.0000; \quad -3.8197$$

The response of the interconnected system to the setpoint sequence from above, and with the same $\Omega\tau = 10.0$, is shown in Fig. 8.11 (left).

Re-tuning the local controllers by increasing the k_{i_j} and k_{a_j} twofold yields

$$-20.000; \quad -10.000 \pm j10.000; \quad -10.000;$$

and speeds up the internal equilibration; Fig. 8.11 (right). Note that such re-tuning will lead to weakly damped response in isolated mode (from $2\zeta = 2.0$ to $2\zeta = 1.0$).

---

[2] Determined numerically from the Simulink model; Fig. 8.10.

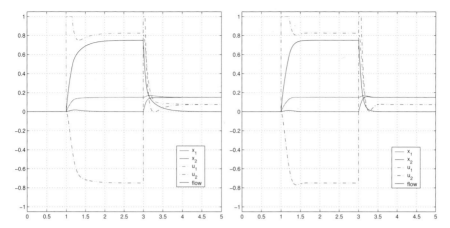

Fig. 8.11. Closed-loop responses for the system in Fig. 8.6, with PI(aw) controllers tuned for isolated mode (left) and re-tuned for interconnected mode (right)

8.5.4 Stability Analysis

To cover almost all of the phase plane trajectories, the multivariable form of the circle criterion must be used, as both saturations will be met. However, for part of the transients, only one of the saturations will be active; see Figs. 8.9 and 8.11. Then, the test for the single input saturation can be applied. One may conjecture that, if the test uncovers a stability problem in the single saturation case, then the full multivariable case will also have a problem. The inverse need not hold, *i.e.* a system with no single saturation stability problems may have low stability properties in the combined case.

For brevity, only the one-saturation case with P controllers shall be investigated here. Then the general result of Chapter 2 applies directly:

$$F + 1 = \frac{\text{``open''-loop denominator polynomial}}{\text{closed-loop denominator polynomial}} \tag{8.40}$$

where the "open" loop is in fact semi-closed, as $k_1 := 0$, but $k_2 \neq 0$. Therefore

$$F + 1 = \frac{\left(s\frac{\tau}{c}\right)^2 + \left(s\frac{\tau}{c}\right)\left[2 + \frac{a_1}{c} + \frac{k_2 + a_2}{c}\right] + \left[\frac{a_1}{c} + \frac{k_2 + a_2}{c} + \frac{a_1}{c}\frac{k_2 + a_2}{c}\right]}{\left(s\frac{\tau}{c}\right)^2 + \left(s\frac{\tau}{c}\right)\left[2 + \frac{a_1 + k_1}{c} + \frac{k_2 + a_2}{c}\right] + \left[\frac{a_1 + k_1}{c} + \frac{k_2 + a_2}{c} + \frac{k_1 + a_1}{c}\frac{a_2 + k_2}{c}\right]} \tag{8.41}$$

Note that this implies $b_j = 1$. If not, replace k_j by $b_j k_j$.

Inserting the numerical values for the benchmark

$$\tau = 1.0; \quad c = 5.0; \quad \frac{a_1}{c} = 0.10; \quad \frac{a_1 + k_1}{c} = \frac{a_2 + k_2}{c} = 2.0$$

yields

$$F + 1 = \frac{\left(\frac{s}{5}\right)^2 + 4.10\left(\frac{s}{5}\right) + 2.30}{\left(\frac{s}{5}\right)^2 + 6.00\left(\frac{s}{5}\right) + 8.00}$$

and the Nyquist contour in Fig. 8.12, which evolves in the right half plane for all ω values. This indicates excellent stability properties, and thus confirms the observation from the simulation.

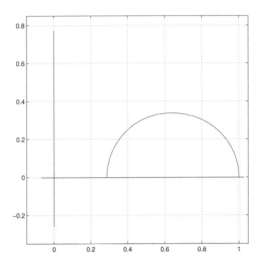

Fig. 8.12. Nyquist contour for the single saturation stability test for the system from Fig. 8.8

Exercise
Consider the full stability problem with both saturations being active.

8.6 Interacting Control Loops

As has been explained in Sect. 8.2.4, the critical windup situation is connected to an ill-conditioned plant ($|A_o| \approx 0$), *i.e.* strong directionality ($S \approx +1$).

The first step is to select a suitable benchmark system. The system used in Sect. 8.5 has a strong directionality. It is based on a physical plant model, its state variables have a direct physical meaning, but it is not yet in the standard form of a transfer function matrix; see Eq. 8.12.

An *alternative* would be the system in Eq. 8.42:

$$\begin{bmatrix} y_1(s) \\ y_2(s) \end{bmatrix} = \frac{1}{sT+1} \begin{bmatrix} 4 & 3 \\ 5 & 4 \end{bmatrix} \begin{bmatrix} u_1(s) \\ u_2(s) \end{bmatrix} \tag{8.42}$$

As

$$S = \frac{G_{12}G_{21}}{G_{11}G_{22}} = \frac{3 \cdot 5}{4 \cdot 4} = \frac{15}{16}; \quad \text{and thus} \quad 1 - S = +\frac{1}{16} \qquad (8.43)$$

it is ill-conditioned and thus also exhibits strong directionality. It has been used very often in this context (*e.g.* see [63, 64, 53]), and thus has become a *de facto* standard.

However, the model of Eq. 8.6 shall be pursued here, also for having been investigated with state feedback control in Sect. 8.5, as a contrast to the design approach used next.

8.6.1 The Plant

The state space representation, denoted as *plant v.1*

$$\begin{bmatrix} s\tau & 0 \\ 0 & s\tau \end{bmatrix} \begin{bmatrix} x_1(s) \\ x_2(s) \end{bmatrix} = \begin{bmatrix} -(a+c) & +c \\ +c & -(a+c) \end{bmatrix} \begin{bmatrix} x_1(s) \\ x_2(s) \end{bmatrix} + \begin{bmatrix} +1 & 0 \\ 0 & +1 \end{bmatrix} \begin{bmatrix} u_1(s) \\ u_2(s) \end{bmatrix} \tag{8.44}$$

and

$$\begin{bmatrix} y_1(s) \\ y_2(s) \end{bmatrix} = \begin{bmatrix} x_1(s) \\ x_2(s) \end{bmatrix} \tag{8.45}$$

has to be transformed into the input-output representation with the four transfer functions $G_{j,k}$. Note that again $\tau_1 = \tau_2 = \tau$ and $c_1 = c_2 = 1$ and that now both u_1 and u_2 act on the inflow.

Using Cramer's rule results in

$$\begin{bmatrix} y_1(s) \\ y_2(s) \end{bmatrix} = \frac{1}{(s\tau + a)(s\tau + 2c + a)} \begin{bmatrix} s\tau + a + c & +c \\ +c & s\tau + a + c \end{bmatrix} \begin{bmatrix} u_1(s) \\ u_2(s) \end{bmatrix} \tag{8.46}$$

with

- one pole at $s = -a/\tau = -0.5$
- one pole at $s = -(2c+a)/\tau = -10.5$
- and zeros at $s = -(a+c)/\tau = -5.5$.

This suggests rewriting Eq. 8.46 as

$$\begin{bmatrix} y_1(s) \\ y_2(s) \end{bmatrix} = \frac{1}{s\tau + a} \begin{bmatrix} \frac{s\tau+a+c}{s\tau+2c+a} & \frac{+c}{s\tau+2c+a} \\ \frac{+c}{s\tau+2c+a} & \frac{s\tau+a+c}{s\tau+2c+a} \end{bmatrix} \begin{bmatrix} u_1(s) \\ u_2(s) \end{bmatrix} \tag{8.47}$$

where the zeros and poles in the matrix elements are approximatively a factor $c/a = 10$ and $2c/a = 20$ further to the left than the pole of the dominant first-order lag.
This suggests an *order reduction* by using only the static gains in the matrix

$$\begin{bmatrix} y_1(s) \\ y_2(s) \end{bmatrix} \approx \frac{1}{s\tau + a} \begin{bmatrix} \frac{a+c}{2c+a} & \frac{+c}{2c+a} \\ \frac{+c}{2c+a} & \frac{a+c}{2c+a} \end{bmatrix} \begin{bmatrix} u_1(s) \\ u_2(s) \end{bmatrix} \rightarrow \frac{1}{sT + 1} \begin{bmatrix} b_{11} & b_{12} \\ b_{21} & b_{22} \end{bmatrix} \begin{bmatrix} u_1(s) \\ u_2(s) \end{bmatrix}$$
(8.48)

with the numerical values for the benchmark

$$T = \tau/a = 2.0; \quad b_{11} = b_{22} = 1.10/1.05; \quad \text{and} \quad b_{12} = b_{21} = 1.00/1.05$$

As the system will be used next with local feedback control of y_1 to u_1 and y_2 to u_2, it is rewritten as

$$y_1 = \frac{1}{sT + 1}(b_{11}u_1 + b_{12}u_2) \quad \text{and} \quad y_2 = \frac{1}{sT + 1}(b_{21}u_1 + b_{22}u_2) \quad (8.49)$$

This is denoted as *plant v.2*.

Fig. 8.13 shows the Bode plots of the four transfer functions for the plant v.1 and for their approximations, plant v.2

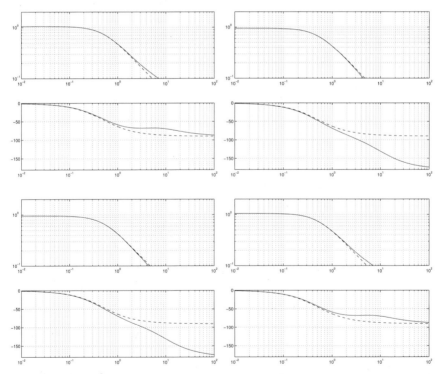

Fig. 8.13. The four frequency responses for the original *plant v.1* (full line) and for the approximations, *plant v.2* (broken line); G_{12} shown at top right

There are two poles at $s = -1/T = -0.5$. And the gain matrix B

$$|B| = b_{11}b_{22} - b_{12}b_{21} = (1/1.05)^2 \left(1.10^2 - 1.0^2\right) \approx +0.1574 \qquad (8.50)$$

indicates both strong directionality and correct input–output pairing.

8.6.2 Local P Controllers

Two local P controllers are used on plant v.2, see Fig. 8.14, similar to what has been used for plant v.1 from Fig. 8.8. They are designed in the same way, i.e.

$$b_{jj}k_j = \Omega T - 1; \quad b_{jj}kr_j = \Omega T; \quad j = 1, \, 2 \qquad (8.51)$$

Then, with the symmetry property of plant v.2 and of the controllers, the characteristic equation (from Eq. 8.16) is

$$0 = 1 + 2RG_{11} + (RG_{11})^2 \left(1 - \frac{1}{1.1^2}\right) \quad \text{with} \quad RG_{11} = \frac{kb_{11}}{sT+1} = \frac{\Omega T - 1}{sT+1} \qquad (8.52)$$

that is

$$0 = (sT+1)^2 + 2(\Omega T - 1)(sT+1) + (\Omega T - 1)^2 \left(\frac{0.21}{1.21}\right) \qquad (8.53)$$

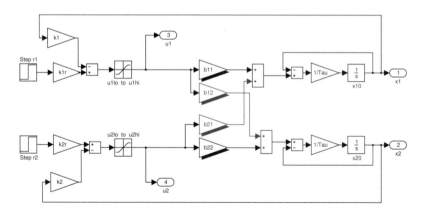

Fig. 8.14. Plant (v.2) with two local P controllers from Fig. 8.8

This leads to the two roots s_1, s_2 at

$$s_1 = -\frac{1}{1.1T}(0.1\Omega T + 1) = -\frac{0.1}{1.1}\Omega - \frac{1}{1.1T} \qquad (8.54)$$

and at

$$s_2 = -\frac{1}{1.1T}\,(2.1\Omega T - 1) \;=\; -\frac{2.1}{1.1}\Omega + \frac{1}{1.1T} \tag{8.55}$$

Note that

$$s_1 + s_2 = -2\Omega$$

and also for

$$\Omega T := 1 \quad \rightarrow \quad s_1 = s_2 = -0.5; \quad \text{resulting for} \quad k = 0$$

In contrast to the case with plant v.1, the slow closed-loop mode associated with s_1 cannot be significantly accelerated, if the value of k is to stay in a practicable region:
let $\Omega := 10 \rightarrow k \approx 18.136$, $s_1 \approx 1.364$ (slow mode), $s_2 \approx 18.636$ (fast mode).

Transient Response

This is visible in the simulation Fig. 8.15 (compare with Fig.8.9).

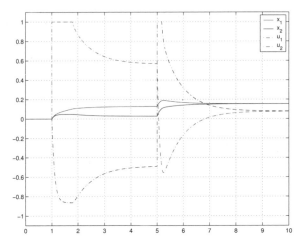

Fig. 8.15. Response of the control system Fig. 8.14

Stability Properties

Again for brevity, only situations are checked where one of the u_j saturates. For the Nyquist contour, applying the general result Eq. 8.40 produces (with $k_2 = 0$ in the "semi-open system" characteristic polynomial for the numerator)

$$F + 1 = \frac{\left(s + \frac{1}{T}\right)^2}{(s - s_1)(s - s_2)}\,[(1 + R_1 G_{11})\,(1 + 0) + 0]$$

$$= \frac{\left(s + \frac{1}{T}\right)(s + \Omega)}{\left(s + \frac{0.1}{1.1}\Omega + \frac{1}{1.1T}\right)\left(s + \frac{2.1}{1.1}\Omega - \frac{1}{1.1T}\right)} \tag{8.56}$$

The Nyquist contour of $F + 1$ evolves in the right half plane, see Fig. 8.16, and thus indicates global asymptotic stability.

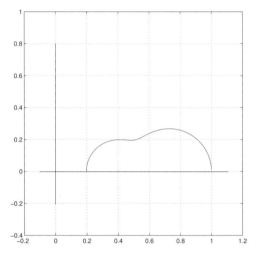

Fig. 8.16. The Nyquist contour for the stability test of the control system in Fig. 8.14 with only the saturation on u_2 being active

8.6.3 Decoupling Control

The basic idea is to feed the current output of the local controller R_1 through a transfer function K_{12} to the input u_2 at the other local controller site. If K_{12} is selected appropriately, then the coupling effect through G_{12} may be greatly reduced or even nullified. It has been mentioned above that, although the concept is intuitively simple, looking at it in detail reveals numerous problems. Nevertheless, this concept shall be applied here. The situation in Fig. 8.14 is particularly suitable, as the coupling paths in plant v.2 are simple gains with no dynamics and are at the input side of the dominant (first-order lag) dynamics; see Fig. 8.17.

Then, for decoupling both loops:

$$K_{12}^* b_{22} + b_{12} \overset{!}{=} 0 \quad \text{that is} \quad K_{12}^* = -\frac{b_{12}}{b_{22}}; \quad \text{and } vice versa \text{ for } K_{21}^* \qquad (8.57)$$

As shown above, such decoupling also affects the gains b_{ii}, $i = 1, 2$, of the local plant transfer functions $u_i \to y_i$

$$\tilde{b}_i = b_{ii}\left(1 - \frac{b_{12}b_{21}}{b_{11}b_{22}}\right) \quad \to \quad \approx 0.1736\, b_{ii} \, ; \quad \text{for} \quad i = 1, 2 \qquad (8.58)$$

In other words, the low plant gain $\tilde{b}_i$ requires large $\overline{u_i}$ for a given setpoint step size, and thus makes the control loop sensitive to saturations. Also, the local controller gain $\tilde{k}_i$ must be increased to attain a given bandwidth $\tilde{\Omega}\ 2\tau$ in the local (decoupled) loops.

$$\tilde{b}_i \tilde{k}_i = \left(\tilde{\Omega}\ T - 1 \right) \qquad \text{for}\ \ i = 1, 2 \tag{8.59}$$

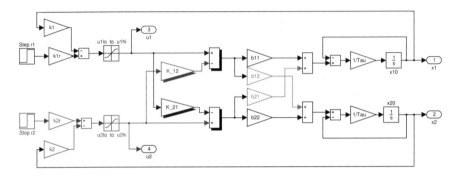

Fig. 8.17. Plant (v.2) with two local P controllers from Fig. 8.14 and decouplers K_{21}, K_{12}

Transient Response

This is illustrated by the simulations in Fig. 8.18.

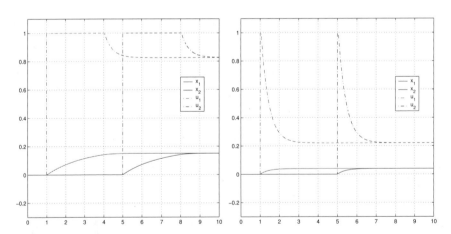

Fig. 8.18. Transient response for the control system with exact compensations K_{jk}^*, with $\tilde{\Omega} = 2.5 \rightarrow \tilde{k} = 22.0$:   (left) with $r = 0.15$;   (right) with $r = 0.04$

Fig. 8.19 illustrates the effect of non-exact compensation, specifically for $K_{jk} = 0.90K_{jk}^*$. The eigenvalues move to $\tilde{s}_1 = -3.2097; \tilde{s}_2 = -1.7903$.

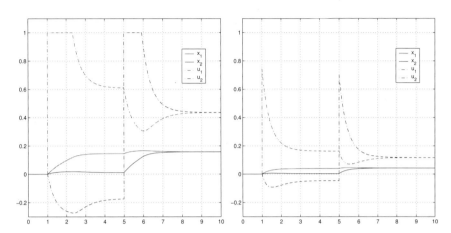

Fig. 8.19. Transient response for the control system with approximate compensations $K_{jk} = 0.9K_{jk}^*$:   (left) with $r = 0.15$;   (right) with $r = 0.04$

Stability Properties

Exact decoupling results in two separate loops, consisting of the plant $\tilde{G}$ and the controller R. This leads to

$$F + 1 = \frac{s + \frac{1}{\tilde{T}}}{s + \tilde{\Omega}} \tag{8.60}$$

Its Nyquist contour is in the right half plane for all ω, and indicates global asymptotic stability.[3]

8.6.4 A Different Control Concept

Another way of accelerating the slow eigenmode from Eq. 8.54 of the system with local controllers in Fig. 8.14 is as follows; see Fig. 8.20:

(a) Cross over the input–output pairing used previously, *i.e.* connect y_2 to controller R_1 and to u_1 and *vice versa* for y_1,
(b) change the negative feedback in the R_2-path to a positive one (note the gain -1 in Fig. 8.20),
(c) and finally reduce the gain k_2 in R_2.

---

[3] As long as the control problem is "well posed".

This strategy has been investigated for instance in [66].

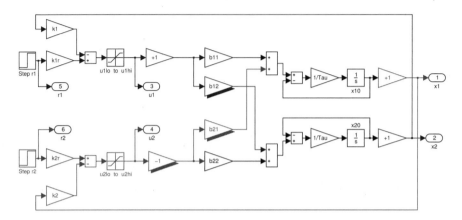

Fig. 8.20. Plant v.2 with two local P-controllers, using "crossed over and inverted" control

Linear Controller Design

Rename

$$H_{11} := G_{21}; \quad H_{22} := G_{12}; \quad H_{12} := G_{11}; \quad H_{21} := G_{22} \tag{8.61}$$

Then the characteristic equation following item (a) is

$$(1 + R_1 H_{11})(1 + R_2 H_{22}) - R_1 H_{12} R_2 H_{21} = 0 \tag{8.62}$$

and following (b) and (c) insert

$$k_2 = -\alpha k_1 \quad \text{with} \quad 0 < \alpha < 1.0 \tag{8.63}$$

That is

$$0 = (sT + 1 + k_1 h_{11})(sT + 1 - \alpha k_1 h_{22}) + \alpha k_1 h_{12} k_1 h_{21}$$

$$= (sT + 1)^2 + (sT + 1) k_1 h_{11} [1 - \alpha] + \alpha k_1^2 h_{11} h_{22} \left[\frac{h_{12} h_{21}}{h_{11} h_{22}} - 1 \right]$$

Inserting $h_{11} = h_{22} = \dfrac{1.0}{1.05}$ and $\dfrac{h_{12} h_{21}}{h_{11} h_{22}} - 1 = 0.210$

yields $\quad 0 = (sT + 1)^2 + (sT + 1) k_1 h_{11} [1 - \alpha] + 0.21 \alpha k_1^2 h_{11}^2$

$$\rightarrow \quad (sT + 1) = -\left(\frac{1 - \alpha}{2} k_1 h_{11}\right) \pm \sqrt{\left(\frac{1 - \alpha}{2} k_1 h_{11}\right)^2 - 0.21 \alpha k_1^2 h_{11}^2}$$

$$\text{finally} \quad (sT+1) = k_1 h_{11} \left[-\frac{1-\alpha}{2} \pm \frac{1}{2}\sqrt{1 - (2+4 \times 0.21)\,\alpha + \alpha^2} \right] (8.64)$$

The next design decision is to choose α such that two equal real poles result, *i.e.*

$$0 = 1 - (2 + 4 \times 0.21)\,\alpha + \alpha^2 \quad \rightarrow \quad \alpha_1 = 2.4282; \quad \alpha_2 = 0.4118 \qquad (8.65)$$

with α_2 being the valid solution; see Eq. 8.63. This leads to

$$s_1 = s_2 = -\left(0.2941\Omega_1 + \frac{1 - 0.2941}{T} \right) \quad \text{with} \quad \Omega_1 = 10 \quad \rightarrow \quad s_{1,2} = -3.2938$$
$$(8.66)$$

to be compared with Eq. 8.54. Also

$$\Omega_2 = \alpha\Omega_1 + \frac{1-\alpha}{T} \quad \rightarrow \quad \Omega_2 = 4.118 + 0.2941 = 4.412 \qquad (8.67)$$

and

$$k_1 = (\Omega_1 T - 1)/h_{21} \rightarrow 19.95; \quad k_2 = (\Omega_2 T - 1)/h_{12} \rightarrow 8.2161 \qquad (8.68)$$

In other words, a considerable acceleration of the slow mode is feasible with reasonably sized controller gains.

Transient Response

This is visualized in the simulations in Fig. 8.21. Note the large transient excursions of the $u_j(t)$, and the non-minimum phase response in the $y_j(t)$.
In the lower plot, the setpoint step size is increased to $r_1 = r_2 = +0.2$. Then both u_j's get stuck on their upper limit. This "stuck" condition *continues* after $t = 6.0$, when the second setpoint is stepped up to the value of the first one. There, the crossflow should be zero again, *i.e.* the system should be operating in the linear range again. In other words, the system response in this case is *unstable*.
As, the open-loop plant response is stable, y_1 and y_2 converge to steady states at $\overline{y_1} \approx 0.0952$ and $\overline{y_2} \approx -0.0952$.

Exercise
Note that Ω_1 and Ω_2 are no longer the same.
- What are the consequences, if the design specification is $\Omega_1 := \Omega_2$?
- And what is the effect of $a_1 = a_2 = a \to 0$?

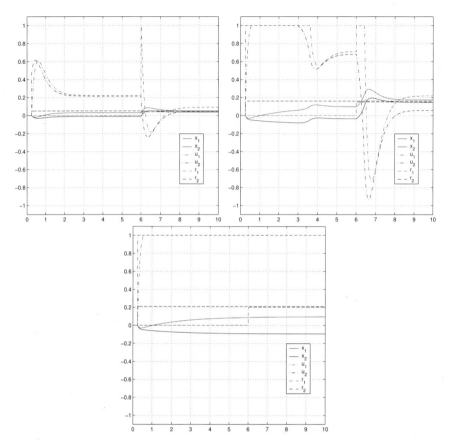

Fig. 8.21. Transient response for the system Fig. 8.20 with values from Eqs. 8.67, 8.68:
(left) for small steps $r_1, r_2 = 0.04$;   (right) for larger steps $r_1, r_2 = 0.15$;
          (bottom) for step size $r_1, r_2 = 0.20$ ("stuck" condition, *unstable*)

Stability Properties

Again for brevity, only the two cases with one saturation being active are investigated.

(a) Consider first the case where only u_2 meets its saturations. Then for the linear subsystem

$$F + 1 = \frac{(sT + 1)(sT + 1 + k_1 b_{21})}{(sT + 1)^2 + (sT + 1)[k_1 b_{21} - k_2 b_{12}] + k_1 k_2 [b_{11} b_{22} - b_{12} b_{21}]}$$

$$= \frac{\left(s + \frac{1}{T}\right)\left[s + (1 + k_1 b_{21})\frac{1}{T}\right]}{(s + \Omega)^2} = \frac{\left(s + \frac{1}{T}\right)\left(s + \Omega_1 - \frac{1}{T}\right)}{(s + \Omega)^2} \qquad (8.69)$$

with Ω from the closed-loop pole assignment, see Eqs. 8.62 to 8.66, and with $k_1 b_{21} := \Omega_1 T - 1$ from the loop design.

For $\Omega_1 T > 1$ this starts and evolves in the right half plane. In other words, this case has good stability properties, and does not weaken the overall stability properties unduly.

Exercise
Check both Eqs. 8.69 and 8.70 with Fig. 8.20.

(b) Consider next the case where u_1 saturates only. Then

$$F + 1 = \frac{(sT + 1)(sT + 1 - k_2 b_{12})}{(sT + 1)^2 + (sT + 1)[k_1 b_{21} - k_2 b_{12}] + k_1 k_2 [b_{11} b_{22} - b_{12} b_{21}]}$$

$$= \frac{\left(s + \frac{1}{T}\right)\left[s + (1 - k_2 b_{12})\frac{1}{T}\right]}{(s + \Omega)^2} = \frac{\left(s + \frac{1}{T}\right)\left(s - \Omega_2 + \frac{2}{T}\right)}{(s + \Omega)^2} \quad (8.70)$$

As $\Omega_2 > 0$ and as also $\Omega_2 \gg \frac{2}{T}$, the Nyquist contour starts from the negative real axis in the downward direction,[4] which indicates a small radius of convergence. In other words, this case is critical for the overall stability properties.

Fig. 8.22 shows the Nyquist contours for the two cases (a) and (b).

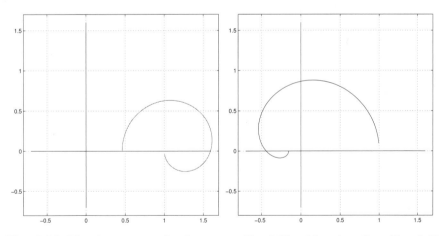

Fig. 8.22. Nyquist contours for the system Fig. 8.20, with values from Eqs. 8.67 and 8.68:   (left) case(a): for u_2 saturating, but not u_1;   (right) case (b): for u_1 saturating, but not u_2

---

[4] See also the stability charts in Sect. 6.4.2.

Introducing Antiwindup Feedback

The unstable response in Fig. 8.21 is the consequence of r_2 being large. This generates a large u_2, which will drive u_1 beyond its saturation. This is equivalent to having R_1 in open-loop, or alternatively to setting $k_1 := 0$. Introducing this into the characteristic equation Eq.8.64 leads to

$$0 = (sT + 1) [sT + (1 - k_2 b_{12})] \tag{8.71}$$

where $1 - k_2 b_{12} \ll 0$, *i.e.* to a highly unstable response of the system. This then will move u_2 quickly to its own saturation. Then R_2 is in open-loop, or equivalently $k_2 := 0$. And the characteristic equation reduces to

$$0 = (sT + 1)(sT + 1) \tag{8.72}$$

and thus the response will finally stabilize. However, u_1 will end up so far beyond its saturation that the setpoint step on r_1 is insufficient to bring it back into its linear range, and thus to case (a) from above. In other words u_1 "winds up".

This suggests inserting an awf from u_1 to u_2; see Fig. 8.23. This is to keep u_1 close to its saturation, such that the linear control feedbacks are reestablished, when the operating conditions are moved back into the range of "well-posed" situations.

The awf gain k_a shall be designed by nonlinear stability analysis here. Another approach is from a Robust Control formulation, *e.g.* see [53, 67] and references therein.

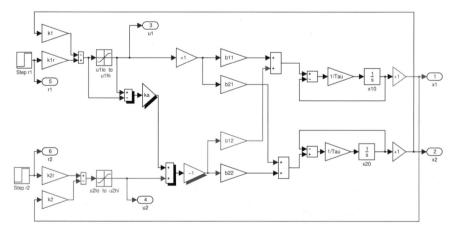

Fig. 8.23. The control structure from Fig. 8.20 with awf from u_1 to u_2, and with gain k_{a_1} as design parameter

Stability Analysis

Denote as v_1 the input to the saturation in the R_1 loop, and as u_1 the output of this saturation, and as w_{21} the output of the awf to u_2, *i.e.*

$$w_{21} := k_{a_1}(u_1 - v_1) \tag{8.73}$$

Then the linear subsystem has one output v_1 and two inputs u_1 and w_{21}. Denote the individual responses as

$$v_{1_1} = L_{1_u} u_1 \quad \text{and} \quad v_{1_2} = L_{1_w} w_{21} \tag{8.74}$$

Then by superposition of the individual responses

$$v_1 = v_{1_1} + v_{1_2} = L_{1_u} u_1 + L_{1_w} w_2 = L_{1_u} u_1 + L_{1_w} k_{a_1}(u_1 - v_1) \tag{8.75}$$

i.e.

$$v_1 = \frac{L_{1_u} + k_{a_1} L_{1_w}}{1 + k_{a_1} L_{1_w}} u_1 = L u_1 \tag{8.76}$$

which defines L.

Again, the saturation nonlinearity shall be replaced by the deadspan nonlinearity and a unity gain in parallel. Then, for the transfer function F of the linear subsystem to be used with the deadspan nonlinearity only:

$$F = \frac{L}{1 - L} = -1 + \frac{1}{1 - L} = \frac{1 + k_{a_1} L_{1_w}}{1 - L_{1_u}} \tag{8.77}$$

The next step is to determine L_{1_u} and L_{1_w}. From Fig. 8.23 read

$$
\begin{aligned}
L_{1_u} &= -k_1 \frac{1}{sT + 1} \left[b_{21} + b_{11} b_{22} \frac{-(-k_2)}{(sT + 1) - k_2 b_{12}} \right] \\
&= -\frac{k_1 b_{21}(sT + 1) + k_1 k_2 [b_{11} b_{22} - b_{12} b_{21}]}{(sT + 1)[(sT + 1) - k_2 b_{12}]}
\end{aligned} \tag{8.78}
$$

and

$$L_{1_w} = -\left[-\frac{1}{1 - \left(+\frac{k_2 b_{21}}{sT + 1} \right)} b_{22} \frac{1}{sT + 1} k_1 \right] = \frac{k_1 b_{22}}{(sT + 1) - k_2 b_{21}} \tag{8.79}$$

into Eq. 8.77 finally yields

$$
\begin{aligned}
F + 1 &= \frac{1}{1 - L_{1_u}} (1 + k_{a_1} L_{1_w}) \\
&= \frac{(sT + 1)\left[(sT + 1) - k_2 b_{12} + k_a k_1 b_{22}\right]}{(sT + 1)^2 + (sT + 1)\left[k_1 b_{21} - k_2 b_{12}\right] + k_1 k_2 \left[b_{11} b_{22} - b_{12} b_{21}\right]} \\
&= \frac{\left(s + \frac{1}{T}\right)\left[\left(s + \frac{1}{T}\right) - \frac{1}{T}(k_2 b_{12} - k_a k_1 b_{22})\right]}{(s + \Omega)^2} \\
&= \frac{\left(s + \frac{1}{T}\right)\left[s + \frac{\beta_1}{T}\right]}{(s + \Omega)^2}
\end{aligned}
\tag{8.80}
$$

where $\quad \beta_1 := +1 - k_2 b_{21} + k_{a_1} k_1 b_{22}$

Comparing this with the result for the no-awf case in Eq. 8.69 reveals that in the square bracket factor of the numerator, a positive term $k_{a_1} k_1 b_{22}$ is added. In other words, β_1 increases for increasing k_{a_1}. Therefore the starting point of the Nyquist contour (at $\omega = 0$) moves monotonously to the right, and thus the radius of attraction increases.

A special case is $\beta_1 = 0$. Then

$$
F + 1 = \frac{\left(s + \frac{1}{T}\right) s}{(s + \Omega)^2}
\tag{8.81}
$$

and the Nyquist contour starts from the origin in the vertical direction. This lets the radius of attraction grow to near infinity.

The corresponding awf gain value shall be denoted as $k_{a_1}^*$. It has a similar role as the compensating awf gain in Chapter 2.

$$
\beta_1 = 0 = +1 - k_2 b_{12} + k_{a_1}^* k_1 b_{22}; \quad i.e. \quad k_{a_1}^* = \frac{1 - k_2 b_{12}}{-k_1 b_{22}}
\tag{8.82}
$$

Inserting from the design

$$
k_2 b_{21} = \Omega_2 T - 1 \quad \text{and} \quad \frac{b_{22}}{b_{12}} k_1 b_{12} = \frac{b_{22}}{b_{12}} (\Omega_1 T - 1)
\tag{8.83}
$$

yields

$$
k_{a_1}^* = \frac{b_{12}}{b_{22}} \frac{\Omega_2 T - 2}{\Omega_1 T - 1} \quad \rightarrow \quad \approx 0.3421
\tag{8.84}
$$

In Fig. 8.24 the Nyquist contours are plotted from the Simulink model corresponding to the control structure in Fig. 8.23 for

$$
v k_{a_1} = [0, \ 0.5, \ 1.0, \ 2.0] \, k_{a_1}^*
\tag{8.85}
$$

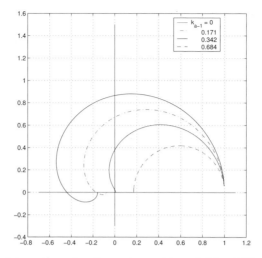

Fig. 8.24. Nyquist contours for the system with awf, Fig. 8.23, with values from Eqs. 8.67) and 8.68 for saturating u_1

8.6.5 Transient Response

Fig. 8.25 shows the response of the system with awf in Fig. 8.23, with the awf gain at the same values as in Fig. 8.24. The setpoint steps applied here ($r_1 = r_2 = 0.50$) are far above the stability limit for the no-awf case ($r_1 = r_2 \lesssim 0.20$ from Fig. 8.21).
This confirms the effect of the awf, and the sizing of the awf gain by the stability-based design method.

To *summarize*: a solution has been found for making this design choice feasible with saturating controls u_1, u_2. However, it still is a dangerous proposition, as the local R_2 loop has been made voluntarily highly unstable by design. There will be grave consequences if the R_1 loop is opened inadvertently by some equipment failure, or if it is switched to manual. Commissioning of the control system will also be rather difficult.

Exercise
- Investigate the effect of the opposite awf (from u_2, added to u_1) on stability and on transient response,
- and also of having both awf loops implemented jointly.
- Are both awf loops needed to ascertain stability?
- And: use local PI(aw) controllers instead of the P controllers.

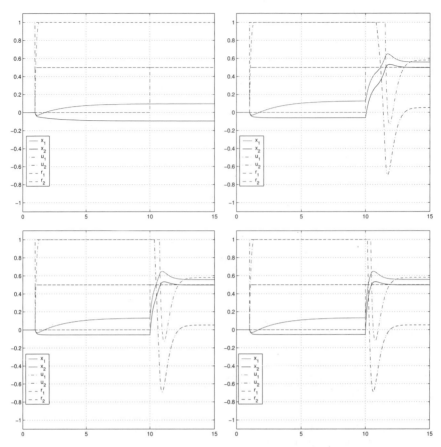

Fig. 8.25. Responses of the system in Fig. 8.23, with values from Eqs. 8.67 and 8.68

with awf from u_2, added to u_1:

(top left) $k_{a_1} = 0$;   (top right) $k_{a_1} = 0.5\ k_{a_1}^*$

(bottom left) $k_{a_1} = k_{a_1}^*$;   (bottom right) $k_{a_1} = 2\ k_{a_1}^*$

8.7 Summary

The orientation towards control system design has motivated the investigation of control systems, where there is one main measured variable y_1, but multiple physical actuators. They may be operated in parallel or sequentially by one main controller with appropriate awf. Or there may be more than one controller for y_1, again with appropriate awf, but with shifted setpoints such that the actuators are operated in sequence. The parameters of those controllers may differ to account for different plant response. Such control structures are used very often in practice.

Obviously these are not of the usual single-input type, nor are they of the fully multivariable type. From their physical background, they have been attributed to the multivariable area here. However, again assuming that the awf has been designed properly, the stability properties can be determined either from fusing the multiple physical actuators into one logical actuator of corresponding work span, or from fusing the sequentially operating control loops into one loop. In other words, the methods developed so far for the scalar u case can be applied. Or such systems are a subclass of the scalar u class, and clearly belong there from a more theoretical point of view.

From the multivariable control systems *per se*, only one (small) part has been selected, where saturating control variables are known to cause performance degradation.[5] The benchmark problem investigated here is based on a simple, transparent physical plant. It leads to a symmetric 2×2 plant model with two dominant state variables, and where standard local P control (without decoupling add-on) is already the full state feedback control.
The system is subject to strong directionality in the control variables. This is also associated with a comparatively slow equilibration transient. Linear decoupling is not an effective design option here, because the coupling reduces the apparent plant gain (as seen from the local control inputs to the local measured outputs, and with the other control loop being closed).

Another approach is to invert the sign in one of the two loops, *i.e.* to make it voluntarily unstable. However, the controller gain of this loop must be lower than the gain of the other loop with traditional negative feedback, in order to conserve overall stability. This speeds up the slow equilibration transient. But stability will be lost if the stable loop is no longer functional. If it is voluntarily switched to manual mode, then the positive feedback can be reverted at once to negative feedback again. But what if its control variable saturates transiently? If this situation persists long enough, then the two loops will not recover, when the cause for the saturating u is removed. They will stay mutually "locked out", or "wind up" persistently.

It has been shown that this behavior can be suppressed by adding an awf from the actuator in the negative feedback loop crossing over to the control input of the positive feedback loop. This re-establishes nonlinear stability. The radius of attraction can be made very large, if the awf gain exceeds a (finite and rather low) threshold value. However, from a more distant perspective, this seems "a patch (awf) on a patch (positive feedback)", and a rather risky proposition. Therefore, an alternative is needed.
One is hinted at by the observation on the benchmark, that state feedback design yields a less slow equilibration than designing local feedback on the traditional 2×2 transfer function matrix. And response may be further improved by appropriate feedforward paths to both control inputs as well. MPC design again supports this in a systematic way.

---

[5] Note that this does not exclude that other such problem areas may exist.

9

Conclusion

Some remarks shall conclude this book on the approach, the main achievements, the limitations, and some open areas for further work.

9.1 Approach

The presentation is oriented to control engineers interested in improving their design methods, to students preparing their professional career in control design, and to researchers interested in applications rather than the control theoretician. An inductive, bottom-up approach has been opted for, rather than a deductive top-down one, starting with the simple (and most frequent) cases and progressing to more complex design tasks. A median level of abstraction is used, where typical industrial control design tasks are still visible, where control theory is used with restraint and as a tool (to improve analysis and design, and not as an end on itself), and where ease of implementation is a key issue. This has led to separating the "standard methods" (where all modules for implementation are available in current process control systems and know-how is the only limitation), from the "advanced methods'. And the focus is on control systems with one manipulated variable (scalar u), with some extension to the multivariable area.

9.2 Achievements

A *systematic design procedure* has been proposed and used throughout. It consists of a balanced application of four elements:

- *Structure* in alternative forms, where functionality and equivalence is a key issue.
- *Transient responses* on benchmarks extracted from industrial control design tasks, serving as a screening tool for structures, and for a qualitative

measure of performance. A more quantitative technique is available from Robust Control, but it is not included here, mainly for brevity.

- *Nonlinear stability analysis* in the form of the sector criteria, which (after some preparation) is a powerful tool for understanding the effects of parameter variations, and also for choosing design parameters well.
- *Links to* more formal methods of *optimal control*, such as minimum-time systems and numerical optimization of trajectories (Explicit MPC).

The application of this systematic design process has shown it to be an effective and efficient approach, much more so than focusing on any one of these elements alone would be.

Input constraints have been handled by *antiwindup feedback (awf)*. The basic situation with a plant of dominant first order, actuator saturation as the input constraint and PI control has been discussed in detail. A generic structure has been used with dynamic awf, from which the most current control structures are instantiated. This has been extended to other typical situations in this context, such as actuators with both slew and stroke constraints, the effect of a derivative action and of high-frequency measurement disturbances. Then awf has been generalized to control of plants with higher dominant order, again with input constraints from actuator stroke saturation and state feedback. Observers have not been considered, for brevity. Stability analysis has been prepared and standardized as "stability charts". They allow one to discern quickly, whether the given control problem is a difficult one, *i.e.* sensitive to such saturation or not.

Again several current and some novel awf structures have been investigated. One of those, the "selection for lower-bandwidth control", performs particularly well, is comparatively simple to implement, and is closely related to the corresponding solution of the Explicit MPC method.

In the field of Multivariable Control, only one case has been investigated, one which is known to be sensitive to saturations and where awf is helpful. A more complete investigation would be outside the scope of this book, as it must include many other relevant design techniques.

Output constraints are handled by the *override* technique. It consists of feedback control along the constraints with standard linear controllers including integral action, and subsequent Max-Min-selection on the controls. In contrast to the awf field, this is not yet an established research area.

Some inroads have been made in this book. Basic elements are presented about currently used alternatives for structures, their functionality and closed-loop performance on the benchmark, a suitable generic structure as their common root, and the preparation of the stability analysis. It has been demonstrated, that the override structure is closely related to what the Explicit MPC method produces as the result of the optimization process. It has also been shown that the input-constraint case is a special case of the output-constraint one.

Again, the basic concept is generalized into implementing alternative control strategies, override control for plants with higher dominant order, and other override paradigms, such as acting on the reference r of the main loop instead on the control u, or resorting to a controlled compromise between performance of the main and of the constrained variable.

9.3 Limitations

As with any other method, the design methods of antiwindup feedback and override control have their limitations. The broad investigations in this book have outlined them.

First, both design methods have been developed originally for single input plants of dominant low order, for which PI control is well suited. Fortunately, this covers most of the design tasks in industrial control design. And the results are comparable to what optimal control would provide. But it has also been shown that the stability properties, and thus performance, will deteriorate progressively as the problem becomes situated farther away from this simple setting. In short, loops with plants of dominant first order are not sensitive to constraints. Loops with plants of dominant second order have a tendency to overshoot. And plants of dominant third order may lead to unstable loop response for large inputs. Also, Hurwitz plants are less sensitive than stable plants with poles on the imaginary axis, and unstable plants are very sensitive to constraints. Here, the design must be very careful, such as providing sufficient actuator working span, other constraints being shifted outward as far as possible, well-fitted reference trajectories, *etc.*

Second, both methods build on a well-designed control system for the linear range. In other words, a mistuned linear control will negatively affect the performance of antiwindup or overrides. And override control may not be a good stand-in for a missing coordinating control in a multivariable situation with strong interactions.

And finally, both methods are iterative and driven by the intuitive insight of the designer into the specific needs for the design task at hand. This insight has to be built up, which requires effort and time, and may be slow going. If the design task is very complex, then this can get out of hand. A numerical optimization technique, such as MPC, will surely be faster, but can have other implementation problems, such as unfeasible trajectories, *etc.*

9.4 Some Further Research Areas

Obviously there is a strong interaction (and competition) with the numerical optimization methods. The question is whether the antiwindup and override designs are going to be eliminated soon by these more general methods.

For the many small-sized design tasks in industry, a well-designed antiwindup or override is a cost-effective solution. Furthermore, it seems to be closely related to the corresponding Explicit MPC solution. On the other hand, numerical optimization is the main method for large-sized problems, where the selection techniques may be helpful in the implementation phase. Further research in this overlapping area will surely be productive for both sides.

Another promising area is two-level control, which from practical experience is a wise approach to large complex control problems. Here, local single-input control loops with high bandwidth reduce the effects of disturbances and parameter variations as seen from the next higher level. This improves the operating conditions for the higher level optimization, which then coordinates the lower level loops through more complex, but slower evolving trajectories, *e.g.* by an MPC control algorithm with its slow updating rate. The local high bandwidth loops will still need the functionality for local input and output constraints, *i.e.* antiwindup and overrides, at least as a precaution, and of course they must communicate current constraints to the upper-level algorithm.

In any case antiwindup and override controls are here to stay in design practice for the foreseeable future, together with the indefatigable PID controller . . . .

A

Nonlinear Stability Tests

The aim of this appendix is to present the main nonlinear stability tests used in this text. The assumptions are listed and discussed in verbal rather than abstract mathematical form. No proofs are given.

Readers interested in the mathematical background are referred to the references.

A.1 The Canonical Loop Structure and Stability of Motion

The canonical system Fig. A.1 is a standard configuration used by most nonlinear stability tests. It consists of two subsystems N and L connected in a negative feedback loop. Any nonlinear system to be tested must be brought to this form first.

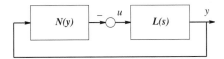

Fig. A.1. the canonical loop structure

Note that both u and y may be vectors for MIMO cases. In this text, only the scalar (SISO) case is needed and therefore considered.

The negative feedback is accounted for separately in Fig. A.1 and is excluded from the following discussion of the properties of N and L.

The subsystem N contains the nonlinear characteristic (a nonlinear function $u = -N(y)$), and the subsystem L contains all the dynamics. It has to be linear and time invariant, and thus can be described by its transfer function $L(s)$. In other words, the nonlinear and the linear dynamic properties must be neatly separable into these two blocks. Further assumptions on both subsystems are test-specific and shall be given there.

Note that no persistent inputs are considered here, *i.e.* both reference and disturbance inputs are set to zero. In other words, the focus is on the *initial condition response* of the closed loop (this is in contrast to classic control theory, where initial conditions are usually set to zero, and the inputs are nonzero, such as steps).

The *undisturbed motion* of the nonlinear system in Fig. A.1 is considered as a reference. This may be a periodic motion of all variables in the loop (such as in a stationary limit cycle), or it may be a steady state with a zero motion of all variables.

Then an initial condition is applied with arbitrary size.

The *stability tests* now consider the initial condition response. Therefore, the term "stability of motion" is often used.[1]

For *asymptotic stability* the initial condition response must decay to the (undisturbed) reference motion, whereas for *stability* it must stay in a bounded neighborhood to the reference motion, where the bounds are determined by the size of the initial condition. Note that this definition of stability allows one to investigate stability of both limit cycles (which would be classified as non-stable in linear control theory anyhow) and steady state.

An important parameter is the size of the initial conditions. One therefore distinguishes between "stability in the small" or "local stability" for infinitesimal size at one end, and "stability in the whole" or "global stability" for the size tending to infinity.

It is useful for applications to insert "stability in the large" for intermediate sizes. Then the initial conditions responses form a compact "region of attraction" of finite size around the reference motion. For initial conditions outside this border the responses diverge, *i.e.* they are not stable.

A.2 Final Steady State and the Sector Criteria

This is a first group of stability tests where the reference is the steady state. This makes them particularly useful for the class of problems considered in this text, where the response should settle to a steady state and not to a limit cycle.

The *linear subsystem* $L(s)$ must be asymptotically stable itself (without the feedback through N): It must be Hurwitz, that is all poles must have nonzero negative parts. Non-rational stable elements, such as a pure delay,

---

[1] Note that, in linear control theory, stability is a system property for short, but not here.

are admissible. Also, $L(s)$ has to be a finite bandwidth system,[2] *i.e.* its gain $|L(j\omega)|$ shall decay to zero for $\omega \to \infty$ (strictly proper). This restriction may be relaxed to "proper", *i.e.* to a finite, bounded gain. In other words, a proportional feedthrough is admissible.

Finally, the mathematical model must be such that the steady state to be investigated is at zero (at the origin). In practice, this requires introducing appropriate offsets to suppress any nonzero steady state values for the particular operating condition (working point) to be investigated.

The *nonlinear subsystem* is the nonlinear characteristic $N(y)$. It must evolve in the first and third quadrant, but need not be symmetric to the origin. It must run through the origin and must be single-valued in a finite region around the origin (no hysteresis is allowed, which includes the origin). It may contain hysteresis outside the origin however, but it shall not contain any integrator elements. The plot of $N(y)$ shall be contained in a "sector" (hence the name of these tests) with upper slope b and lower slope a, where $|b|, |a|$ shall be bounded (they may be large but finite). Note that $a < 0$ is admissible, but not needed and not considered here (see references). Within this sector $N(y)$ may also be time varying. This need not be so slow such as to be quasi-stationary, but may be on the same time scale as the variables $u(t)$, $y(t)$.

Under these conditions the circle test in its graphic form consists of drawing
– for the linear subsystem the Nyquist contour of $L(j\omega)$ for $\omega \geq 0$
– and for the nonlinear subsystem the circle with its center on the negative real axis and running through the points $(-1/a,\ j0)$ and $(-1/b,\ j0)$.

In Fig. A.2

$$L(s) = \frac{6}{(s+1)^2 (0.1s+1)^2} \tag{A.1}$$

is used as a numerical example.

If there is no intersection, as in Fig.A.2 (a), with $1/a_1 = 4$, $1/a_2 = 8$, $1/b = 1$, then the closed loop initial conditions response will be asymptotically stable to the origin as the reference motion.

If there is an intersection as in Fig. A.2 (b), $1/a_1 = 4$, $1/a_2 = 8$, $1/b = 1/2$, then no stability result is delivered. The actual system response may converge or diverge. This is due to the inherent conservativeness of the test, which checks only "necessary but not sufficient" conditions (see references).

In applications, this may seem to be a severe deficiency. However, from experience, stability results from sector tests are not as conservative as other, more general, nonlinear stability tests may be. In any case, if the nonlinear loop is designed such that the test shows asymptotic stability, then (from our experience) the real system will show an acceptably well-damped reference

---

[2] Which all physical systems in fact are, but not necessarily their mathematical models.

step response.

Notice that for $a = b$, N will become a linear gain element. Then the circle degenerates into a point at $-1/b$ on the negative real axis and the circle test turns into the well known Nyquist stability test for linear loops. And for $a = 0$ the circle will degenerate into a vertical straight line at $-1/b$, which makes the graphic test even easier to perform.

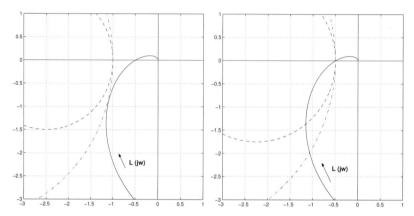

Fig. A.2. The graphic stability test with the "on-axis" circle test:
(left) case (a) indicating asymptotic stability of the initial condition response;
(right) case (b) allowing no conclusion about stability (the circle and $L(j\omega)$ intersect)

An important sub-case is N time-invariant and without hysteresis; Fig. A.3.

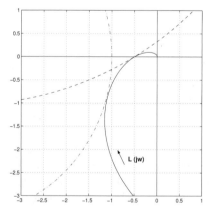

Fig. A.3. The graphic stability test with the "off-axis" $(--)$ compared with the "on-axis" $(\cdot - \cdot)$ circle test with $L(j\omega)$ $(—)$

Then the "on-axis" circle test can be relaxed into the "off-axis" circle test, which allows a larger sector: an upper slope of 2 is admissible instead of 1 from before. Note that $a = b = 2$ is also on the linear stability border.

If, additionally, $a = 0$, then the Popov test can be used instead. It allows one to use an arbitrary inclined straight line, as in Fig. A.4, together with the Popov plot $P(j\omega)$ instead of its Nyquist plot $L(j\omega)$. P is generated from L by

$$P(j\omega) = \Re[L(j\omega)] + j\omega\Im[L(j\omega)] \qquad (A.2)$$

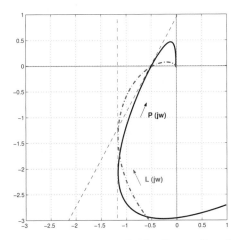

Fig. A.4. The graphic stability test with the Popov line (– –) and the Popov contour P (—) compared to the "on-axis" circle test with $L(j\omega)$ (· · ·) and the vertical straight line at $-1/b_v \approx -1/0.8621 = -1.165$

In this case, the Popov test produces the same upper bound for b as the off-axis circle test[3] and a higher upper bound than the on-axis circle test for $a = 0$, which produces the vertical straight line · · · at $(-1/b_v)$ in Fig. A.4.

A.3 Limit Cycles and the Describing Function Method

This method is best suited to check for limit cycles and their stability in loops such as in Fig. A.1. It is mathematically not very sound, but effective for engineering analysis and design. One should be careful not to overextend its usage.

In the context of the systems considered here, phenomena similar to limit cycles are known to appear for constrained systems of dominant order three

---

[3] This need not always be the case.

or higher. In constrained systems of dominant lower order, look for other causes than the constraints.

The linear subsystem $L(j\omega)$ in Fig. A.1 need not be asymptotically stable by itself, but it has to be a sufficient low-pass filter for the higher harmonic components of the limit cycle in the loop.

The nonlinear subsystem characteristic is to be non-time-varying. If it is not symmetric to the origin, a steady state offset is generated. The nonlinear characteristic may now contain a hysteresis around the origin and very steep slopes (such as for a relay). It is now described by the pair of gain $|B|$ and phase $\arg B$ of the first harmonic component of $u(t)$, if its input is a pure sinus oscillation. Both gain and phase are functions of the input amplitude $\hat{y}$. This is known as harmonic linearization of N.

Now the loop is closed with B instead of N. The elements in the loop are linear and the Nyquist test applies

$$1 + B(\hat{y}_{LC})L(j\omega_{LC}) = 0 \quad \rightarrow \quad L(j\omega_{LC}) = -\frac{1}{B(\hat{y}_{LC})} \qquad \text{(A.3)}$$

This equation holds at the intersection of the plots of the Nyquist contour $L(j\omega)$ and of the negative inverse of $B(\hat{y})$ in the complex plane. There, the linearized model loop is at its oscillatory stability border with frequency ω_{LC} and amplitude $\hat{y}_{LC}$ of the limit cycle, where ω_{LC} is to be read from $L(j\omega)$ and $\hat{y}_{LC}$ from the negative inverse of B. Fig. A.5 depicts this situation for the numerical case of L from above and for a relay with no hysteresis.

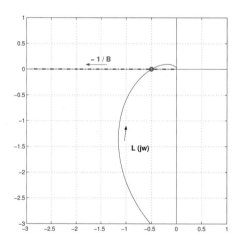

Fig. A.5. The describing function test for L from the previous example and for a relay with no hysteresis as the nonlinear element

Stability of this motion may be checked by increasing $\hat{y}_{LC}$ by a small amount, and locating the corresponding point on the plot of $-1/B$. If this point is on the left side of the Nyquist contour L for increasing ω, then this is a stable motion, otherwise it is not. From Fig. A.5 the motion is stable. A second case with the deadspan nonlinearity is shown in Fig. A.6 with[4]

$$L = -1 + \frac{(s+1)^3}{(s+\Omega)^3} \quad \text{with} \quad \Omega = 7.0 \tag{A.4}$$

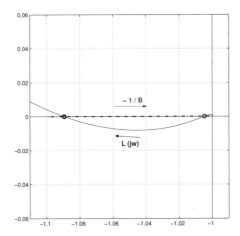

Fig. A.6. The describing function test for L from Eq. A.4 and for the deadspan nonlinearity

Here the leftmost intersection point indicates an unstable equilibrium, *i.e.* the motion will decay to zero for smaller $\hat{y}$ (represented by points farther to the left), and diverge for larger $\hat{y}$ (represented by points farther to the right). The right-hand intersection indicates a stable motion. Here, a diverging motion from the leftmost intersection will be stabilized.

A.4 Summary

Both the sector and describing function tests for stability of motion are useful tools in this context of constrained systems. They are complementary and not antagonistic. The sector criteria are useful as long as the Nyquist contour of $F + 1$ stays well within the second quadrant and away from the negative real axis. They fail, however, if the Nyquist contour crosses over the negative real axis, as then there will always be an intersection. Clearly, the describing function test is better suited for such cases. But it should not be used if the Nyquist plot stays far off the negative real axis, and thus far from any intersection points.

---

[4] Note that this corresponds to Peter's case, Sect. 6.2.

References

1. Aström K.J., B. Wittenmark, (1984)
   Computer Controlled Systems – Theory and Design, Prentice Hall
2. Aström K.J., T. Hägglund, (1995)
   PID Controllers: Theory, Design, and Tuning, 2nd ed., ISA (Instrument Society of America)

Theoretical Basis
3. Bellman R., (1957)
   Dynamic Programming, Princeton University Press, Princeton, N.J.
4. Pontrjagin L.S., V. Boltyanskii, R. Gamkredlidze, E. Mishchenko, (1962)
   The Mathematical Theory of Optimal Processes, Interscience Publishers Inc. New York
5. Narendra K.S., R.M. Goldwyn, (1964)
   A geometrical criterion for the stability of certain nonlinear nonautonomous systems, *IEEE Transactions on Circuit Theory*, vol. CT-11, 406–407
6. Zames G., (1966)
   On the input-output stability of time-varying nonlinear feedback systems, *IEEE Transactions on Automatic Control*, vol. AC-11, no 3, 465–476
7. Hsu J.C., A.U. Meyer, (1968)
   Modern Control Principles and Applications, McGraw-Hill, New York
8. Cho Y.S., K.S. Narendra, (1968)
   An off-axis criterion for the stability of feedback systems with a monotonic nonlinearity, *IEEE Transactions on Automatic Control*, vol. AC-33, 413–416
9. Narendra K.S., J.H. Taylor, (1973)
   Frequency Domain Criteria for Absolute Stability, Academic Press, New York
10. Utkin V., (1977)
    Variable structure systems with sliding modes, a survey, *IEEE Transactions on Automatic Control*, vol. AC–23, 212–222
11. Ackermann J., (1985)
    Sampled-Data Control Systems, Springer
12. Khalil H.K., (1996)
    Nonlinear Systems, 2nd ed., Prentice Hall
13. Saberi A., A.A. Stoorvogel, P. Sannuti, (1999)
    Control of Linear Systems with Regulation and Input Constraints, Springer

496   References

Early Publications

14. Oppelt W., (1958)
    Steuerung und Regelung bei absatzweisem Betrieb, *Regelungstechnik*, Olden-
    bourg, München-Wien, vol. 6, no. 2, 52–59
15. Pahl G., W. Reitze, M. Salm, (1964)
    Ueberwachungseinrichtung für zulässige Temperaturänderungen an Dampftur-
    binen, *BBC Mitteilungen*, vol. 46, no. 3, 139–147
16. Fertik H.A., C.W. Ross, (1967)
    Direct digital control algorithms with anti-windup feature, *ISA Transactions*,
    vol 6, no. 4, 317–328
17. Berlemont V., J. Debelle, (1968)
    Quelques aspects du démarrage automatique d'une chaudière à ballon, *Proceed-
    ings IFAC-Symposium on multivariable control,* VDI-Verlag, Düsseldorf
18. Stoll A., E. Hähle, A. Fischer, (1968)
    Materialgerechter Betrieb grosser Dampferzeuger mit Hilfe eines Freilast-
    rechners, *Siemens-Zeitschrift*, vol. 42, no. 2, 112–116
19. Bloch H. (1969)
    Festprogrammierte elektronische Einrichtung zur automatischen Steuerung von
    Dampfturbogruppen, *BBC Mitteilungen*, vol. 51, 156–164
20. Latzel W., (1969)
    Hybrider Turbinenregler mit Regelungs-, Überwachungs- und Schutzfunktion,
    BBC-Nachrichten, vol. 17, 71–74
21. Buckley P.S., (1971)
    Designing override and feedforward controls, *Control Engineering*, vol. 18, no.
    8, 48–51, and no. 9, 82–85
22. Sarantsev L.B., (1971)
    "Analysis of the dynamics of gas turbines and their control systems", *Transac-
    tions ASME, Journal of Engineering for Power*, no. 7, 300–306
23. Glattfelder A.H., (1974)
    Regelungssysteme mit Begrenzungen, Oldenbourg, München-Wien
24. Wührer W., (1976)
    The governing of water turbines using electronic modules, *Escher Wyss News*,
    Zürich, vol. 49, 3–14

Research from the Last Two Decades

25. Krikelis N.J., (1980)
    State feedback integral control with "intelligent" integrators, *International Jour-
    nal of Control*, vol. 32, no. 3, 465–473
26. Foss M.A., (1981)
    Criterion to assess stability of a 'lowest wins' control strategy, *Proceedings IEE*,
    vol. 128, part 1, 1–8
27. Glattfelder A.H., W. Schaufelberger, (1983)
    Stability analysis of single loop control systems with saturation and anti-reset-
    windup circuits, *IEEE Transactions on Automatic Control*, vol. 28, no. 12, 1074–
    1081
28. Glattfelder A.H., H.P. Fässler, W. Schaufelberger, (1983)
    Stability of override control systems, *International Journal of Control*, vol. 37,
    no. 5, 1023–1037

29. Krikelis N.J., S.K. Barkas, (1984)
    Design of tracking systems subject to actuator saturation and integrator windup,
    International Journal of Control, vol. 39, no. 4, 667–682

30. Wu Q.D., (1986)
    Stabilitätsanalyse von Regelsystemen mit Begrenzungen, PhD thesis, ETH
    Zürich, no. 7930

31. Glattfelder A.H., W. Schaufelberger, (1986)
    Start-up performance of different proportional–integral anti-windup regulators,
    International Journal of Control, vol. 44, no. 2, 493–505

32. Hanus R., M. Kinnaert, J.-L. Henrotte, (1987)
    Conditioning technique, a general anti-windup and bumpless transfer method,
    Automatica, vol. 23, 729–739

33. Noisser R., (1987)
    Anti-reset windup Massnahmen für Eingrössenregelungen mit digitalen Reglern,
    Automatisierungstechnik at, Oldenbourg München-Wien, vol. 35, no. 12, 499–
    504

34. Glattfelder A.H., W. Schaufelberger (1988),
    Stability of discrete override and cascade-limiter single loop control systems,
    IEEE Transactions on Automatic Control, vol. 33, no. 6, 532–540

35. Glattfelder A.H., L. Guzzella, W. Schaufelberger, (1988)
    Bumpless transfer, anti-reset-windup, saturating and override controls - a status
    report on self selecting regulators, *Proceedings IMACS 88, 12th World Congress
    on scientific computation*, vol. 2, 66–72

36. Guzzella L., A.H. Glattfelder, (1989)
    Stability of multivariable systems with saturating power amplifiers, *Proceedings
    American Control Conference*, Pittsburgh, 1687–1692

37. Aström K.J., L. Rundqwist, (1989)
    Integrator windup and how to avoid it, *Proceedings American Control Confer-
    ence*, Pittsburgh, 1693–1698

38. Campo P.C., M. Morari, C.N. Nett, (1989)
    Multivariable anti-windup and bumpless transfer: a general theory, *Proceedings
    American Control Conference, Pittsburgh*, p. 1706–1711

39. Hanus R., M. Kinnaert, (1989)
    Control of constrained multivariable systems using the conditioning technique,
    Proceedings American Control Conference, Pittsburgh, 1712–1718

40. Glattfelder A.H., L. Guzzella, W. Schaufelberger, (1990)
    Stability and design of single loop state control systems with actuator satura-
    tions, *Proceedings 11th IFAC World Congress*, Tallinn 1990, vol. 6, 81–89

41. Wurmthaler Ch., P. Hippe, (1991)
    Systematic compensator design in the presence of input saturations, *Proceedings
    1st European Control Conference*, 1268–1273

42. Rundqwist L., (1991)
    Anti-reset windup for PID controllers, PhD thesis, Department of Automatic
    Control, Lund Institute of Technology

43. Glattfelder A.H., L. Huser, (1993)
    "Hydropower reservoir level control – a case study", *Automatica*, vol. 29, no. 5,
    1203–1214

44. Glattfelder A.H., W.Schaufelberger, (1993)
    Generalized anti-windup, *Proceedings 2nd European Control Conference*, Gron-
    ingen, vol. 2, 1078–1083

498   References

45. Glattfelder A.H., Chr. Eck, W.Schaufelberger, (1995)
    Stability analysis of two different anti-windup systems, *Proceedings 3rd European Control Conference*, Rome, vol. 2, 1492–1496
46. Rönnbäck S., (1996)
    Nonlinear dynamic windup detection in anti-windup compensators *Proceedings CESA'96 Multiconference, Symposium on Control, Optimization and Supervision*, vol. 2, 1014–1019
47. Glattfelder A.H., W. Schaufelberger, (1996)
    Stability of trajectory generator control systems with anti-windup, *Proceedings CESA'96 Multiconference, Symposium on Control, Optimization and Supervision*, vol. 2, 1020–1025
48. Zirn, O., A.H. Glattfelder, (1997)
    Stability analysis for the design of fast axis feed drives, *Proceedings 4th European Control Conference*, Bruxelles, vol. 6, part B
49. Teel A., N. Kapoor, (1997)
    The L_2 anti-windup problem: its definition and solution, *Proceedings 4th European Control Conference*, Brussels, paper 494
50. Odenthal D., T. Bünte, J. Ackermann, (1999)
    Nonlinear steering and braking control for vehicle rollover avoidance, *Proceedings 5th European Contol Conference*, Karlsruhe, Paper F0241
51. Bühler H., (2000)
    Regelkreise mit Begrenzungen, Fortschritt-Berichte VDI, Nr. 828, Düsseldorf
52. Glattfelder A.H., J. Tödtli, W.Schaufelberger, (2001)
    Stability properties and effects of measurement disturbances on anti windup PI- and PID-control", *European Journal of Control*, vol. 7, no. 5, 435–448
53. Mulder E.F., M.V. Kothare, M. Morari (2001)
    Multivariable anti-windup controller synthesis using bilinear matrix inequalities, *European Journal of Control*, vol. 7, no. 5, 455–464
54. Güvenc, B.A., L. Güvenc, D. Odenthal, T. Bünte, (2001)
    Robust two degree of freedom vehicle steering control satisfying mixed sensitivity constraint, *Proceedings 6th European Contol Conference*, Porto, 1198–1203
55. Turner M.C., D.G. Bates, I. Postlethwaite, (2001)
    A conditioning strategy for performance improvement of an integrated flight and propulsion control system for a V/STOL aircraft, *Proceedings 6th European Contol Conference*, Porto, 3050–3055
56. Grimm G., I. Postlethwaite, A.R. Teel, M.C. Turner, L. Zaccarian, (2001)
    Case studies using linear matrix inequalities for optimal anti-windup synthesis, *Proceedings 6th European Contol Conference*, Porto, 3794–3799

Links to Model Predictive Control
57. Camacho E.F., C. Bordons, (1999)
    Model Predictive Control, Springer
58. Maciejowski J.M., (2002)
    Predictive Control with Constraints, Pearson Education
59. Bemproad A., M. Morari, V. Dua, E.N. Pistikopoulos, (2002)
    The explicit linear quadratic regulator for constrained systems, *Automatica*, vol. 38 no. 1, 3–20
60. Borrelli F., (2002)
    Discrete time constrained optimal control, PhD thesis, ETH Zürich, no. 14666

Multivariable Control

61. Maciejowski J.M., (1989)
    Multivariable Feedback Design, Addison-Wesley
62. Skogestad S., I. Postlethwaite, (1996)
    Multivariable Feedback Control, Analysis and Design, John Wiley & Son
63. Doyle J.C., R.S. Smith, D.F. Enns, (1987)
    Control of plants with input saturation nonlinearities, *Proceedings Americal Control Conference*, Minneapolis, 1034–1039
64. Zheng A., M.V. Kothare, M. Morari, (1994)
    Antiwindup design for internal model control, *International Journal of Control*, vol. 60, no. 5, 1015–1024
65. Campo P.C., M. Morari, (1994)
    Achievable closed-loop properties of systems under decentralized control: conditions involving the steady state gain, *IEEE Transactions on Automatic Control*, vol. 39, no. 5, 932–943
66. Kapoor N., A.R. Teel, P. Daoutidis, (1998)
    An anti-windup design for linear systems with input saturation, *Automatica*, vol. 34, no. 5, 559–574
67. Turner M.C., I. Postlethwaite, (2002)
    Local anti-windup compensation using output limiting, *Proceedings 15th IFAC World Congress*, Barcelona, paper 278

Index